IEE ELECTROMAGNETIC WAVES SERIES 33

Series Editors: Professor P. J. B. Clarricoats
Professor Y. Rahmat-Samii
Professor J. R. Wait

RADIO DIRECTION FINDING AND SUPERRESOLUTION

Other volumes in this series:

Volume 1 **Geometrical theory of diffraction for electromagnetic waves** G. L. James
Volume 2 **Electromagnetic waves and curved structures** L. Lewin, D. C. Chang and E. F. Kuester
Volume 3 **Microwave homodyne systems** R. J. King
Volume 4 **Radio direction-finding** P. J. D. Gething
Volume 5 **ELF communications antennas** M. L. Burrows
Volume 6 **Waveguide tapers, transitions and couplers** F. Sporleder and H. G. Unger
Volume 7 **Reflector antenna analysis and design** P. J. Wood
Volume 8 **Effects of the troposphere on radio communications** M. P. M. Hall
Volume 9 **Schumann resonances in the earth-ionosphere cavity** P. V. Bliokh, A. P. Nikolaenko and Y. F. Flippov
Volume 10 **Aperture antennas and diffraction theory** E. V. Jull
Volume 11 **Adaptive array principles** J. E. Hudson
Volume 12 **Microstrip antenna theory and design** J. R. James, P. S. Hall and C. Wood
Volume 13 **Energy in electromagnetism** H. G. Booker
Volume 14 **Leaky feeders and subsurface radio communications** P. Delogne
Volume 15 **The handbook of antenna design, Volume 1** A. W. Rudge, K. Milne, A. D. Olver, P. Knight (Editors)
Volume 16 **The handbook of antenna design, Volume 2** A. W. Rudge, K. Milne, A. D. Olver, P. Knight (Editors)
Volume 17 **Surveillance radar performance prediction** P. Rohan
Volume 18 **Corrugated horns for microwave antennas** P. J. B. Clarricoats and A. D. Olver
Volume 19 **Microwave antenna theory and design** S. Silver (Editor)
Volume 20 **Advances in radar techniques** J. Clarke (Editor)
Volume 21 **Waveguide handbook** N. Marcuvitz
Volume 22 **Target adaptive matched illumination radar** D. T. Gjessing
Volume 23 **Ferrites at microwave frequencies** A. J. Baden Fuller
Volume 24 **Propagation of short radio waves** D. E. Kerr (Editor)
Volume 25 **Principles of microwave circuits** C. G. Montgomery, R. H. Dicke, E. M. Purcell (Editors)
Volume 26 **Spherical near-field antenna measurements** J. E. Hansen (Editor)
Volume 27 **Electromagnetic radiation from cylindrical structures** J. R. Wait
Volume 28 **Handbook of microstrip antennas** J. R. James and P. S. Hall (Editors)
Volume 29 **Satellite-to-ground radiowave propagation** J. E. Allnutt
Volume 30 **Radiowave propagation** M. P. M. Hall and L. W. Barclay (Editors)
Volume 31 **Ionospheric radio** K. Davies
Volume 32 **Electromagnetic waveguides: theory and application** S. F. Mahmoud
Volume 33 **Radio direction finding and superresolution** P. J. D. Gething

RADIO DIRECTION FINDING AND SUPERRESOLUTION

P J D GETHING

Peter Peregrinus Ltd. on behalf of the Institution of Electrical Engineers

Published by: Peter Peregrinus Ltd., London, United Kingdom

Reprinted 1986, 1989.
Second Edition 1991.

Peter Peregrinus Ltd.,
Michael Faraday House,
Six Hills Way, Stevenage,
Herts. SG1 2AY, United Kingdom

British Library Cataloguing in Publication Data
A CIP catalogue record for this book is available from the British Library.

ISBN 0 86341 238 6

Printed in England by Short Run Press Ltd., Exeter

Contents

Preface

My aim in writing this book has been to present a systematic and coherent account of modern DF principles, suitable for use by research students. The treatment is therefore at postgraduate level and the reader is assumed to possess at least a slight acquaintance with the basic features of radio propagation and antenna-array theory. The main emphasis is on direction finding in the HF band (2–30 MHz), but the underlying principles should also be of interest to readers working in the fields of radar, sonar, or radioastronomy.

The major challenge in current DF research is to find reliable methods, suitable for routine operational use, of resolving a multiray wave-field into constituent rays, the directions of arrival of which can be determined. The attack advocated here is by wavefront analysis (WFA), in which measurements are made at a number of antenna elements, recorded in digital form and subsequently analysed in a computer. I have found it convenient to treat WFA as an extension of classical array theory, concerned with the computer synthesis of array-radiation patterns containing nulls in desired directions. The advantages of building on array theory are that the physical significance of mathematical results can be explained and that the generality of certain results becomes apparent.

Research workers are also greatly interested in the small deviations of ray paths produced by 'tilts' and other effects in the ionosphere. The wide-aperture direction finder has become an important tool of ionospheric and propagation research in recent years. This book therefore contains chapters dealing with the measurement of ionospheric parameters, ray-tracing, tilt-correction schemes, plotting methods, and the statistical analysis of bearings and plots.

The results obtained from some of the sample WFA calculations are presented in tabular rather than graphical form. The tables contain a very abbreviated version of the output from computer programs written to solve the type of problem described in Chapters 10 and 11. The reader who studies these tables will become familiar with the kind of output he will have to interpret if he develops similar programs; he may also be able to use some of the results to test his programs. Elsewhere in the book I have preferred to present results graphically as far as possible.

Preface to second edition

In updating the contents of this book, I have again aimed to match the level of treatment to the needs of a postgraduate research student. Over the last decade there has been a tremendous growth of interest in the application of superresolution processing techniques to problems in radio direction finding, radar and sonar. The determination of the angular spectrum of signals arriving at an array of sensors is also closely related to the separation of frequency components in conventional spectrum analysis and time-series analysis. The applications of such analyses are legion and include, for example, the interpretation of seismological records and of electro-encephalograms. A very extensive research literature of several hundred papers, mainly of a theoretical and mathematical nature, has been built up on these subjects and their inter-relationships since the first edition of 'Radio direction finding and the resolution of multicomponent wave-fields' was published in 1978.

I have therefore added Chapters 12 and 13 to deal with these developments. The main emphasis is on the underlying principles (which follow naturally from the material in earlier chapters) and the results that have been achieved; it is clearly neither possible nor desirable, in a textbook of this nature, to reproduce several pages of mathematics from every significant paper. I have exercised some selectivity in quoting references, with a preference for accessible and recent papers from which earlier work can usually be traced rather than the first paper on a new topic, and for lucid survey papers likely to be useful to the new research worker. Slight changes have been made to Chapter 11 in order to lead smoothly into the new chapters, and a new Appendix 5 has been added to define the notation used.

Most of the terms that have become widely accepted in the superresolution literature have also been adopted here. Whenever the context is sufficiently general I have referred to sensors rather than antenna elements. However, the term ray is preferred to signal, because the former corresponds more clearly to the wavefield model and the latter is ambiguous: the signal from one transmitting source could give rise to several arriving rays. In so far as any distinction exists between the terms wavefront analysis and superresolution, I have tended to use the former as a general term for the collection and analysis of digital data from many sensors and the latter for the detailed processing techniques.

The progress on topics other than superresolution has been steady rather than spectacular. However, it has been necessary to revise each of the original chapters; I have therefore added new material and references at a number of places throughout the text.

Acknowledgements

I have benefited from the advice and assistance of many people since I first became interested in DF some thirty years ago. I am particularly grateful to A. N. Beresford, who directed my attention to the theoretical problem of angle measurements in multi-ray conditions and did much to encourage the parallel development of practical techniques. Many others have participated in the development of WFA theory. D. Cawsey and J. H. Ellis have made particularly significant contributions and are responsible for some of the numerical examples used in Chapters 10 and 11. The following sections are based largely on their work: one-dimensional theory (10.5), matrix inversion (11.2.3), eigenvalues (11.3), angle-free linear relationships (11.6), masquerading (11.7) and resolving power (11.8).

The treatment of wavefront curvature in Section 3.2 is based on an unpublished note by Dr. R. Benjamin. In particular, he showed that the range of the transmitter can, in suitable circumstances, be deduced without any knowledge of arrival angles. Many of the results in Chapter 4 were first obtained by Dr. S. M. S. A. Shirazi (now Dr. M. J. Archer) when he was a research student at the University of Birmingham, working under the supervision of Dr. D. C. Cooper.

Dr. M. Mellors first drew my attention to the distinction between electronic and mechanical scanning of the multiplicative array discussed in Section 5.8 and Appendix 2. The treatment in Chapter 6 of the effects of site re-radiators is based partly on numerical calculations undertaken by J. S. B. Reynolds and partly on unpublished notes by Dr. W. C. Bain and Dr. E. N. Bramley. In particular, the theory given in Section 6.4.4 is due to Dr. Bramley. The results embodied in eqns. 6 and 7 of Chapter 6 were suggested to me by J. H. Ellis.

R. M. Chamberlain provided the proof in Chapter 9 that the envelope of ellipses seen on an interferometer display, when two rays arrive from constant directions, is a parallelogram. The brief treatment in Section 16.4.4 of the plotting problems that arise with non-normal distributions is based partly on an investigation by A. D. V. Ridlington. P. J. Meade originally provided the proof of the result in Appendix 9.

E. G. Shepherd and J. Hockley lent me photographs and drawings from which some of the diagrams have been prepared. Permission to copy a number of diagrams from published journals has been generously granted by the IEE (Figs. 2.3, 2.5 and 7.4), the IERE (Figs. 5.5 and 5.6) and the American Geophysical Union (Figs. 2.6, 9.1 and 14.3); full details of the authors and sources are given in the main text.

To all of the above, and to the many colleagues who participated in the development of hardware and in the collection and analysis of experimental results, I record my grateful thanks.

Additional acknowledgements for second edition

In preparing material for the second edition, I have taken account of a number of mathematical and statistical points made by K. C. Bowen in a generous review of the first edition in the *Journal of Naval Science*. D. S. C. Paine drew my attention to many important publications when I returned to this subject after an interval. Friendly contacts with Dr. D. N. Travers and his team at the Southwest Research Institute, San Antonio, Texas, have been a source of much help and encouragement over a period of many years.

I am grateful to Professor T. B. Jones and D. S. C. Paine for helpful comments on draft material for Chapters 12 and 13 and Appendix 5.

The index for the first edition was compiled by J. E. Holstrom; it has now been expanded to include new material.

I am again pleased to acknowledge the assistance of the IEE and IEEE in granting permission to copy diagrams from published papers as follows: Fig. 12.4, 14.5 and 14.6 (IEE) and Figs. 12.5, 12.6 and 13.2 (IEEE). The cover design is based on Fig. 2 in a paper by R. L. Johnson and G. E. Miner (*IEEE Trans.*, **AES-22**, July 1986, pp. 432–442), with the agreement of the IEEE.

List of abbreviations and acronyms

AAR	adaptive angular response
AIC	an information criterion suggested by Akaike
AOA	angle(s) of arrival
AP	alternating projection
AR	auto regressive
ARMA	auto regressive moving average
ASAT	adaptive spatial averaging technique
ASRP	augmented smoothed rank profile
BERF	bearing error reduction factor
BFN	beam-forming network
BPE	best point estimate
BS	backscatter
CA	(locus of) constant amplitude
CADF	commutated antenna direction finder
CDAA	circularly disposed antenna array
c.d.f.	cumulative (probability) distribution function
CEP	circular error probability (region)
CP	(locus of) constant phase
CRC	Communications Research Centre (Canada)
CRLB	Cramer–Rao lower bound
CT	check transmitter
CW	continuous wave
DF	direction finder or direction finding
DOA	direction(s) of arrival
ESPRIT	estimation of signal parameters via rotational invariance techniques
FBLP	forward-backward linear prediction
g.c.	great circle
HAM	horizontal angle measurement
HF	high frequency
IF	intermediate frequency

JG	Johnson–de Graaf method
KT	Kumaresan–Tufts method
LAT	large-area tilt
MADF	medium aperture direction finder
MESA	maximum entropy spectral analysis
MID	magneto-ionic deviation
MJG	modified Johnson–de Graaf method
MLE	maximum likelihood estimator
MLM	maximum likelihood method
MODE	method of direction estimation
MOF	maximum observed frequency
MS	modified Schmidt method
MUF	maximum usable frequency
MUSIC	multiple signal characterisation
NADF	narrow aperture direction finder
NR	Nakagami–Rice (distribution)
NSP	noise space projection
OISR	output interference to signal ratio
PC	phase coincidence
PCA	principal component analysis
p.d.	phase difference
p.d.f.	probability density function
PHD	Pisarenko harmonic decomposition
PRO-ESPRIT	Procrustes version of ESPRIT algorithm
PR	probability region
PTMF	parametric target model fitting
QP	quasi-parabolic
QUMP	quasi unimoded propagation
RF	radio frequency
r.m.s.	root mean square
SAB	scanned array beam
SABTF	SAB with triangular factorisation
s.d.	standard deviation
SE	systematic error
SFB	scanned fixed beam
SIT	systematic ionospheric tilt
SLC	sidelobe canceller method
SNR	signal to noise ratio
SSL	single station location
TAM	Toeplitz approximation method
TDDF	time difference direction finding
TID	travelling ionospheric disturbance
TLS-ESPRIT	total least squares version of ESPRIT
TNA	thermal noise algorithm
TT	target transmitter
UHF	ultra high frequency
VAM	vertical angle measurement
VHF	very high frequency

VI	vertical incidence
VWADF	very wide aperture direction finder
WADF	wide aperture direction finder
WFA	wavefront analysis
WI	wave interference

Chapter 1
Introduction

1.1 Some preliminary definitions

The aim in this chapter is to review briefly the physical principles involved in radio-direction finding in the HF band. We shall consider only direction finding from fixed land-based stations, on signals propagated via the ionosphere. The abbreviation DF will be used here, as in most published literature, for both direction finder and direction finding; the correct choice should always be obvious from the context.

The ideal direction finder would be capable of working over a wide frequency band, 360° of azimuth and 90° of elevation, of dealing with all forms of modulation, and of giving accurate and reliable bearings, even on comparatively brief transmissions. A signal such as a distress call from a ship may be brief and signal parameters such as frequency may not be optimum at all receiving points. The transmissions on which bearings are to be taken in normal applications of civil and military DF must therefore be regarded as nonco-operative.

The terminology to be used in this book conforms as far as possible with the relevant British Standard (1971) and with the definitions recommended by the IEEE (1983). These publications can be consulted for definitions of terms such as DF sensitivity, pick-up factor, pick-up ratio and standard-wave error, which are sometimes used (but not here) in comparisons of the merits of various types of DF.

The symbol α will be used for the azimuth angle of arrival of a ray, measured clockwise from North, and Δ for the elevation angle. These two angles, or the corresponding direction cosines, provide a complete measurement of the direction of arrival (DOA) of the ray. A measured azimuth angle α may differ from the true great-circle bearing β of the transmitter from the receiver for a number of reasons to be described. The term 'bearing' is sometimes used in published literature for both angles but will generally be reserved here for the true bearing. Nevertheless, the small angle $(\alpha-\beta)$ will be referred to as the bearing error; $(\beta-\alpha)$ is the ideal bearing correction to add to α in order to obtain β. Some forms of DF instrument are capable of providing elevation-angle measurements as well as azimuth-angle measurements and are sometimes referred to as three-dimensional direction finders (Adcock, 1959). In any case the DF should be regarded as a device that determines the direction of arrival of a signal, not the direction of the transmitter.

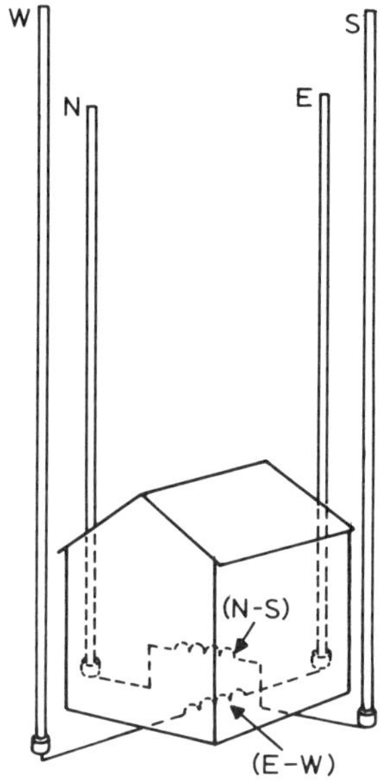

Fig. 1.1 Four-element Adcock direction finder

A wide-aperture direction finder (WADF) is defined in the British Standard as one 'in which the arriving wavefront is sampled simultaneously (or nearly so) by an array of aerials extending over a distance comparable to or greater than one wavelength.' Narrow-aperture systems (NADF) such as the Adcock have apertures of a few tenths of a wavelength in the frequency band for which they are designed. Linear arrays with apertures of more than ten wavelengths have been built by some research groups and are often referred to as very-wide-aperture systems (VWADF). Less commonly, arrays intermediate in size between the NADFs and the WADFs are said to be of medium aperture (MADF). In what follows we shall be largely concerned with WADFs.

It is convenient to describe a few of the basic types of DF at this stage and to give the reader some idea of physical dimensions. Typically, the aperture of a four-element Adcock is about 6 m. The vertical monopole elements, about 10 m high, are at the corners of a square, with buried feeders running to a central hut, as shown in Fig. 1.1.

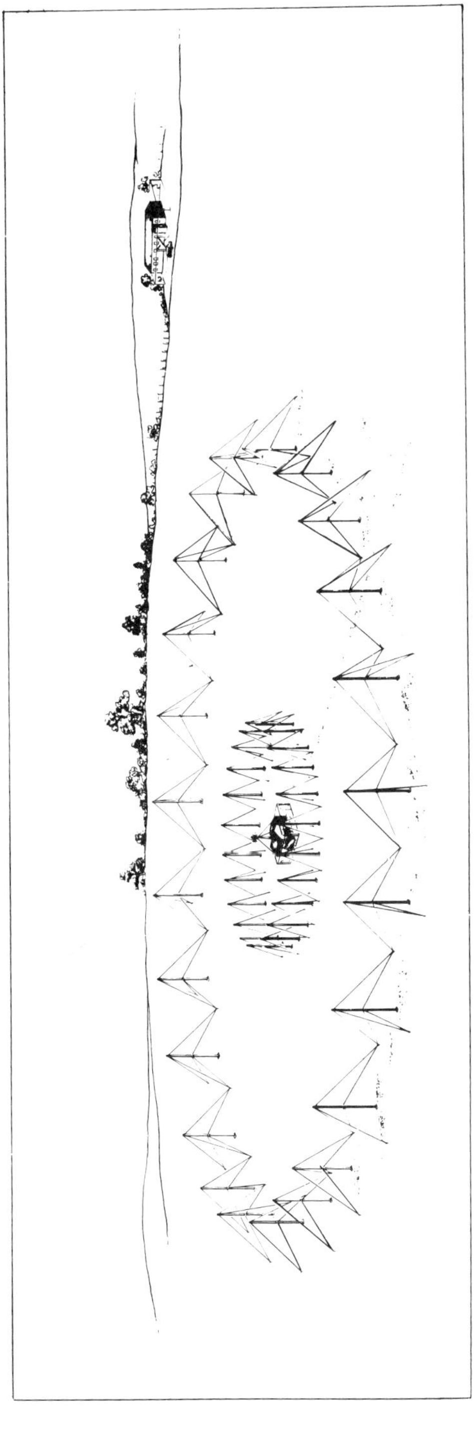

Fig. 1.2 Two-ring DF array

Most WADFs employ circular rings of identical elements, i.e. circularly disposed antenna arrays (CDAA). Because it is difficult to maintain satisfactory radiation patterns from a given ring over more than about a 3:1 frequency range, or 4:1 at most, at least two rings of elements are required to cover the whole 15:1 ratio of the HF band from 2 MHz to 30 MHz. Fig 1.2 shows a sketch of a two-ring system in which the larger ring has a diameter of (say) 150 m and is used for the lower part of the band, while the inner ring, with a diameter of (say) 50 m is used for the upper part of the band.

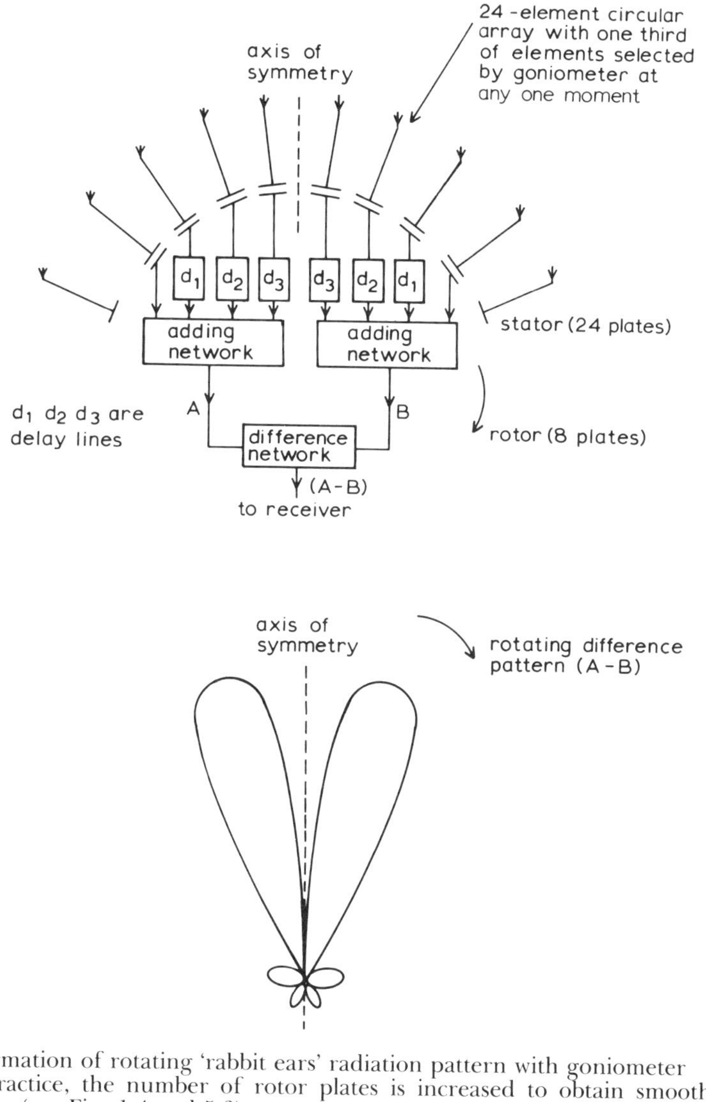

Fig. 1.3 Formation of rotating 'rabbit ears' radiation pattern with goniometer
Note: In practice, the number of rotor plates is increased to obtain smoother commutation (see Figs 1.4 and 5.3)

Three main types of CDAA may be distinguished. In the Wullenweber type (Rindfleisch, 1956) the outputs from a sector of about one-third of the elements in the ring being used are combined to form a directional pattern with desired properties. This pattern is rotated in azimuth by means of a spinning goniometer, effectively a capacitative switch that selects which elements are to be used at any moment. The way in which a 'rabbit-ears' difference pattern can be formed and rotated is shown in Fig 1.3, and the mechanical design of a goniometer is indicated in Fig. 1.4 The azimuth measurement, for a single arriving ray, is made by noting the pointing

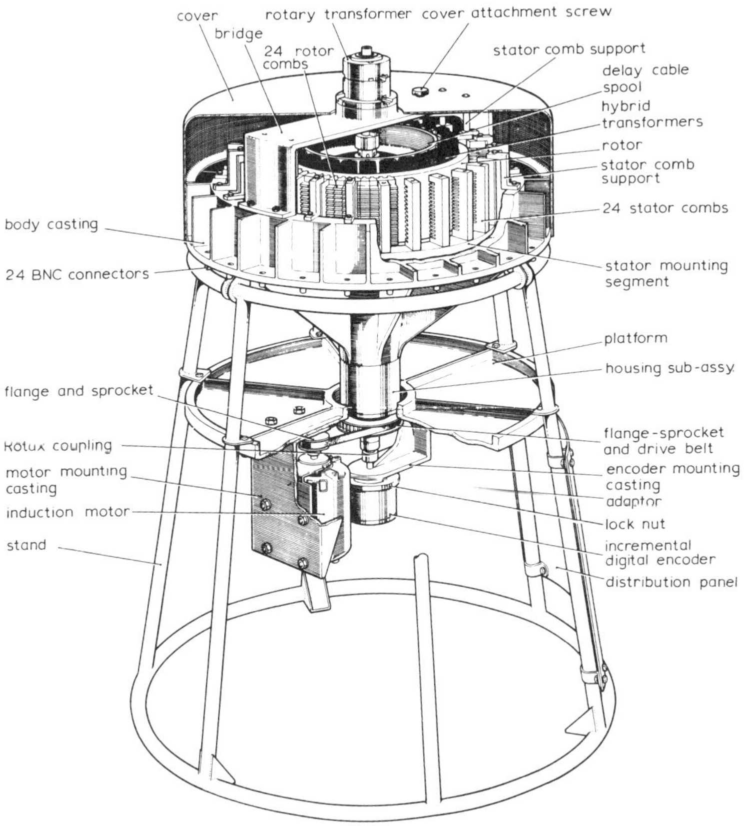

Fig. 1.4 Mechanical design of rotating goniometer

angle of the goniometer when a minimum signal between the two maxima is observed. A Doppler DF is a system in which a frequency modulation is imposed on the signal by moving a single receiving element rapidly round the perimeter of a circle. Whale (1954) used a system of this sort, but it is more usual to obtain a simulated Doppler effect by smooth commutation round a ring of fixed elements. The third type of CDAA is the system suggested by Earp and Godfrey (1947), in which the bearing information is obtained from phase comparisons between successive pairs of elements in the ring. The simulated Doppler and Earp–Godfrey systems are sometimes referred to as commutated antenna direction finders (CADF).

Research workers often take DF measurements on a transmitter of known position, either to study the various sources of DF error or as a means of detecting travelling ionospheric disturbances (TIDs) and tilts. It is convenient to refer to the transmitter as the 'target' in order to distinguish the wanted signal from interfering transmitters and noise, but this term is not intended to suggest some state of hostility between the points of transmission and reception. Indeed, for research purposes a 'co-operative' transmitter under the direct control of the experimenters is often used. Short pulses can then be transmitted in order to achieve time resolution of the various possible modes of propagation. Alternatively, if single-moded signals are required, the frequency can be chosen so that only one mode is received, or one mode is so dominant that the signal is effectively single-moded.

Since rays can arrive at the DF from 'unwanted' transmitters and from re-radiating objects in the vicinity of the DF array, it is better to refer to multiray rather than multimode wave-fields, and to restrict the use of the term 'modes' to direct arrivals from the target transmitter. The principal topic in this book is the performance of WADFs in multiray wave-fields. The term 'complex fading' will be used to describe changes in relative amplitude, or relative phase, or both, when two or more rays 'illuminate' the DF.

Since much of this book is concerned with the measurement of the azimuth angles and elevation angles of arriving rays, it is convenient in places to use the abbreviation HAM for horizontal angle measurement and VAM for vertical angle measurement. HAM, for example, might refer either to a technique for measuring ray azimuths or to the action of taking an azimuth observation; the exact shade of meaning should be clear from the context.

Many DFs of conventional design incorporate a bearing readout device, such as a digital display corresponding to the position of a cursor on a visual display. Clearly, such a device can display only one number at a given moment and manifestly cannot provide simultaneous HAM on several rays. This number will be referred to as the indicated bearing. In single-ray conditions the indicated bearing should equal the azimuth of the ray but will not, in general, coincide exactly with the true great-circle bearing of the transmitter; gradients in ionospheric layers can easily produce differences of a degree or more between the indicated and true bearings. In multiray conditions there is no reason to suppose that the indicated bearing is equal to the azimuth of any one ray. It is important that the distinctions between true bearing (one number), indicated bearing (one number) and the azimuths of arrival of n rays, $\alpha_1, \alpha_2, \alpha_3, \ldots, \alpha_n$ (n numbers) should be clearly appreciated.

It will frequently be assumed in numerical simulations that the arriving rays fluctuate with time in amplitude and phase (complex fading) but that the ray DOAs remain sensibly constant. In these circumstances, the indicated bearing varies with time, whereas the true bearing and the ray azimuths do not vary. Such fluctuations in indicated bearing are often described in the literature as arising from wave interference (WI). It will be shown that WI fluctuations provide some information about the incident rays.

Surfaces of constant phase at the DF will be referred to as wavefronts. Because DF measurements are often based on the determination of phase difference between signals in two or more antenna elements, we are particularly interested in the characteristics of wavefronts. For a single mode from a distant transmitter, the wavefronts can generally be regarded as plane, and the normal to these planes is defined (in free space) as the ray direction. For two or more rays, the resultant wavefronts are no longer plane. Wave-fields can be economically described in terms of the ray parameters, but it is sometimes helpful to use the picture of distorted wavefronts, particularly in comparing the merits of DFs with different apertures. If the correct ray parameters can be calculated from measurements with a DF, the DF will be said to have resolved the wave-field.

1.2 Review of DF techniques

A brief description of the highlights in the evolution of the HF direction finder may be found helpful to an understanding of later chapters.

Because no antenna element has a truly isotropic radiation pattern in free space, any element has directional properties that can, in principle, be exploited. A simple vertical loop, for example, can be rotated about a vertical axis to find a position giving a signal minimum in a receiver connected to it; provided the waves from the target transmitter have travelled in a horizontal plane, the normal to the loop should then 'point' to the transmitter (or in the reciprocal direction, 180° away). It was realised early in the history of DF, however, that the simple rotatable loop is quite unsuitable for use on sky wave signals arriving from unknown elevation angles and of mixed polarisation after their passage through the ionosphere. The position of a signal minimum can give an indicated bearing grossly in error (e.g. by 30°) because of what would now be called polarisation error. This difficulty became generally recognised after the publication of a paper by Smith-Rose and Barfield (1926); they pointed out that the direction finder patented by Adcock (1919) would be very much less susceptible than the loop to polarisation error.

The Adcock DF was therefore increasingly studied during the 1930s and was used extensively during the Second World War. The importance of DF in wartime gave a stimulus to research, in which the various sources of site error and instrumental error were gradually identified and reduced. Some DFs had to be erected on sites that were known to be poor in terms of conductivity and freedom from obstacles; suitably designed earth mats were shown by Smith-Rose and Ross (1947) to be very valuable on sites of low or irregular conductivity.

The occurrence of polarisation error in the bearings from a single loop does not prevent the use of loop elements in various array configurations, provided the error is effectively balanced out in pairs of identical elements. Loops have some attractive properties as array elements (e.g. see Karavassilis and Davies 1985; Hipp 1988), although it is more usual to use vertical monopoles in HF DF arrays.

As instrumental errors were reduced, the importance of errors arising from site imperfections and wave-interference effects became greater, and interest grew in systems working on a larger portion of the arriving wavefront. The combination of two or more Adcocks in a scheme of group working was tried in Britain (Crampton, 1947). The Wullenweber was introduced in Germany as a single-ring system with an inner-reflecting screen to improve the radiation pattern. After the war, the University of Illinois built a system of this type for research purposes (Hayden, 1961a). There are 120 elements in a circle of diameter 291 m, with a wire screen of height approximately 20 m. A research programme can be planned to match the limited frequency coverage of a single-ring system, whereas a fuller coverage of the HF band may be required in an operational system.

British research in the post-war years was increasingly devoted to the study of wavefront behaviour. The Wullenweber was not immediately adopted as a research tool, mainly because of its expense; CADF systems were tried, but were found to suffer from a capture effect which was a disadvantage when dealing with comparatively weak signals in the presence of stronger interfering signals. Research workers therefore chose to experiment with interferometer systems such as that used by Ross, Bramley and Ashwell (1951), which had the added advantage of providing some information about elevation angles.

In recent years, however, the Wullenweber has attracted increased attention because it seems to offer the closest approach to the ideal operational direction finder (for azimuth determinations only) that can be suggested at present. The expense of a circular reflecting screen can now be avoided by adopting instead a doublet of elements in place of each single element (Byatt, 1967)*. The cardioid pattern in azimuth of a doublet is very similar to that of an element in front of a reflecting screen.

A Wullenweber with the best modern instrumentation, installed on a good site and well maintained, should measure the azimuth of arrival of a single ray to an accuracy of 0·1° or better. This is an order of magnitude smaller than the typical deviations of the ray path (~1°) produced by the ionosphere. The emphasis in current DF research is therefore very much on the elucidation and correction of these propagation effects. By a closely related but slightly different argument, the ideal DF for use in such research ought to have an instrumental accuracy no worse than 0·1° if effects of the order of 1° are to be studied quantitatively. It should also be

* The doublet arrangement was described earlier by K. Fischer, in literature issued by Telefunken (1958).

capable of resolving modes that may be less than 1° apart in azimuth and of making at least approximate measurements of elevation angles, as an aid to mode identification. Mode identification is essential; there is little point in measuring a tilt if we do not know whether it is in the E-region or F-region of the ionosphere. It is likely to be used, in conjunction with auxiliary sounding equipment, to take a series of observations on a controlled propagation path.

It is easily shown that an aperture of nearly 100 km is required to provide 'instant' resolving power of 0·1° at the lower end of the HF band. By choosing to work in the middle or upper part of the band we can reduce the requirement to a few tens of kilometres, and by making use of time averaging in conditions of good signal/noise ratio we can cut the aperture requirement still further. Two research groups have constructed very wide aperture linear arrays for DF research, with the transmitter on or near the broadside direction. An array belonging to Stanford University, described by Sweeney (1970), consists of 256 vertical elements over an aperture of 2·5 km. Rice (1971) described an array constructed for the Communications Research Centre, Canada; the main arm is 1·2 km long and there is a shorter crossarm for elevation measurements.

It is virtually impossible to build a circular array large enough to provide 'instant' resolution of the order of 0·1° at the lower end of the HF band. Those research groups unable to erect a long linear array, or wishing to make measurements in any direction on a variety of transmitters, must therefore manage with much smaller apertures, not usually larger than about 300 m, i.e. the size of the Illinois system. A useful rule of thumb is that, compared with the Adcock, a system with an aperture of about one wavelength (1λ) produces a useful suppression of site errors, but that it is necessary to go to about 4–5λ before any worthwhile reduction of wave interference effects occurs. Physically this distinction arises because site re-radiators may produce unwanted rays well separated in azimuth from the wanted rays, and therefore outside the main beam of an array of quite modest aperture, whereas wave interference usually involves comparatively closely spaced modes.

The full benefits of time averaging to reduce the effects of wave interference are not available on any signals of brief duration. Modern techniques of spread-spectrum communications and frequency hopping also pose problems to a conventional DF system tuned to a fixed frequency, because of the limited or brief bursts of energy within the passband. Wide aperture systems are particularly valuable when there are restrictions of this nature on time averaging.

We shall be particularly concerned here with WADFs of the order of 300 m in diameter. A very important current aim is to improve the classical resolving power of the circular WADF substantially, by making optimum use of time averaging, by the exploitation of relative fading between modes, and by the development of techniques to separate modes in elevation. Attempts have been made to measure elevation angles with the Wullenweber; see, for example, Jones, Schlicht and Ernst (1966). An alternative is to use a separate vertical array for this purpose (Gething *et al.* 1969).

1.3 The impact of new techniques

The advent of multichannel receivers and the availability of digital computers have opened up new possibilities in the processing of signals received at the elements of a DF array. These may be thought of as probes in the wave field; simultaneous measurements of amplitude and phase at each probe provide a space sample of the field at one instant of time. One of the effects of relative fading between two or more incident modes is to produce movements of the wave-field pattern relative to the array. Thus some of the advantages of spatial exploration over an area larger than the array can be obtained by taking successive samples (sometimes called frames of data or 'snapshots') at suitable time intervals.

The general technique of measurement and calculation described above has come to be known as wavefront analysis (WFA). It provides a convenient tool for the analysis of multiray wave-fields; various signal-processing schemes can be tried out in the computer at leisure, using recorded data. WFA also provides physical insight into the behaviour of different types of DF in multiray wave-fields. The origins of this approach can be traced to a pioneering paper by Bain (1956).

The best tactics to employ in the resolution of multiray wave-fields are still imperfectly understood at present. It cannot be claimed that WFA provides a dramatic increase in classical resolving power in all circumstances. It does, however, have many attractions: it provides a rational approach to the exploitation of fading; it allows the power of digital signal processing methods giving 'superresolution' to be applied to the DF problem; and it provides the only sound method known to the author of measuring elevation angles of CW signals in multiray conditions. Although it may be many years before WFA techniques and superresolution are widely used in operational DF, it is difficult to see how further progress can be made in research activities without this attack on the problem of mode resolution.

A succinct statement of the need for a study of the performance of DF systems in multiray fields can be found in a paper by Hayden (1961b). He wrote:

> 'The development of the detailed configuration of any direction-finding system is based on some concept of the character of the incident signal field. This character is described in terms of a mathematical model, the parameters of which are the quantities to be evaluated by the system. To this point, the incident signal-field models which have been used as bases for direction-finding systems development have been rather simple, far too simple in fact to satisfactorily describe the actual signal fields. One of the frontiers in direction-finding system development lies, at the moment, in the area of improving the incident field model and of developing more sophisticated systems based on the improved signal-field models.'

Considerable progress has been made since those words were published in 1961, but it cannot be claimed that all the problems have been solved. It is hoped that the treatment given here will provide a useful summary of the present state of understanding and will help to expose areas in which further work is required.

1.4 The value of computer simulations

It is not, of course, necessary to build and test various types of direction-finder in order to study their capabilities; computer studies can be used for at least a preliminary assessment and should give a clearer understanding of the relative merits of different array configurations and signal processing schemes, in situations of increasing complexity. Various mock 'battles' can be staged between the wave-field and the DF system, in which the system is regarded as the winner if the ray parameters are correctly recovered within some suitable tolerance.

It is important that the model of the wave-field should not be too idealised. In a single-ray situation with plane surfaces of constant phase, any direction finder worthy of the name should give the correct answer; such problems provide no discrimination in comparisons of apertures or processing schemes. A problem so difficult that it defeats all reasonable DF systems is also of little value; such a problem might involve many closely-spaced rays, each surrounded by a 'cone' of energy, plus a relatively large amount of external noise. It is with problems of intermediate complexity that the value of a large array and sophisticated processing can be most convincingly demonstrated.

1.5 Other applications of DF techniques

Interest in the peformance of antenna arrays illuminated by two or more rays is not, of course, confined to those engaged in DF research. Directive-antenna systems are used in a wide variety of applications as a means of discriminating against unwanted rays. The commonly quoted 'worst sidelobe level' is a measure, when superposition applies, of the highest response to a second ray when the wanted ray arrives from the direction of maximum response. Shaw and Davies (1964) pointed out that comparisons of arrays using multiplicative processing with those using conventional 'additive' processing are misleading unless multiple-ray situations are considered. The ability to discriminate between two or more targets is a very desirable feature in radar and sonar systems. Cooper (1973) suggested that, when the weaker of two rays is the wanted signal, a form of automatic DF system can be used to improve the signal/noise ratio.

More generally, it seems probable that ray selection and rejection techniques could be used to obtain improvements in the quality of reception of signals affected by multipath distortion and interference. The aim is to place a null in the array-radiation pattern on each unwanted arrival. Lim (1977) showed how a null could be generated in the pattern of a circular array and steered electronically to any selected azimuth. There is a growth of interest in the use of adaptive arrays, in which the nulls are steered automatically on to the required directions; see, for example, papers by Gabriel (1976), L. J. Griffiths (1977), J. W. R. Griffiths (1983) and a book by Hudson (1981).

It is worth noting at this point that circular arrays can be used to provide fixed beams for reception, as well as DF facilities. Starbuck (1969) described a circular array of 24 guyed elements on a diameter of 153 m, similar to the outer ring of the array sketched in Fig. 1.2. An arrangement of multicouplers and beam-forming networks produced 24 fixed beams at azimuth intervals of 15°. Limited reception tests suggested that the performance, in the band 1·5–10·0 MHz, was as good as that of a set of rhombic antennas occupying a much larger area of ground.

In summary, the distinction between arrays for reception and DF is becoming more blurred. However, because directional measurements are not required in the former case, it is not necessary to know the exact angle to which a null has been steered and it is not necessary to know the exact radiation pattern of the array on a real site. Nevertheless, mathematical methods for the analysis of multicomponent wave-fields, reviewed by Baur (1976a, b), may have applications in the quest for improved reception. Wavefront analysis and superresolution, as discussed in Chapters 10–13, provide a new approach to a number of aspects of array design, particularly those involving the control of nulls (Gething, 1973; Gething and Haseler, 1974; Cawsey, 1974). The theory provides a simple mathematical basis for the real-time digital control of null-steering systems with minicomputers, in which sets of simultaneous linear equations are solved.

1.6 Additional position-fixing techniques

In time-difference direction finding (TDDF), the position of a target transmitter is determined from the times of arrival of some recognisable signal feature at three or more receiving sites, measured against a common time standard. The time difference for each pair of sites places the transmitter on a hyperboloid of revolution with the line joining the sites as axis. The intersection of the hyperboloid with the surface of the earth is a position line of approximately (but only approximately) hyperbolic shape (see Fig. 1.5). The intersection of the hyperboloid with a sphere representing the earth is referred to as a spherical hyperbola in the IEEE standard. The intersections of two or more position lines give possible positions for the transmitter.

For signals that have travelled via the ionosphere, a complication is that each time difference (expressed as a distance) is the difference of the effective oblique paths from the transmitter to the two receiving sites, and is not in general equal to the difference of ground distances. Because it is difficult to allow for this effect, termed 'ionospheric height error' in the IEEE standard, TDDF does not seem to have found much application in the HF band. The technique is used in some microwave systems, with a

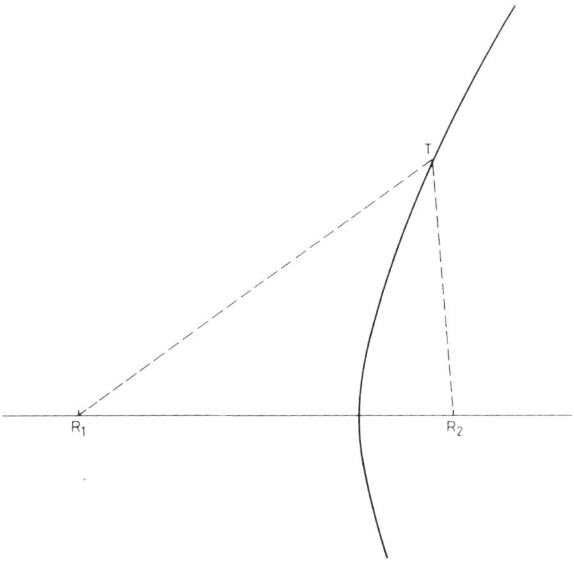

Fig. 1.5 Hyperbolic position line

Locus of possible transmitter positions T determined from time difference measured at two receiving sites R_1 and R_2

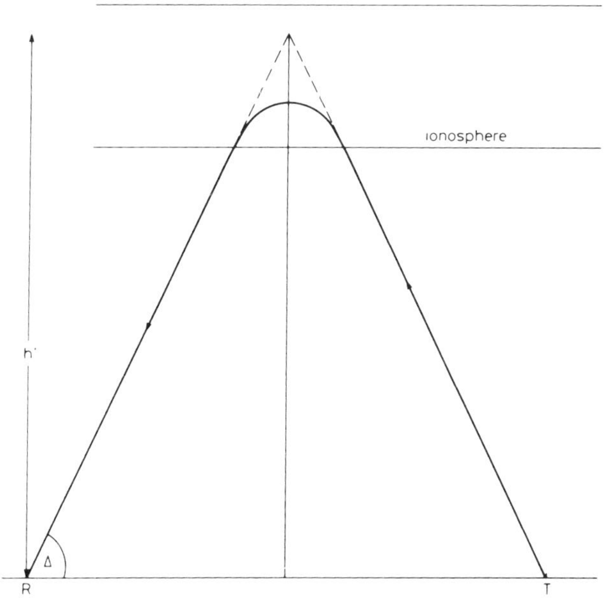

Fig. 1.6 Range by vertical triangulation
$RT = 2\ h' \cot \Delta$

typical separation between receiving sites of a few tens of kilometres. Cusdin and Mallinson (1988), however, used a baseline of a few metres in a system which might therefore be regarded as a single piece of equipment determining a form of bearing.

In a single-station location (SSL), both the bearing and estimated range of the transmitter are determined. The range is estimated from a triangulation in the vertical plane, using one or more measured elevation angles and the apparent height of reflection. Fig. 1.6. shows a very simplified version of the calculation. It is again necessary to have some knowledge of the properties of the ionosphere at one or more 'reflection' points, more properly regarded as refraction regions. The functions of bearing measurement and range estimation could, in principle, be carried out at two distinct sites, but the use of a single site reduces the amount of real estate required and avoids the need for intersite communications.

The terms TDDF and SSL have been defined so that occasional references can be made to these techniques in later chapters, but neither forms a major theme of the developments to be described.

References

ADOCK, F. (1919): 'Improvements in means for determining the direction of a distant source of electro-magnetic radiation', British patent 130490

ADCOCK, F. (1959): 'Radio direction finding in three dimensions', *Proc. Inst. Radio Eng. Australia*, **20,** pp. 7–11

BAIN, W.C. (1956): 'The theoretical design of direction-finding systems for high frequencies', *Proc. IEE.*, **103B,** pp. 113–119

BAUR, K. (1976a): 'Die Interferenzanalyse als Mittel der Peiltechnik (Teil I)', *Frequenz*, **30,** pp. 9–12

BAUR, K. (1976b): 'Die Interferenzanalyse als Mittel der Peiltechnik (Teil II)', *Frequenz*, **30,** pp. 26–31

BRITISH STANDARDS INSTITUTION (1971): 'Radio location and navigation terminology', BS 4727: Part 3: Group 06

BYATT, D.W.G. (1967): 'Wullenweber arrays using doublet aerials', *Marconi Rev.*, **30,** pp. 1–16

CAWSEY, D.C. (1974): 'Linear antenna arrays with broadened nulls', *Electron. Lett.* (GB)., **10,** pp. 291–292

COOPER, D.C. (1973): 'Zero-steering antenna system for receiving a signal close to the direction of strong interference', *Electron. Lett.* (GB)., **9,** pp. 140–141

CRAMPTON, C. (1947): 'Naval radio direction-finding', *J. IEE.*, **94,** part IIIA pp. 132–153

CUSDIN, A.R., and MALLINSON, P. (1988): 'Time of arrival direction finding for ESM', Proc. Conf. Military Microwaves '88, London 5–7 July, pp. 20–24 (Tunbridge Wells, Microwave Exhibitions and Publishers)

EARP, C.W., and GODFREY, R.M. (1947): 'Radio direction-finding by the cyclical differential measurement of phase', *J. IEE.*, **94,** part IIIA, pp. 705–721

GABRIEL, W.F. (1976): 'Adaptive arrays – an introduction', *Proc. IEEE.*, **64,** pp. 239–272

GETHING, P.J.D. (1973): 'Concept of phase centre of an array, applied to elevation–angle measurements', *Proc. IEE.*, **120,** pp. 541–543

GETHING, P.J.D., and HASELER, J.B. (1974): 'Linear antenna arrays with broadened nulls', *Proc. IEE.*, **121,** pp. 165–168

GETHING, P.J.D., MORRIS, J.G., SHEPHERD, E.G., and TIBBLE, D.V. (1969): 'Measurement of elevation angles of h.f. waves', *Proc. IEE.*, **116,** pp. 185–193

GRIFFITHS, J.W.R. (1983): 'Adaptive array processing – a tutorial', *IEE. Proc.*, Pt. F **130,** pp. 3–16

GRIFFITHS, L.J. (1977): 'A comparison of quadrature and single-channel receiver processing in adaptive beamforming', *IEEE Trans.*, **AP–25,** pp. 209–218

HAYDEN, E.C. (1961a): 'Propagation studies using direction-finding techniques', *J. Res. Natl. Bur. Stand.*, **65D,** pp. 197–212

HAYDEN, E.C. (1961b): 'Instrumentation for propagation and direction-finding measurements', *J. Res. Natl. Bur. Stand.*, **65D,** p. 253

HIPP, J.E. (1988): 'Experimental comparisons of sky wave DF algorithms using a small circular array of loop antennas', Proc. 4th int. conf. on HF radio systems and techniques, IEE. Conf. Publn. 284, pp. 215–220

HUDSON, J.E. (1981): 'Adaptive array principles' (Peter Peregrinus)

IEEE (1983): 'Standard definitions of navigation aid terms', IEEE. Standard 172 (New York, IEEE.)

JONES, E.C. Jr., SCHLICHT, H.C., and ERNST, E.W. (1966): 'Elevation angle of arrival determination with circular antenna arrays', Proc. Nat. Elec. Conf. Chicago, **22,** pp. 7–12

KARAVASSILIS, N., and DAVIES, D.E.N. (1985): 'An adaptive circular array for HF direction finding and null steering', ICAP 85, IEE Conf. Publn. **248,** pp. 295–299

LIM, J.C. (1977): 'Introduction of a sharp null response in an otherwise omnidirectional pattern using a circular array', *Radio & Electron. Eng.*, **47,** pp. 30–32

RICE, D.W. (1971): 'The CRC high frequency direction finding research facility', Conference Digest, 1971 International electrical and electronics conference, Toronto. (IEEE, New York)

RINDFLEISCH, H. (1956): 'The Wullenweber wide aperture direction finder', *Nachrichtenech. Z.*, **9,** pp. 119–123

ROSS, W., BRAMLEY, E.N., and ASHWELL, G.E. (1951). 'A phase-comparison method of measuring the direction of arrival of ionospheric radio waves', *Proc. IEE.*, **98,** pp. 294–302

SHAW, E. and DAVIES, D.E.N. (1964): 'Theoretical and experimental studies of the resolution performance of multiplicative and additive arrays', *Radio & Electron. Eng.*, **28,** pp. 279–291

SMITH-ROSE, R.L., and BARFIELD, R.H. (1926): 'The cause and elimination of night errors in radio direction-finding', *J. IEE.*, **64,** pp. 831–838

SMITH-ROSE, R.L., and ROSS, W. (1947): 'The use of earth-mats to reduce the polarization error of U-type Adcock direction finders', *J. IEE.*, **94,** part III, pp. 91–98

STARBUCK, J.T. (1969): 'A multiple-beam high-frequency receiving aerial system', *Radio & Electron. Eng.*, **37,** pp. 229–235

SWEENEY, L.F. Jr. (1970): 'Spatial properties of ionospheric radio propagation as determined with half degree azimuthal resolution', Thesis, Stanford University

WHALE, H.A. (1954): 'A rotating interferometer for the measurement of the direction of arrival of short radio waves', *Proc. Phys. Soc.*, **67B,** pp. 553–562

Chapter 2

Ionospheric modes

2.1 General description

Both single-hop and multiple-hop echoes from the ionosphere are frequently seen in records made with vertical-incidence sounders (ionosondes) radiating short pulses. It is convenient to regard the region of the ionosphere in which reflection occurs as a well-marked layer; that is to say, the D-, E-, and F-regions of ionospheric physics tend to be visualised as discrete layers in ray-tracing applications. The surface of the earth and the reflecting layer act very much like a pair of parallel mirrors in optics, and the separate returns correspond to multiple images of the transmitter.

Ionospheric modes may be described in the standard notation recommended by Davies (1967). The principal features are as follows:

(*a*) A symbol such as 2*F*2 means a two-hop path with reflections in the F2-layer and an intermediate ground reflection. 3*E* means a three-hop *E* path with two intermediate ground reflections.

(*b*) For propagation paths involving reflections by different layers, the reflections (or hops) should be specified in order of their position with respect to the transmitter. However, this order will not always be known.

(*c*) A dash is used to represent a ground reflection in mixed modes. For example, a one-hop *F*2 reflection followed by a one-hop reflection from sporadic *E*-layer ionisation (*E*s) is written *F*2-*E*s (sometimes called an *N*-type reflection). *F*2*E*s*F*2 must be interpreted as a reflection from the F2-layer followed by a reflection from the *top* of the Es-layer and another *F*2 hope (*M*-type reflection).

(*d*) When it is necessary to distinguish between the ordinary and extraordinary raypaths, an *o* or *x* may follow in parentheses. Thus *F*2 *MOF*(x) means the maximum observed frequency of the extraordinary wave which is reflected once at the F2-layer.

Variations of critical frequency, height of maximum ionisation and other ionospheric parameters with geographical position are sometimes referred to as ionospheric tilts or gradients; the former term will be used here for brevity, but it must be appreciated that the effective tilt of an ionospheric layer for an incident ray depends on the depth of penetration of the ray into the layer. For this reason, a tilted layer should *not* be visualised (except possibly in the case of a very thin and intense sporadic E-layer) as a tilted 'copper sheet'.

The time delays for vertical reflection from an effective height of (say) 150 km are 1, 2, 3 ... millisec for 1, 2, 3 ... hops. Provided the transmitted pulse is sufficiently short, these echoes are resolved in time. Magneto-ionic splitting into ordinary and extraordinary characteristic waves, a consequence of the existence of the earth's magnetic field, can also be seen in suitable circumstances, particularly near the critical (penetration) frequency for each layer.

Similarly, resolved echoes may also be received over oblique-incidence paths from a distance pulsed transmitter. Three new features now appear:

(*a*) E-layer and F-layer modes can both be open at the same time on the same frequency,
(*b*) Mixed modes involving reflections from two or more layers can occur,
(*c*) In a layer of finite thickness, a high-angle path is open, in addition to the low-angle path, provided the frequency exceeds the critical frequency.

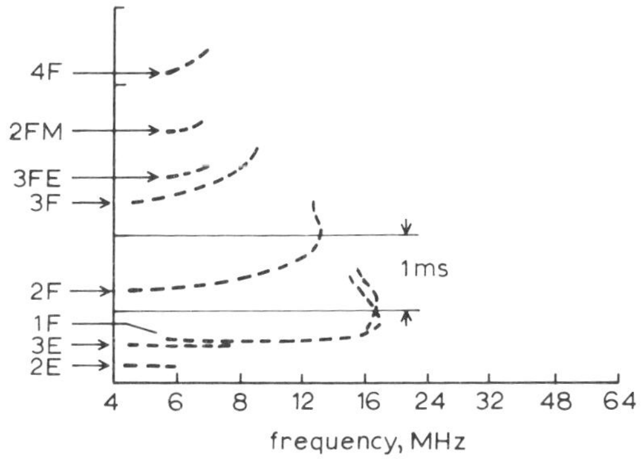

Fig. 2.1 Sketch of oblique ionogram

The standard oblique ionogram is a plot of time delay as ordinate against frequency as abscissa, for a fixed point-to-point link. The existence of two or more branches to the ionogram at a particular frequency corresponds to the reception of two or more resolved echoes (modes) at that frequency. Fig. 2.1, based on an ionogram in a paper by Lomax (1969), provides an example of the large number of modes that can sometimes be seen, particularly at the lower end of the frequency band. Below the critical frequency of a given layer there is, in principle, no limit to the number of possible multihop paths. Magneto-ionic splitting is often seen on high-angle rays, particularly for the F-layer. In practice, absorption places a practical limit on the number of modes exceeding the noise level, and on the detectability of very high-angle paths. Fig. 2.2 shows the sort of amplitude record that might be expected on a fixed frequency at which three modes exceed the noise level. A total time spread of about one or two milliseconds between the first and last detectable arrivals is fairly typical, although time delays of 5 ms or more can occur between the arrival of (say) a 1*E*-mode and a 3*F*2-mode.

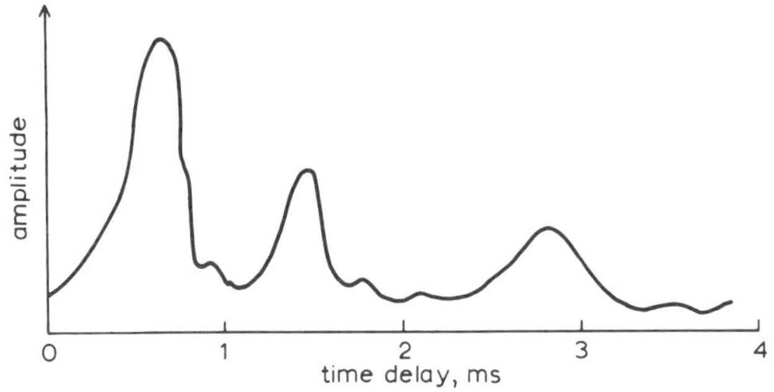

Fig. 2.2 Pulse amplitude against time (A-display) for fixed frequency

An oblique ionogram is an excellent guide to the selection of a frequency on which only one mode of significant strength is present; if such a frequency or range of frequencies exists, it is commonly found (as in Fig. 2.1) between about 60% and 80% of the maximum observed frequency (MOF). Lynch, Fenwick and Villard (1972) found minimum-time dispersion in this frequency range.

2.2. Mode identification

The labels attached to various branches of the oblique ionogram in Fig. 2.1 show the nature of the modes as inferred from the record and from ray-tracing studies, although labels such as 2FM and 3FE do not necessarily show the true order in which the hops occurred. In DF experiments, mode identification is certainly desirable. Complete identification requires the specification for each ray of the layer or layers involved, the number of hops, the magneto-ionic component and whether the path is low-angle or high-angle.

It is also desirable to know something about mode amplitudes. Suppose that a strong *E*-mode and a weak *F*-mode arrive at direction-finder *A*; the mean azimuth measured over a period will approximate to that of the *E*-mode. If a weak *E*-mode and a strong *F*-mode arrive at another direction-finder, *B*, the mean measured azimuth there will approximate to that of the *F*-mode. It would then be misleading to say that the mode structures at *A* and *B* are identical. An approximate measurement of average mode strength, not necessarily on an absolute scale, is therefore valuable.

Various crude sorting schemes can be used in the analysis of measurements, e.g.

(*a*) single mode conditions compared with multimode conditions,
(*b*) observations on one-hop modes compared with observations on multi-hop modes,
(*c*) observations on *E*- or *Es*-modes compared with observations on *F*-modes.

It must be emphasised that these classification schemes are *not* identical (*a* and *b* are often confused), nor are they based on full mode identification including amplitude information. However, classification schemes of this sort can be valuable when full mode identification is not possible.

Because of the difficulties of separating *o* and *x* components and because amplitudes change as a result of fading, full mode identification is rarely possible in practice. Nevertheless, a correct explanation of each discrete echo in terms of the reflecting layer or layers involved and the number of hops can be very valuable; the corresponding label for each echo will be referred to as the 'path geometry'. Deductions about path geometry may be difficult to make if the echo structure is known only on one frequency at one time, but become much easier if records from a swept-frequency (or stepped-frequency) oblique sounder are available, and if the diurnal changes of mode structure can be studied. Approximate elevation-angle measurements are also very helpful.

2.3 Fading

2.3.1. Amplitude distributions

An excellent survey of literature on amplitude fading is given in a CCIR report (1970). The main physical causes of fading are listed as

(*a*) variations of absorption,
(*b*) movements of ionospheric irregularities producing focusing and defocusing,
(*c*) changes of path length among component signals propagated via multiple paths or multiple modes,
(*d*) changes of polarisation involving multiple paths or multiple modes.

The fading range is defined as the difference (in dB) between the signal levels exceeded 10% and 90% of the time. Values of 13 ± 3.2 dB and 16.6 ± 3.2 dB are quoted for two long-distance HF paths.

Amplitude distributions obtained practically over short analysis intervals of a few minutes approximate fairly closely to the Rayleigh distribution or to the Nakagami–Rice (NR) distribution. A very good mathematical simulation of a Rayleigh distribution can be obtained by adding together *n* unit vectors with a rectangular probability distribution of phase from 0 to 2π (i.e. random phase); Norton *et al.* (1955) showed that any value of n greater than about four gives a satisfactory approximation. If a steady ('specular') component is now added to the vectors of random phase, the NR distribution is obtained. The probability $p(A)$ that the amplitude exceeds A is then given by

$$p(A) = \frac{2}{A_r^2} \int_A^\infty A \exp\{-(A^2 + A_s^2)/A_r^2\} I_0\left(\frac{2AA_s}{A_r^2}\right) dA$$

where

> A_r is the r.m.s. value of the random component (i.e. the vector sum of the random vectors),
> A_s is the amplitude of the steady component,
> I_o is the modified Bessel function of order zero.

For $A_s \ll A_r$, the NR distribution approaches the Rayleigh law

$$p(A) = \frac{2}{A_r^2} \int_A^\infty \exp(-A^2/A_r^2)\, dA$$

For $A_s \gg A_r$, it approaches a normal distribution with mean A_s and standard deviation 0.707 A_r.

The probability of occurrence of deep fades is of some interest, both in communications (where transmission errors are likely to occur) and in DF (where large errors in azimuth measurements are found to correspond to times of low amplitude). Let P be the probability of the signal amplitude lying below some small level L, $N(L)$ be the number of fades per unit time below L, and $\bar{t}(L)$ be the average duration of fades below L. Lin (1951) stated and proved three theorems for a wide variety of fading models:

(*a*) $P(A \leq L) \propto L^2$
(*b*) $N(L) \propto L$
(*c*) $\bar{t}(L) \propto L$

He also stated that most practical results seem to fit these power laws extremely well.

2.3.2 Fading rates

The rapidity of fading can be defined in various ways. If the autocorrelation function R of an amplitude record against time t is approximately represented by a Gaussian curve

$$R(t) = R(0) \exp(-t^2/2t_0^2)$$

the value of t_o giving the best fit to the observations is a useful measure of the fading period. Values in the range 0·5 to 2·5 s are quoted in the CCIR report for some HF paths, presumably in multimode conditions.

Experiments in which a true single mode consisting of one magneto-ionic component has been isolated for study are comparatively rare. Balser and Smith (1962) found some examples of very slow fading in such circumstances, with several minutes per fading cycle. More commonly, however, polarisation rotation when both *o* and *x* waves are present leads to fading rates for a 'single' mode of several seconds per cycle. When two or more such modes are present, the rate of change of relative phase is such that fading rates of several cycles per second are typical. The following rough guide is therefore suggested:

(*a*) Single *m-i* component: minutes per cycle,
(*b*) Single mode: seconds per cycle,
(*c*) Two or more modes: cycles per second.

Differences of Doppler shift between different modes provide some information about the rate of change of relative phase. For example, a difference of 1 Hz (i.e. 360° of phase per second) between two modes would produce one amplitude maximum and one minimum per second.

2.3.3 Diversity distances

Generally, a strong specular component and a slow fading rate can be visualised as corresponding to a nearly-plane wavefront. The diversity distance at which the correlation coefficient of the amplitude variations on two separated antennas drops to (say) 0·5, then tends to be comparatively large: Balser and Smith found distances of the order of 40λ for single-hop paths and about 10λ for multihop paths. Diversity distances may drop to a few wavelengths in multimode conditions.

The diversity distance and fading rate are approximately related in practice by a drift speed for ionospheric irregularities of the order of 50–100 metres per second. Bearing in mind that the amplitude pattern on the ground moves with twice the drift speed, we can write diversity distance (m) × fading rate (Hz) ≃ 2 × drift speed.

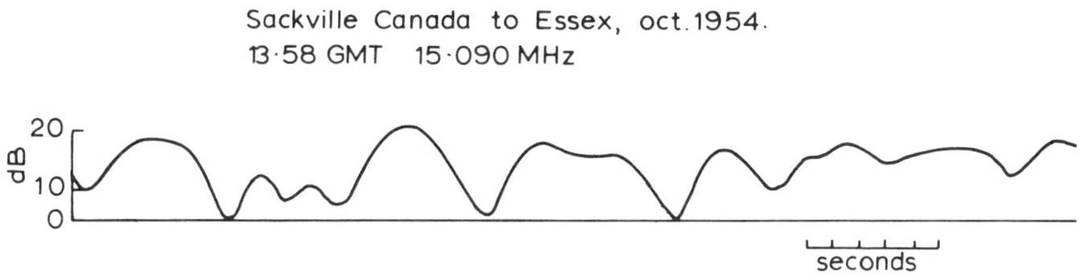

Fig. 2.3 Amplitude fading record

2.3.4 Fading models

It is comparatively easy to generate a simulated amplitude-fading record for a single antenna, with the aid of a computer incorporating a random-number generator. Fig. 2.3 shows an actual fading record, copied from a paper by Grisdale, Morris and Palmer (1957), while Fig 2.4 shows a simple computer approximation to an NR distribution, for comparison. It was obtained by adding together 16 unit vectors of variable phase and one specular vector of length four units and constant phase. The unit vectors were given random phases initially and these phases were then allowed to change randomly, but by limited amounts (maximum ± 50°) from point to point. The time scale of this graph is arbitrary and has been adjusted so that 100 points in Fig. 2.4 correspond to about 37 s in Fig. 2.3. Note that the vertical scales are not identical.

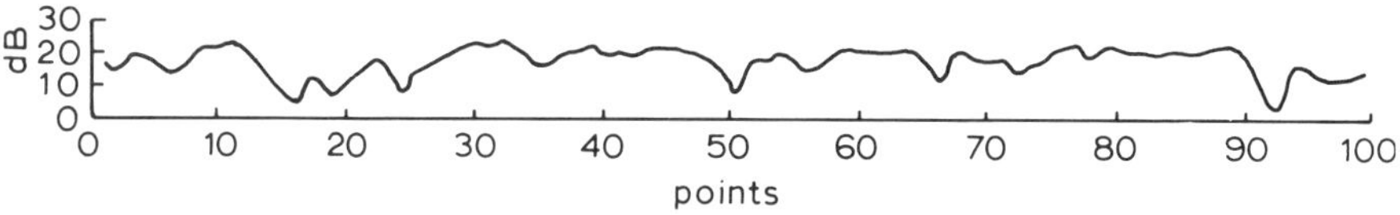

Fig. 2.4 Computer simulation of amplitude fading

The two patterns show fair agreement in general characteristics, although there are naturally differences of detail. Various other parameters were tried before a satisfactory model was obtained. These numerical experiments suggested that very shallow fading is obtained unless the specular component is made approximately equal to the square root of the number of unit vectors, i.e. about half the total power should be in the specular component. Very irregular fading is obtained unless some restriction is placed on the speed at which phase can change. However, if the phase changes are reduced too much, long 'drifts' of signal level can occur.

In the published literature, models and measurements are often compared by means of their cumulative (probability) distribution function (c.d.f). Two general points are worth making. Firstly, perfect agreement cannot be expected between c.d.f.s. calculated for finite lengths of record; all that can be expected is agreement to within some acceptable tolerance that has to be defined. Secondly, good agreement between c.d.f.s. is a necessary, but not a sufficient, test of whether a model is a satisfactory representation of a live signal. The c.d.f. is not affected by the time-scale, that is to say, either Fig. 2.3 or Fig 2.4 could be stretched or compressed in the x co-ordinate without affecting the c.d.f. Some test of fading rate, e.g. by means of an autocorrelation function, is therefore essential.

When a satisfactory model of single-mode fading has been obtained it is reasonable to ask whether the combination of two such models gives the more rapid fading characteristics of multimode conditions. For example, we could combine (say) 10 unit vectors to give a near-Rayleigh model A, then add another near-Rayleigh model B generated from (say) 12 unit vectors. However, the combination of these models A and B will simply give a third model C, which is a near-Rayleigh distribution generated from 22 unit vectors. There is no reason why the fading rate of model C should be any different from the fading rates of models A and B *unless* we make up additional rules about the phases of the unit vectors. As noted in Section

2.3.2, measured Doppler shifts (typically a few tenths of a Hz) give some information about the rate of change of phase difference between modes.

For the communicator interested in (say) error rates due to fades and the potential value of diversity, comparatively simple models of amplitude fading may be adequate. However, a much more sophisticated model would be desirable for a full study of the action of a large DF array, perhaps deployed in three dimensions. The ultimate aim in this case would be to model fluctuations of amplitude, phase, DOA, and polarisation in space as well as in time. Note that these parameters cannot be treated as independent variables: the partial correlations of (for example) amplitude fades with DOA fluctuations are vital features of the model.

The construction of such a model would be a major undertaking and it is by no means certain that available information about the behaviour of real wave-fields is sufficient for realistic models to be constructed. However, there is a growing interest in computer simulations of fading across an array due to travelling ionospheric disturbances and it is probable that more papers on this topic will be published. A few key references to existing literature, additional to those on coning noted in the next Section, are given here.

Amplitude patterns across the ground, i.e. in two dimensions, are discussed by Ames (1964). Information about polarisation variations can be found in papers by Hedlund and Edwards (1958), Porter (1966) and Epstein (1969). Whale and Delves (1958) defined a correlation ratio connecting the changes in amplitude and the changes in bearing of a fading radiowave made up of a number of randomly phased components. In a later paper (Delves and Whale (1962)), they discussed the joint distribution of the bearing, elevation and amplitude of a fading radiowave for several situations. This paper, and one by Pitteway, Wright and Fedor (1971), who attempted a simulation of fading records for spaced-antenna drift experiments, seem particularly relevant to the construction of fading models.

2.4 Specular and diffracted components

A single ionospheric mode is often pictured as a single ray specularly reflected by a 'smooth-copper-sheet' ionosphere, surrounded by a cone of rays produced by the roughness of the ionosphere; the former is referred to as the specular component and the latter as the scattered, diffracted, or noise component. The specular component is assumed to lie on the axis of the (symmetrical) cone.

The origins of this model can be traced to early work on drifting irregularities in the ionosphere and to studies of amplitude fading at a single antenna element. In a classic paper, Booker, Ratcliffe and Shinn (1950) treated the ionosphere as a diffraction grating and showed that the angular-power spectrum of the scattered radiation should be proportional to the Fourier transform of the autocorrelation function of the complex amplitude of the emergent wave. Their one-dimensional analysis was extended to two dimensions by Briggs and Phillips (1950). Clearly, the concept of specular and diffracted components in the angular-power spectrum links up neatly with the 'steady' and 'random' phasors of the Nakagami–Rice fading model.

Coning, as it is called, has important implications in the design of antenna rays. It places a practical limit on the useful aperture of large arrays; if the beamwidth of a receiving array is narrower than the angular width of the cone, the array rejects some of the incident energy. It affects DF accuracy. It broadens the width of narrow pulses because of the corresponding spread in path length. It is closely related to the diversity distance discussed in Section 2.3.3. For these reasons, many investigators have endeavoured to confirm the model and to determine three basic parameters. These are:

(*a*) The ratio B of power in the specular component to power in the diffracted component,
(*b*) The angular standard deviation in azimuth, σ_α, of the diffracted component,
(*c*) The angular standard deviation in elevation, σ_Δ, of the diffracted component.

B is often called the coherence ratio; the angular-power spectrum is usually assumed to have a Gaussian form.

There is an extensive literature on this subject; measurements have been made on short paths (one hop) and on long paths (multihop), at vertical or oblique incidence, and on various frequencies. Only the most important theoretical and practical results can be summarised here.

A few general points are worth making before these results are presented. Firstly, the values of B, σ_α and σ_Δ may vary very much from day to day and from path to path. The question: 'How rough is the ionosphere?' must be answered in statistical terms. Nevertheless, the best way of comparing results taken on different frequencies, or at different elevation angles, but at about the same time, on the same layer and in the same geographical area, is to interpret the results in terms of the physical size of typical 'irregularities' in the ionospheric layer concerned.

It is well known that an infinitely narrow antenna beam can be produced only by an infinitely large antenna. It may be helpful to picture a circular, horizontal disk of finite size in the ionosphere as the 'antenna' responsible for the cone. The size of the disk falls as ionospheric 'roughness' increases. This model correctly predicts the variation of σ_α with frequency and of σ_Δ with frequency and elevation angle. Typically, the diameter of the disk for the F-region might be a few hundred metres, giving cone angles of a few degrees.

2.5 Experiments on fine structure

Experimental methods to investigate the fine structure of ionospheric modes can be classified under the following headings:

(*a*) Measurements of amplitude fluctuations at one antenna,

(*b*) Measurements of amplitude fluctuations at two antennas, e.g. of the amplitude difference,
(*c*) Measurements of the fluctuations in phase difference at two antennas.
(*d*) Measurements involving more than two antennas.

Several authors have pointed out that method *a* is poor; the shape of the NR distribution is not very sensitive to the value of *B*. Fig. 2.5, copied from a paper by Hughes and Morris (1963), illustrates this point.

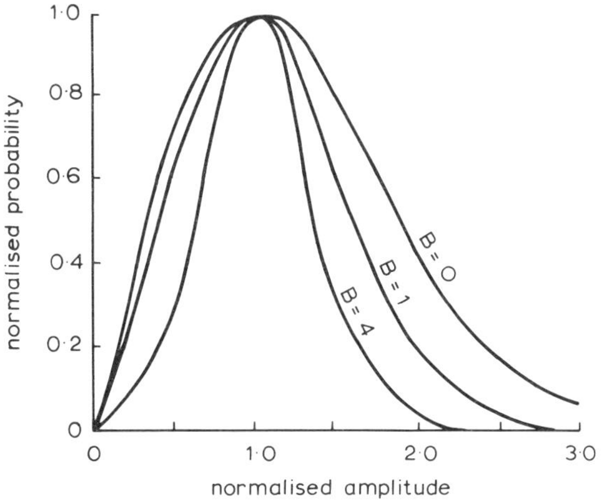

Fig. 2.5 Nakagami–Rice amplitude distribution

With two antennas, it is usual to arrange them broadside to the vertical plane containing the mode in order to determine σ_α and 'in line' with the mode when determining σ_Δ.

In any comparison of the results of various experiments described in the literature, it is helpful to apply the following tests to each paper:

(*a*) Has a single mode been isolated for study, e.g. by the use of short pulses?
(*b*) Have both the broadside and in-line antenna configurations been used?
(*c*) Have measurements been made at various antenna separations?
(*d*) Have both amplitude and phase measurements been collected, and a respectable sample of data analysed?
(*e*) What assumptions or approximations have been used in the analysis?

Item *d* is not intended to imply that it is essential to analyse both amplitude and phase measurements. Broadly, phase measurements seem to be the more useful, particularly if the specular component is rather small ($B < 1$); see Morris and Hughes (1959) and Hughes and Morris (1963). Since phase difference measurements are closely related to variations in apparent DOA, they are particularly relevant in DF theory and are therefore considered in more detail below.

2.6 Theory of phase measurements

2.6.1 General survey

Suppose we wish to calculate the azimuth angle of arrival α of a mode from phase difference measurements φ between two antennas spaced d metres apart in the horizontal plane. If α is measured from the broadside direction, and Δ is the elevation angle,

$$\varphi = \frac{2\pi d}{\lambda} \sin \alpha \cos \Delta$$

Thus fluctuations in φ from moment to moment produce corresponding fluctuations in α, with Δ assumed constant. However, these fluctuations cannot be directly equated with σ_α, which does *not* represent the variations in direction of the axis of the cone.

At very wide spacings of many λ, DF is only possible if a specular component is present; a diffracted component alone would give a rectangular distribution of phase difference from 0 to 2π with no recognisable mean value. As a specular component is added and becomes stronger, so a recognisable 'hump' should develop and become more marked. Roughly speaking, the specular component provides the bearing, while the diffracted component contributes noise. The ratio B can be estimated from the shape of the phase-difference distribution.

At much closer spacings, DF is possible even if there is no specular component. This is because the phases of the diffracted component at two antennas, although apparently random (rectangular distribution from 0 to 2π) are correlated for close ionospheric paths; thus the phase *difference* can show a recognisable hump. Several authors introduce a correlation function R *for the diffracted component alone*. For an exact definition the individual papers must be consulted, but R is a function of spacing, falling off from 1 towards 0 as spacing increases from zero. Suppose R falls to a value of $1/e$ in a distance of (say) 300 m. This distance, a, tells us something about the scale of irregularities in the ionosphere (or the size of the disk mentioned earlier).

The standard deviation (s.d.) of a phase-difference distribution at a spacing d is a function of B and of $R(d)$. It is small when B is large and/or when R is near 1. It is large when B is small and R is near zero.

The general procedure might therefore be summarised as follows:

(*a*) Determine phase-difference s.d. at very wide spacing and hence extimate B.

(*b*) Now determine s.d. at narrower spacing. With B known, R can now be estimated.

(*c*) From measurements of $R(d)$ at a few different spacings d estimate a.

(*d*) From a calculate σ_α and σ_Δ.

2.6.2 Full theory

The first authors to give reliable curves for the s.d. of the phase-difference distribution as a function of *R* and *B* seem to have been Whale and Gardiner (1966). Their Fig. 1, copied here in Fig. 2.6, differs in detail from approximate curves derived by Hughes and Morris (1963), which are therefore suspect.

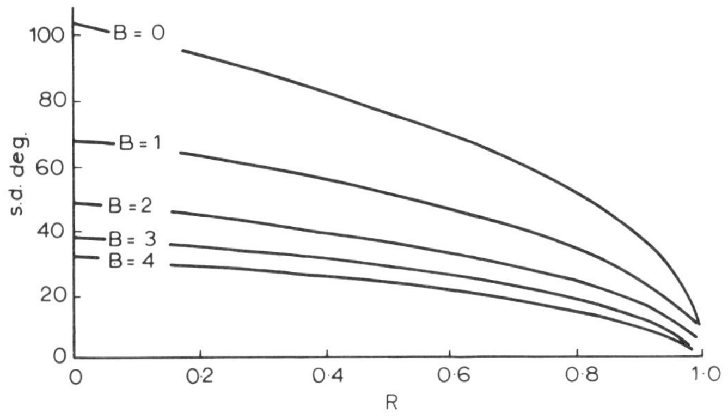

Fig. 2.6 Standard deviation of phase difference between two spaced antennas. Based on Fig. 1 in a paper by Whale and Gardiner, Radio Sci., **1**, pp. 557–570, May 1966, copyrighted by American Geophysical Union.

Whale and Gardiner started from the basic equations defining the NR distribution. They first obtained a joint probability distribution for A_1, A_2, φ_1, φ_2, the amplitudes and phases at two antennas (their eqn. 12). They then integrated over all values of the amplitudes, changed the variable to phase difference and obtained a probability distribution expressed as an infinite series of Laguerre polynomials (their eqn. 17). This series was presumably evaluated numerically for various values of *B* and *R* in order to plot Fig. 1. They also derived an interesting and simple result, that the probability *p* of the phase difference falling in the range $-\pi/2$ through 0 to $+\pi/2$ for a broadside configuration is

$$p = 1 - \left(\frac{1 - R}{2}\right) \exp\,(-B)$$

(their eqn. 21).

2.6.3 Approximations

Before the full theory was developed, various authors derived results for special cases, e.g. very large or very small specular component, very wide spacings etc. Some of these apply to amplitudes, some to phases.

Because most of these results seem to be consistent with Whale–Gardiner theory, they will not be reviewed in detail here.

2.7 Experimental results

Original papers must be consulted for details of paths, frequencies etc. B seems to be small or zero for multihop paths. Briggs and Phillips (1950) found B to be less than 1. Bramley (1951) quoted $b(=\sqrt{B}) = 2.5$ as fairly typical for the paths he measured, and an angular spread of about 1°. Hughes and Morris (1963) found values of $b(=\sqrt{B})$ ranging from 0.4 to 1.9 and standard deviations of several degrees: σ_α ranges from 1·5° to 4·5°, σ_Δ from 2·5° to 8°. However, their results should be treated with some caution, partly because their theoretical approximations are suspect (as noted above) and partly because the measurements were taken on CW signals; the evidence quoted in support of a dominant single mode hypothesis is not entirely convincing.

Some later investigations have cast doubt on the basic model of a specular component plus cone. Whale and Boys (1968) made phase measurements on pairs of antennas selected from a three-arm array and concluded from their analysis that

> 'In many cases, especially when the propagation distance was only a few thousand kilometres, two or three distinct specular components could be identified'.

Boys (1968) regarded the true coherence ratio as zero. Results obtained by D. W. Rice (1973) with a broadside array 1.18 km long showed nearly plane wavefronts for some modes; from this and other evidence the author concluded that 'the signal is usually resolvable into a relatively small number of discrete signals' and that 'any random component must be at least 23 dB smaller than the discrete components. Thus a ray interference model better represents the data than a diffracting-screen model'. Elevation angle measurements made by Clarke and Tibble (1978) under carefully-monitored conditions lend support to the hypothesis that each mode is composed of a small number of close components.

Using a V-shaped array with a greatest dimension of 1.5 km, Warrington, Thomas and Jones (1990) measured values of B for periods identified as 'approximately single moded propagation'. On the assumption of one specular component arriving from a constant direction plus a cone of random components, the calculated values of the coherence ratio were between 0 and 7; however, the same data interpreted in terms of a specular component with a variable direction of arrival plus a cone of random components yielded values of B greater than 10 and often greater than 40. The results obtained under all conditions suggested that the wavefield can be considered as the sum of a small number of components (corresponding to the different modes present in the signal) with large coherence ratios (i.e. essentially planar wavefronts) and directions of arrival that vary slowly by about 1–2° over several minutes.

Ionospheric reflections can occur under some conditions of very oblique incidence at frequencies above the HF band. Schoffel and Landstorfer (1987) included one direct component and at least six scattered components in their modelling of VHF/UHF wavefields, the scattered components including those produced by ionospheric reflections. Measurements of the statistical properties of the phase difference at two points λ/2 apart showed fair agreement with the results expected from the model.

2.8 Comment

Even allowing for the expected variability of the ionosphere from day to day, results obtained by various investigators do not seem to be entirely consistent. The hypothesis of a specular component of constant amplitude and constant DOA must clearly be an over-simplification in practice; it carries an implication of a smooth ionosphere of infinite extent. While it is perhaps true that the ionosphere can be regarded as 'level' over large areas, with more local irregularities superimposed, the model requires justification.

There is scope for further thorough measurements, perhaps involving simultaneous amplitude and phase measurements at several antenna elements with a multichannel receiver.

2.9 Reflection of radiowaves at the ground

2.9.1. Complex reflection coefficient

The reflection coefficient of the ground is a function of the ground constants (conductivity and permittivity), and of the radio frequency, elevation angle and polarisation of the incident waves. The functions for horizontal and vertical polarisation, and curves showing the amplitude of the reflected ray and the phase change on reflection (which together define the complex reflection coefficient), are given for typical ground constants in many textbooks, e.g. Wait (1986).

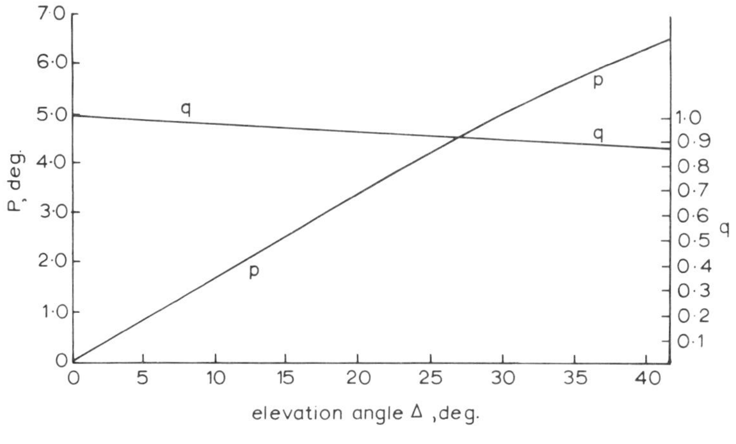

Fig. 2.7 Reflection parameters

For a ray arriving from a low angle the reflection coefficient for horizontally-polarised HF waves on a site of high conductivity is very close to -1, i.e. there is a phase change of π radians without change of amplitude. Let q be the amplitude of the reflected wave, where $q < 1$, and let the phase change on reflection be $(180 - p)$ degrees, where p is positive. Some typical values of p and q are shown in Table 2.1, and plotted in Fig. 2.7, for a site conductivity of 0.035 S/m and permittivity of 10, and for a frequency of 10 MHz.

Table 2.1 Reflection parameters

Δ deg	q	p deg
0	1.00	0.0
5	0.98	0.9
10	0.97	1.7
15	0.95	2.6
20	0.93	3.4
25	0.92	4.2
30	0.91	5.0
35	0.89	5.7
40	0.88	6.4

Examination of these and other results shows that, for ground of comparatively high conductivity, the phase change on reflection can be closely approximated up to quite large values of Δ by the expression $(\pi - \varphi \sin \Delta)$ radians, where φ is a function of frequency, and of the ground constants. This result can be interpreted as a penetration of the incident waves into the ground to a depth $\delta h = \lambda\varphi/4\pi$, where they encounter a perfect reflector. For ground of high conductivity, δh is typically a few tenths of a metre.

For vertically-polarised waves, the values of q and p vary in a more complicated way; q is substantially less than 1 for many frequencies and elevation angles.

2.9.2 Effects of ground reflection

The amplitude and the phase of the signal induced in an element such as a verticle monopole in an antenna array by one or more incident rays depend on the ground reflection coefficient, but provided the ground properties are uniform over the site this factor can be ignored in many types of calculation. For example, the phase difference between two identical elements of a horizontal array is not affected by the reflection coefficient.

However, if elements are arrayed in the vertical plane a knowledge of reflection coefficient will usually be required. For a reflection coefficient of -1, the signal induced in a small horizontal element at height h by a ray from elevation Δ is proportional to sin $(2\pi h \sin \Delta/\lambda)$. Thus the height-gain function plots as a sine wave; the signal is zero at height $h = 0$ and reaches its first maximum at a height given by $h = \lambda/(4 \sin \Delta)$. The signals induced in a vertical array of such elements by a single ray are all in phase, or antiphase for the negative portions of the height-gain sine wave; the changes of phase occur at heights of $\lambda/(2 \sin \Delta)$, $\lambda/(\sin \Delta)$, $3\lambda/(2 \sin \Delta)$ etc. as shown in Fig. 2.8.

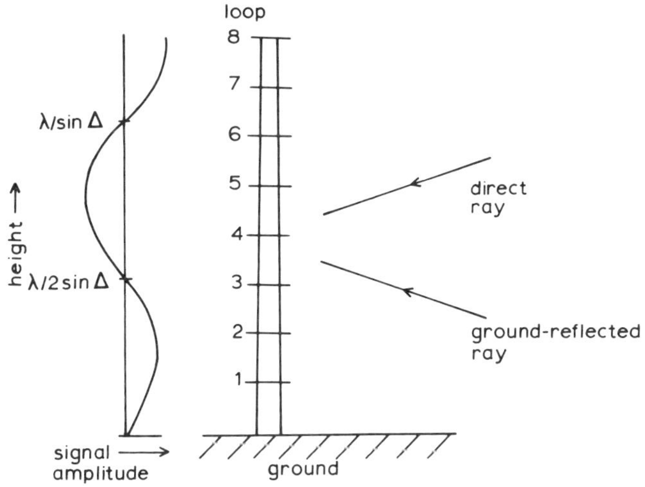

Fig. 2.8 Height-gain function for vertically-stacked array

More generally, in any horizontal plane $z = z_0$ there will be for each direct wave

$$A \exp \{j(\omega t + \varphi)\}$$

a reflected wave

$$A\Gamma(\Delta) \exp \{j(\omega t + \varphi - \theta)\}$$

where $\Gamma(\Delta)$ is the complex reflection coefficient at angle Δ and θ is a geometrical factor of $2\pi z_0 \sin \Delta/\lambda$. The combination of these two rays can be written

$$A\{1 + \Gamma(\Delta) \exp (-2j\pi z_0 \sin \Delta/\lambda)\} \exp \{j(\omega t + \varphi)\}$$

which is equivalent to a new incident ray of modified amplitude and phase. The new amplitude and phase are functions of z_0, i.e. the height above the ground plane.

We shall refer to these parameters as the effective amplitude and the effective phase. If models of wave-fields are constructed for specified ray parameters, but the effects of ground reflection are ignored (equivalent to an assumption of zero reflection coefficient), the ray parameters must be interpreted as the effective values when the direct and reflected rays have been combined.

References

AMES, J. (1964): 'Spatial properties of the amplitude fading of continuous HF radio waves', *Radio Sci.*, **68D,** pp. 1309–1318

BALSER, M., and SMITH, W.B. (1962): 'Some statistical properties of pulsed oblique HF ionospheric transmissions', *J. Res. Natl. Bur. Stand.*, **66D,** pp. 721–730

BOOKER, H.G., RATCLIFFE, J.A. and SHINN, D.H. (1950): 'Diffraction from an irregular screen with applications to ionospheric problems', *Philos. Trans.*, **A242,** pp. 579–609

BOYS, J.T. (1968): 'Statistical variations in the apparent specular component of ionospherically reflected radio waves', *Radio Sci.*, **3,** pp. 984–990

BRAMLEY, E.N. (1951): 'Diversity effects in spaced-aerial reception of ionospheric waves'. *Proc. IEE* Pt III, **98,** pp. 19–25

BRIGGS, B.H., and PHILLIPS, G.J. (1950): 'A study of the horizontal irregularities of the ionosphere', *Proc. Physc. Soc.*, **B63,** pp. 907–923

CCIR (1970): '12th Plenary Assembly, New Delhi,' Vol. II, part 2, pp. 197–213

CLARKE, R.H., and TIBBLE, D.V. (1978): 'Measurement of the elevation angles of arrival of multicomponent HF skywaves', *Proc. IEE.*, **125,** pp. 17–24

DAVIES, K. (1967): 'A nomenclature for oblique ionospheric soundings and ray tracing', *Radio Sci.*, **2,** pp. 1395–1396

DELVES, L.M., and WHALE, H.A. (1962): 'The short-term statistics of a fading radio wave', *J. Atmos. & Terr. Phys.*, **24,** pp. 225–236

EPSTEIN, M. (1969): 'Polarisation of ionospherically propagated HF radio waves with applications to radio communication', *Radio Sci.*, **4,** pp. 53–67

GRISDALE, G.L., MORRIS, J.G., and PALMER, D.S. (1957): 'Fading of long-distance radio signals and a comparison of space- and polarisation-diversity reception in the 6–18 Mc/s range', *Proc. IEE.*, **104B,** pp. 39–51

HEDLUND, D.A., and EDWARDS, L.C. (1958): 'Polarisation fading over an oblique incidence path', *IRE Trans.*, **AP–6,** pp. 21–25

HUGHES, C.J., and MORRIS, D.W. (1963): 'Phase characteristics of HF radio waves received after propagation by the ionosphere', *Proc. IEE.*, **110,** pp. 1720–1734

LIN, S.H. (1951): 'Statistical behaviour of a fading signal', *Bell Syst. Tech. J.*, **50,** pp. 3211–3270

LOMAX, J.B. (1969): 'Automatic analysis of oblique-incidence sounder data', AGARD conference proceedings No. 13, (Technovision Services), pp. 503–524

LYNCH, J.T., FENWICK, R.B., and VILLARD, O.G. Jr. (1972): 'Measurement of the best time-delay resolution obtainable along east-west and north-south ionospheric paths', *Radio Sci.*, **7,** pp. 925–929

MORRIS, D.W., and HUGHES, C.J. (1959): 'Phase characteristics of radio signals received via the ionosphere', *Nature*, **183,** pp. 310–311

NORTON, K.A., VOGLER, L.E., MANSFIELD, W.V., and SHORT, P.J. (1955): 'The probability distribution of the amplitude of a constant vector plus a Rayleigh-distributed vector', *Proc. IRE.*, **43,** pp. 1353–1361

PITTEWAY, M.L.V., WRIGHT, J.W., and FEDOR, L.S. (1971): 'The interpretation of ionospheric radio drift measurements – III Validation of correlation analysis by computer simulation', *J. Atmos. & Terr. Phys.*, **33,** pp. 635–660

PORTER, G.C. (1966): 'Measurements of polarisation statistics of signals received over a short range HF path', *IEEE Trans.*, **COM–14,** pp. 484–494

RICE, D.W. (1973): 'Phase characteristics of ionospherically-propagated radio waves', *Nature Physical Science*, **244,** pp. 86–88

SCHOFFEL, P.H., and LANDSTORFER, F.M. (1987): 'A field model for the evaluation of statistical DF-errors in the UHF/VHF range', *ICAP 87, IEE. Conf. Publn.*, **274, 2,** pp. 23–26

WAIT, J.R. (1986): 'Introduction to antennas and propagation', (Peter Peregrinus)

WARRINGTON, E.M., THOMAS, E.C., and JONES, T.B. (1990): 'Measurements on the wavefronts of ionospherically propagated HF radio waves made with a large aperture antenna array', *Proc. IEE.*, **137H,** pp. 25–30

WHALE, H.A., and BOYS, J.T. (1968): 'Measurements of the coherence ratio of ionospherically propagated radio waves', *Radio Sci.*, **3,** pp. 977–984

WHALE, H.A., and DELVES, L.M. (1958): 'Some relations between the bearing and amplitude of a fading radio wave', *J. Atmos & Terr. Phys.*, **13,** pp. 72–85

WHALE, H.A., and GARDINER, C.W. (1966): 'The effect of a specular component on the correlation between signals received on spaced antennas', *Radio Sci.*, **1,** pp. 557–570

Chapter 3

Wave-field models

3.1 Essential features of wave-field models

Having examined some of the properties of real ionospheric modes and wave-fields, we can now consider the construction of simplified mathematical models. The major justification for using idealised models of a complex real situation is that the physical significance of various factors can be more readily understood. In this chapter we shall describe and illustrate the properties of models for use in computer simulations, as a preliminary to the discussion in later chapters of the circumstances in which resolution of the field into its component rays can be achieved. For the moment we are concerned only with the structure of wave-fields; no assumption is made about the instrument used to measure their properties.

The preparation of a mathematical model of a wave-field, to be used eventually in some comparison of the merits of various forms of direction finder, is analogous to the setting of an examination paper. A paper designed to produce a ranking order of candidates should be neither so easy that they all score full marks, nor so difficult that they all score no marks. The properties of one-ray wave-fields, with plane surfaces of constant phase, are so simple that any well designed direction finder should give a good determination of the ray direction, at least in the azimuth plane. Models involving a large number of rays, plus cones of diffracted energy, plus distributed noise, with time-varying amplitudes, phases and directions of arrival (DOA) are too complicated for our purpose and are not amenable to analysis. If the rays are comparatively closely spaced, resolution is not likely to be achieved with arrays of practicable size.

The essential properties of a useful model between these extremes of simplicity and complexity appear to be as follows. Firstly, at least two rays should be involved, in order to introduce various phenomena of wave interference. Secondly, relative fading in amplitude and phase (complex fading) should be assumed so that the standing-wave pattern, characteristic of wave-interference conditions, drifts relative to the DF array and changes in shape with the passage of time; the exact fading law is probably not important. We do not need to specify absolute amplitudes in units of field strength such as volts per metre (V/m); it is relative amplitudes that determine the pattern.

Such a model is sufficient for initial numerical experiments. When a method of analysis, or signal-processing scheme, has been found that enables a specified DF to resolve the rays, the method should be further tested against problems in which increasing amounts of random noise are added to the measurements made with the DF. In this way, the initial intrinsic assumption of an infinite signal/noise ratio is removed, and the unlimited resolution resulting from this assumption is avoided.

Provided these three features of multiple arrivals, complex fading and noise are included in the model, a much more realistic assessment of the performance of DF systems can be made than in the single-ray situation. New features can then be added to the model, one at a time, in order to study their separate effects by means of controlled numerical experiments. The main emphasis in this book is therefore on wave-fields produced by two or three rays. Coning, as discussed in the previous chapter, will be ignored, on the grounds that DFs with apertures not exceeding a few wavelengths are unlikely to be able to distinguish directly between the specular and diffracted components. It is possible to simulate a cone of diffracted power by means of a bundle of (say) 50 weak rays arriving from directions clustered round a central axis, in order to model the fine structure of a mode; Mathews and Sherrill (1969), for example, used this approach. A model based on a much smaller number of rays might be adequate for some purposes; if each mode is represented by (say) three very closely spaced specular rays, the phenomenon sometimes described as 'within mode wave interference' will certainly be generated as the rays fade relative to each other.

The aim in this chapter is to explore the basic properties of wave-fields produced by a small number of rays. A mathematical approach is used to derive the basic equations defining the fields and the interpretation of the equations is then illustrated by means of computer-produced plots. Some of the results to be presented can be found in published DF literature, but the author knows of no book or paper where the theory is developed systematically. The main emphasis will be on the properties of fields in the ground plane, $z = 0$.

3.2 Wavefronts and wavefront curvature

3.2.1 Phase path

The first step is to define in more detail what we mean by terms such as rays and wavefronts. The concept of a ray is widely used in both optics and radio propagation. A ray path can be briefly defined as a locus of energy flow from the transmitter to the receiver. A large transmitting array can launch energy in a narrow pencil beam, and a ray may be thought of as the limiting case for an infinitely large aperture. In free space the wave normal at any point on the ray path, i.e. the normal to the local surface of constant phase, coincides with the ray direction.

For any point on the ray path, suppose that the time of travel, t, of a surface of constant phase from the transmitter to that point is known. The phase path P is then defined by the equation

$$P = ct$$

where c is the velocity of light in free space. Phase path is therefore essentially a time interval, converted to distance units. A surface of constant phase is a surface on which P is constant. Since, to a first approximation, many types of DF are designed to determine the orientation of surfaces of constant phase, phase path is a key parameter in, for example, ray-tracing studies of the effects of travelling ionospheric disturbances.

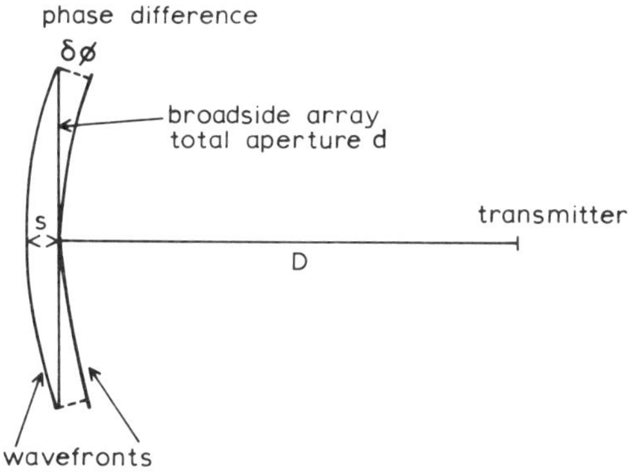

Fig. 3.1 Wavefront curvature

3.2.2. Wavefront curvature

In free space, the surfaces of constant phase are clearly spheres centred on the transmitter. How significant is wavefront curvature at a distant antenna array? Let us consider a broadside array of aperture d at a distance D from the transmitter, as shown in Fig. 3.1. The phase variation $\delta\varphi$ from the centre of the array to each wing is $2\pi s/\lambda$ radians where s is the sagittal distance given by $s = d^2/8D$. All lengths are measured in metres. For $D =$ 1000 km, i.e. 10^6 m and $d = 400$ m we obtain $s = 0{\cdot}02$ m and the phase variation is 0·24° at 10 MHz. Wave-front curvature is therefore entirely negligible in this situation. Reflection in a spherical ionospheric layer adds a slight complication, but the centre of curvature of the wavefronts reaching the DF must be at least as far away as the point of emergence of the ray from the ionosphere. A better approximation to allow for the effects of earth curvature and nonzero elevation angles is

$$s = d^2 \cos \Delta / 8R \tan \theta$$

where Δ is the elevation angle of the (single) mode, R is the radius of the earth and θ is the angle at the centre of the earth between transmitter and receiver. This is very similar to the previous equation if $R \tan \theta/\cos \Delta$ is interpreted as the oblique-path length.

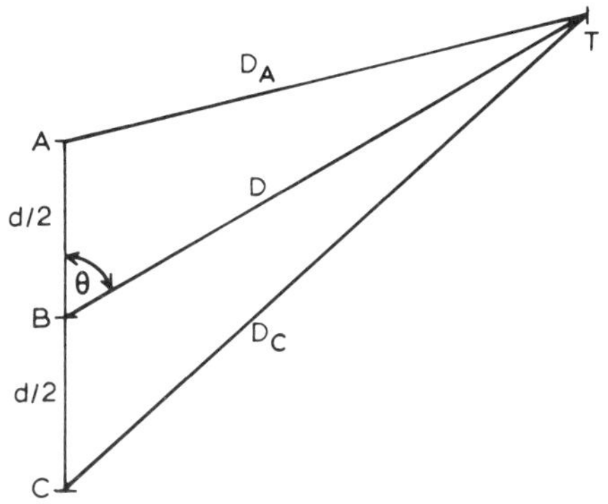

Fig. 3.2 Range determination from wavefront curvature

On low-angle HF oblique paths, therefore, wavefront curvature can be safely neglected for arrays with apertures of a few wavelengths. However, both $\delta\varphi$ and s vary as the square of the aperture and can become significant for a really large array, such as the 2·5 km broadside array at Stanford (Sweeney, 1970), particularly if the transmitter is only a few hundred kilometres away. For example, consider a transmitter at an oblique range of 500 km. The phase difference $\delta\varphi$ for the Stanford array is then 19° at 10 MHz. The transmitter is almost within the near field of the array determined from the well known criterion $2d^2/\lambda$, which corresponds to a phase difference between middle and outer elements of 22·5°.

Experimental measurements of such a phase difference could lead to an approximate estimate of the distance D to the transmitter. The following approach was suggested by Dr Ralph Benjamin. Consider three collinear receiving elements A, B, C, where the spacing $AB = BC = d/2$ (see Fig. 3.2). Let the image T of the transmitter in the ionosphere be at the range D from B and at D_A and D_c from A and C respectively. Let the angle TBA be θ, a function of both α and Δ. It can then be shown that

$$D = \frac{d^2 - (D_C - D_A)^2}{4(D_A + D_C - 2D)}$$

The right-hand side is a function of the known spacing d and of the *diffferences* between ranges, which can be derived from time-difference or phase-difference measurements. It is not necessary that α, Δ or θ should be steady or known. The accuracy of the measurement varies as the square of the 'projected aperture' perpendicular to the direction of arrival, i.e. as $d^2 \sin^2\theta$, and therefore falls off away from the broadside direction.

Webber (1979) used the 1·2 km array at the Communications Research Centre neat Ottawa in attempts to measure the ranges of transmitters at up to 111 km from the array. On ground waves the bias and r.m.s. deviation were low for the nearer transmitters, but were up to 33% and 45% of range respectively for the more distant sites. An unsuccessful attempt was made to compute the range of a signal that had been reflected by the E layer of the ionosphere; the wavefronts may have been distorted by the arrival of two or more modes.

The aperture of a vertical array is unlikely to exceed a few hundred metres at most and wavefront curvature on signals from distant transmitters can be safely neglected. The effect of wavefront curvature when the transmitter is much closer, as in an aircraft calibration of a system for vertical angle measurements, will be discussed in Chapter 6.

3.3 Lines of constant phase for one ray

Having established that, over the sort of aperture we shall normally be concerned with here, the surfaces of constant phase can be regarded as planes, we can now consider the lines of constant phase in the ground plane. These are straight lines, with a spacing of $\lambda/\cos\Delta$ between lines corresponding to $0, 2\pi, 4\pi \ldots$, or indeed to any phase φ, $(2\pi + \varphi)$, $(4\pi + \varphi)$, etc. The wave-field is one of constant amplitude, that is to say the amplitude A should be constant over the whole aperture. At any point, of course, the amplitude fluctuates at radio frequency as $A \sin(\omega t + \varphi)$, where ω is the RF in radians per second and φ is the phase at that point at the arbitrary time origin $t = 0$. The wave crests move across the plane with an apparent phase velocity of $c/\cos\Delta$; note that this velocity is greater than c. In many problems it is convenient to suppress the RF terms and represent the wave at any point by a phasor of length A and phase φ. One advantage of this representation is that a one-mode field is static, that is to say it does not vary with time. In practice the phase of the arriving wave at a reference point may drift slowly relative to the phase of a locally-generated RF wave (because of slight changes of phase path which produce slight frequency variations) but we are usually concerned with phase *differences* between different points in the wave-field and the phasor representation is then very convenient.

The phase difference for two identical vertical monopoles a distance d apart horizontally is $2\pi d \cos\alpha \cos\Delta/\lambda$ where α is now the azimuthal angle between the vertical plane of the ray arrival direction and the line joining the two elements. For directive antennas such as Yagis, the situation is more complicated: in order to calculate the correct phase difference we need to know the separation of the 'phase centres' of the antennas and these phase centres may vary in physical position with frequency *and* with DOA, as pointed out by Dyson (1967). However, for two identical antennas aligned to point in the same direction, the separation of the phase centres must be equal to the physical separation of identical elements.

3.4 Properties of two-ray wave-fields

3.4.1. Notation

Consider two rays represented by

$$A_1 \exp\{j(\omega t + U)\}$$

and

$$A_2 \exp\{j(\omega t + V)\}$$

where A_1 and A_2 are the amplitudes*, ω is the circular frequency in radians per second, j is the conventional symbol for $\sqrt{-1}$ and U and V are given by

$$U = k(x \sin \alpha_1 \cos \Delta_1 + y \cos \alpha_1 \cos \Delta_1 + z \sin \Delta_1)$$

$$V = k(x \sin \alpha_2 \cos \Delta_2 + y \cos \alpha_2 \cos \Delta_2 + z \sin \Delta_2)$$

where k is $2\pi/\lambda$ and λ is the wavelength. The origin of the (x, y, z) axes has been chosen, without loss of generality, such that the waves are in phase at this point. The angles α_1, α_2 represent azimuths relative to the y axis and Δ_1, Δ_2 are elevation angles relative to the horizontal plane. The sign conventions are such that x, y and z must decrease as time increases for a given wave-crest, i.e. the waves are moving from positive x, y and z towards the origin.

The phasor sum of the two waves at any point may be written

$$A_3 \exp\{j(\omega t + \psi)\}$$

where

$$A_3 \sin \psi = A_1 \sin U + A_2 \sin V$$

$$A_3 \cos \psi = A_1 \cos U + A_2 \cos V$$

Hence

$$A_3^2 = A_1^2 + A_2^2 + 2A_1A_2 \cos(U - V) \tag{1}$$

$$\tan \psi = \frac{A_1 \sin U + A_2 \sin V}{A_1 \cos U + A_2 \cos V} \tag{2}$$

The surfaces of constant amplitude are given by A_3 = constant and the surfaces of constant phase by ψ = constant. As expected, A_3 and ψ are functions of position (x, y, z) but not of time t. We therefore have a standing wave pattern and it is, of course, the properties of this pattern that we wish to describe. Complex fading produces changes in the standing wave pattern.

3.4.2 Surfaces of constant amplitude

From eqn. 1 above, the surfaces of constant amplitude at a given moment are surfaces on which the phase *difference* between the rays is constant:

$$U - V = \text{const}$$

i.e.

$$ax + by + cz = d$$

* If sensors are used to measure the properties of the wavefield, the amplitudes of the appropriate polarisation component should be used.

where
$$a = \sin\alpha_1 \cos\Delta_1 - \sin\alpha_2 \cos\Delta_2$$
$$b = \cos\alpha_1 \cos\Delta_1 - \cos\alpha_2 \cos\Delta_2$$
$$c = \sin\Delta_1 - \sin\Delta_2$$

The corresponding amplitude is

$$A_3 = \sqrt{(A_1^2 + A_2^2 + 2A_1A_2 \cos kd)}$$

The surfaces of constant amplitude are clearly parallel planes; they intersect the ground plane in parallel straight lines given by

$$ax + by = d$$

The common slope m of these lines of constant amplitude (CA) is

$$m = \frac{-a}{b}$$
$$= \frac{\sin\alpha_2 \cos\Delta_2 - \sin\alpha_1 \cos\Delta_1}{\cos\alpha_1 \cos\Delta_1 - \cos\alpha_2 \cos\Delta_2}$$

For the special case $\Delta_1 = \Delta_2$ we can write

$$m = \frac{\sin\alpha_2 - \sin\alpha_1}{\cos\alpha_1 - \cos\alpha_2}$$
$$= \frac{2\cos\left(\frac{\alpha_1+\alpha_2}{2}\right)\sin\left(\frac{\alpha_2-\alpha_1}{2}\right)}{-2\sin\left(\frac{\alpha_1+\alpha_2}{2}\right)\sin\left(\frac{\alpha_1-\alpha_2}{2}\right)}$$
$$= \cot\{(\alpha_1 + \alpha_2)/2\}$$

This slope is the tangent of an angle from the x-axis, whereas the azimuths are measured from the y-axis. It follows that the CA bisects the azimuth angle between the two arriving rays. For the special case $\alpha_1 = \alpha_2 = \alpha$ but $\Delta_1 \neq \Delta_2$ we obtain

$$m = -\tan\alpha$$

and the CAs are perpendicular to the arrival azimuth.

Ames (1964) emphasised that the slope can change considerably as a result of small changes in α_1 and α_2 for two close rays. The slopes shown in Table 3.1 were calculated for the case $\alpha_1 = 45°$, $\Delta_1 = 20°$, $\Delta_2 = 30°$.

Table 3.1 Slopes of lines of constant amplitude

α_2	Slope m	$\tan^{-1}(m)$
44°	−1·52	−56·6°
45°	−1·00	−45·0°
46°	−0·66	−33·4°

There is a considerable multiplication factor, in the sense that the CAs change in bearing by 23·2° as the bearing of the second mode changes by 2°.

An amplitude maximum occurs whenever

$$2\pi(ax + by + c)/\lambda = 0,\ 2\pi,\ 4\pi \quad \text{etc}$$

The spacing of the intersections with the y-axis is λ/b and the spacing on the x-axis is λ/a. These spacings are 19.5λ for the above example, with $\alpha_2 = 45°$. The perpendicular distance between successive lines is $19{\cdot}5\lambda/\sqrt{2}$, i.e. 13·6λ. For this special case of two rays on the same azimuth, the perpendicular spacing is $\lambda/(\cos\Delta_1 - \cos\Delta_2)$.

These results are correct for a ground reflection coefficient of zero. With a nonzero coefficient, there will be for each direct ray

$$A \exp\{j(\omega t + U)\}$$

a reflected ray

$$A\Gamma(\Delta) \exp\{j(\omega t + U')\}$$

where $\Gamma(\Delta)$ is the complex reflection coefficient at angle Δ and

$$U' = U - 2kz \sin\Delta$$

The combined effect must be

$$A\{1 + \Gamma(\Delta) \exp(-2jkz \sin\Delta)\} \exp\{j(\omega t + U)\}$$

Thus the amplitude and phase are altered, in any plane z = constant, by constant amounts. The lines of constant amplitude in the plane $z = z_0$ for two incident direct rays can now be written

$$az + by = d(z_0)$$

i.e. the 'constant' d now varies with z. The constants a and b are unaffected, because they depend only on the DOAs of the rays and not on their amplitudes or phases. We therefore have straight lines of constant amplitude in any horizontal plane, but the surfaces are not necessarily plane.

For a ground reflection coefficient of zero, and for the case $A_1 > A_2$, the amplitude A_3 fluctuates between limits of $(A_1 + A_2)$ and $(A_1 - A_2)$. The effective amplitude of each ray must be substituted into these expressions when the coefficient is not zero. However, neither the spacing nor the slope of the CAs is affected.

3.4.3 Surfaces of constant phase

From eqn. 2,

$$A_1 \sin(U - \psi) + A_2 \sin(V - \psi) = 0 \qquad (3)$$

Surfaces of constant phase, ψ = const., can therefore be plotted by calculating U for selected values of V, or vice versa. For any given horizontal plane, e.g. $z = 0$, x and y may be calculated from U and V. Note that pairs of values of U, V that satisfy eqn. 3 can be used to plot curves for various values of arrival angles.

Fig. 3.3 shows a plot prepared in this way for two rays arriving from zero-elevation angle, the stronger along the x-axis and the weaker from an azimuth of 80°. Amplitudes are in arbitrary units. The scale on the y-axis has been compressed relative to the x-axis in order to show more than one kink. The curve of constant phase (CP) passing through the origin never departs by more than about 0.2λ from the y-axis, and a line joining two widely-separated points on any CP is nearly parallel to the y-axis.

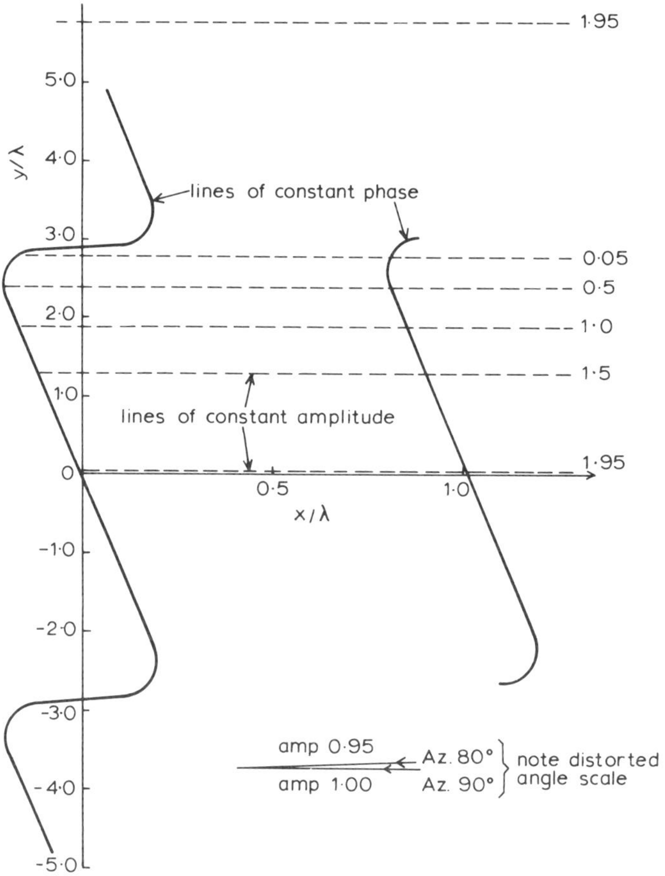

Fig. 3.3 Two-ray wave-field, horizontal plane

For two rays arriving from azimuth angles $+\theta$ and $-\theta$ relative to the x-axis and from zero-elevation angle it may be shown that the equation of a CP is

$$y = \frac{\lambda}{2\pi \sin\theta} \tan^{-1}\left\{\left(\frac{1+h}{1-h}\right) \tan\left(\frac{2\pi x \cos\theta}{\lambda} - \psi\right)\right\} - \frac{n\lambda}{2 \sin\theta}$$

where

$$n = \ldots -3, -2, -1, 0, +1, +2, \text{etc}$$

and

$$h = A_2/A_1 \text{ for } A_2 < A_1.$$

Hayden (1961) gave this expression for the particular case $\psi = 0$ and both he and Miwa (1969) show curves for various values of the amplitude ratio h. In the case $h = 1$ the CPs consist of straight lines with perpendicular kinks.

Two-ray CPs are clearly periodic in nature, the pattern repeating at intervals of $\lambda/(2 \sin \theta)$ along the y-axis in the above example. They are also of identical shape for any value of the phase ψ. Corresponding points on separate CPs are linked by the CAs.

3.5 Three or more rays

When three or more rays are present, some of the simple features of the two-ray wave-field are lost. Let us first consider whether the CP pattern in the ground plane is repetitive. Suppose that lines of zero phase for ray 1 are parallel to the y-axis, those for ray 2 cut the y-axis at intervals of (say) 4λ and those for ray 3 at intervals of (say) 5λ. Then, if all three rays have zero phase at the origin, they will again all have zero phase at $y = 20\lambda$, 40λ etc but not at intermediate points. Thus the wavelength of the 'beats' along the y-axis is 20λ, whereas it is only 4λ for rays 1 and 2 alone. The repetition distance becomes much greater if we replace 4λ by $4{\cdot}1\lambda$ and 5λ by $5{\cdot}1\lambda$. We can be sure that the pattern will repeat after $41 \times 51\lambda$, but perhaps not at any smaller distance. The repetition distance depends on the mathematical property of commensurability. The CAs are no longer straight lines in general. If A_1 is the strongest ray of three and A_3 is the weakest, the maximum amplitude is $(A_1 + A_2 + A_3)$. The minimum is $A_1 - (A_2 + A_3)$ if $A_1 > (A_2 + A_3)$, but zero if $A_1 \leqslant (A_2 + A_3)$. In the latter case, it will be shown by means of computer plots (see Section 3.8) that the phase exhibits sudden 'jumps' of π radians at various points in the wave-field, where the resultant amplitude passes through zero.

In a two-ray wave-field, we know that the loci of maximum amplitude, $(A_1 + A_2)$, in the ground plane are parallel straight lines. The phases of rays 1 and 2 are equal but not constant along these lines. Similarly, the lines of maximum amplitude for rays 2 and 3 alone would be another set of parallel straight lines. In general, these two sets of lines will intersect and produce a regular lattice of points at which the maximum amplitude of $(A_1 + A_2 + A_3)$ is achieved; the phase of the resultant is not the same at all these points. We might regard the points as the peaks of amplitude mountains. The loci of points for a slightly smaller amplitude, say 95% of the maximum, are then likely to be contours enclosing the peaks. If A_3 is very small compared with A_1 and A_2, these loci cannot differ greatly from the straight-line CAs obtained when A_3 is zero. When A_3 is small, therefore, we expect the contours of constant amplitude to delineate parallel mountain ridges with regularly-shaped peaks.

In special circumstances, the CAs for an n-ray wave-field can be a set of parallel straight lines. If a line can be found such that the angle between each ray and the line is the same, the rate of change of phase along this line will be the same for each ray and the line will be one of constant amplitude.

With four rays, there is no reason in general why the maximum possible amplitude ($A_1 + A_2 + A_3 + A_4$) should be achieved anywhere in the wave-field. Suppose we plot the lines of maximum amplitude for rays 1 and 2 alone and for rays 3 and 4 alone. At the intersections of these two sets of lines, there is no reason to suppose that the phases of the two resultants will be equal. Clearly, as the number of rays increases beyond four, so the probability of finding a point at which all the phases are equal becomes remote.

The above arguments apply to the spatial variation of phase and amplitude when the ray parameters do not vary with time. In a real situation, the ray phases at a particular point are likely to vary with time. Suppose that we use the strongest ray as our phase reference and that we have two other rays with steadily changing phases arising from Doppler shifts. If ray 2 changes in phase by 2π every t_2 seconds and ray 3 every t_3 seconds where t_2 and t_3 are integers and *if* the three rays are in phase coincidence at time $t = 0$, they will certainly be in phase coincidence again after $t_2 t_3$ seconds. The time interval could be shorter; for example, if $t_2 =$ 3 s and $t_3 = 6$ s, phase coincidence occurs every 6 s. But, in general, there is no reason to assume commensurability between t_2 and t_3.

Although it is difficult to find general theorems that are valid for three or more rays, there is in principle no difficulty in plotting the wave-field. Eqn. 3 now becomes

$$A_1 \sin (U - \psi) + A_2 \sin (V - \psi) + A_3 \sin (W - \psi) + \ldots = 0$$

For two rays, we are able to calculate x and y in the plane $z = 0$ by first finding V for specified values of A_1, A_2, U and ψ; as noted above, this procedure is attractive because the same pairs of (U, V) values can be used to construct curves for various DOAs. This short-cut is no longer available with three or more rays: it is necessary to start by specifying (x, y, z) and then to calculate ψ by simple phasor addition.

Bain (1956) showed by simple reasoning that the CPs for a multiray wave-field depart by less than $(\lambda/4) \sec \Delta_1$ from the corresponding lines for the strongest component wave, ray 1, provided that this wave is of greater amplitude than all the others combined in phase.

3.6 Wave-fields in the vertical plane

With a horizontal ground plane, the direct rays from elevations Δ_1, Δ_2 etc are associated with reflected rays from $-\Delta_1$, $-\Delta_2$ etc. Thus the basic unit is a pair of rays; we can speak of a one-pair field, a two-pair field, etc. With horizontal polarisation and perfect ground reflection (coefficient -1), the CAs are horizontal and the CPs vertical, with a jump of π in phase at the heights at which the height gain function is zero. These heights, as noted in

Section 2.9.2, are $n\lambda/(2 \sin \Delta)$, where $n = 0, 1, 2$ etc. The phase on each CP at $z = 0$ differs by $\pi/2$ radians from the phase of the direct ray at that point.

In a two-pair wave-field, we have a height gain function of wave-length $\lambda/\sin \Delta_1$ for pair one alone and of wavelength $\lambda/\sin \Delta_2$ for pair two alone. These two sine waves both have amplitude zero at $z = 0$, but there is no reason why they should again pass through zero simultaneously at some other height. The wave-field for two or more direct waves therefore appears to be nonrepetitive vertically within any reasonable distance, except in special cases of commensurability.

3.7 The significance of pattern repetition

For two rays from zero-elevation angle, it has been shown that the shape of a CP in the horizontal plane is a periodic waveform, with a wavelength of $\lambda/2 \sin \theta$ perpendicular to the mean-arrival azimuth. This wavelength is of the right order of magnitude for nonzero but low-elevation angles; the appropriate geometry is shown in Fig. 3.4.

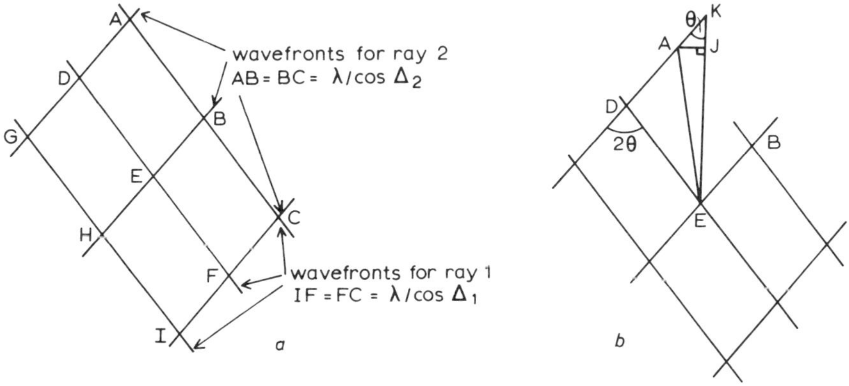

Fig. 3.4 Geometry relating to wavelength of curves of constant phase, two rays, horizontal plane

Produce DA to K, where $DK = DE = \lambda/\cos \Delta_2$ and $A\hat{K}E = \theta$

$$\tan A\hat{E}K = \frac{AJ}{EJ} = \frac{AK \sin \theta}{(DA + DE)\cos\theta} = \tan \theta \left(\frac{\cos \Delta_1 - \cos \Delta_2}{\cos \Delta_1 + \cos \Delta_2} \right)$$

$$AE^2 = AD^2 + DE^2 + 2AD.DE. \cos 2\theta$$

If the CP pattern repeats with a wavelength d, and if we can make amplitude and phase measurements at all points over this aperture, we can then infer the amplitude and phase values at all points on an infinitely extended aperture. Hence, by taking the Fourier transform of the distribution, we can determine the azimuths of arrival of the rays exactly (in the absence of noise). This argument does suggest that the repetitive pattern of the two-ray wave-field is an important property which sets it apart from the patterns for three or more rays; from arguments given above, it seems unlikely that the latter patterns repeat in distances at all comparable with the apertures of typical DFs.

3.8 Computer plots

3.8.1 Method of construction

Approximate plots of the curves of constant amplitude (CA) and constant phase (CP) in a given plane can be readily produced with a computer for various input ray parameters. The computer is programmed to perform phasor addition at a closely-spaced network of points over a specified region of the plane. Several such plots, for a variety of ray parameters, are shown in the next few pages and discussed very briefly. The meaning of various symbols will first be described, using Fig. 3.5 as an example; this is a two-ray pattern in the *xy* (horizontal) plane.

The point O in the bottom left-hand corner is the origin at which the phases of the incoming rays are defined. North is vertically up the sheet and East is to the right. A ray coming in from (say) an azimuth of 90°, measured clockwise from N, therefore moves along the *x*-axis from the right towards the origin. For each point defined by a row-column intersection on standard computer print-out, the resultant phasor was obtained by summing the separate ray contributions. The amplitude of the resultant is always taken to be positive, with phase anywhere in the range 0 to 2π. For Fig. 3.5, the ground-reflection coefficient was taken to be zero; as noted in Section 3.4.2, this assumption does not greatly affect the character of the wave-field.

Fig. 3.5 Approximate computer plot of wave-field, horizontal plane

ray	amp.	az.	elev.	phase
1	1·0	45·0	15·0	0·0
2	0·9	55·0	15·0	0·0

Fig. 3.6 Approximate computer plot of wave-field, horizontal plane

ray	amp.	az.	elev.	phase
1	1·0	45·0	15·0	0·0
2	0·9	55·0	50·0	0·0

The computer was programmed to print certain symbols in appropriate columns for each line of output, as follows:

(*a*) The square of dots in the bottom left-hand corner of Fig. 3.5 shows a 1λ by 1λ area, and serves to define the scale.

(*b*) A plus symbol, +, is printed whenever the resultant phase (measured from 0° to 360°) is closest to 0°. The computer tests for a local minimum compared with neighbouring values and also checks that the phase is within a suitable tolerance of 0°; the tolerance is adjusted according to the scale, which determines the rate of change of phase with position.

(*c*) Similarly, a minus symbol is printed whenever the phase is nearest to 180° and within the same tolerance.

(*d*) A symbol *S*, for strong, is printed whenever the resultant amplitude passes through a local maximum and is within a small (arbitrary) tolerance of the sum of the two-ray amplitudes; in Fig 3.5, *S* means 1·9 ± 0·1.

(*e*) A symbol *W*, for weak, is printed whenever the resultant amplitude passes through a local minimum and is within a small (arbitrary) tolerance of the difference of the two-ray amplitudes; in Fig. 3.5, *W* means 0·1 ± 0·1.

Clearly the plots are approximate because the positions of the +, −, *S* and *W* symbols may be in error by up to half a column (about 0·05 cm.). For example, the lattice point at which the phase is nearer to 0° than at either of

Fig. 3.7 Approximate computer plot of wave-field, horizontal plane

ray	amp.	az.	elev.	phase
1	1·0	45·0	29·6	0·0
2	0·9	55·0	15·0	0·0

the neighbouring points on the same line is not usually a point at which the phase is exactly 0°. In Fig. 3.5, it is not difficult to distinguish the real kinks in the wavefronts, near the *W* symbols, from more minor irregularities arising from the finite width of a column. The computer plots, even though approximate, provide a convenient way of examining many different situations. The reader is asked to visualise smooth curves, similar in character to those shown in Fig. 3.3, passing close to the printed symbols.

Fig. 3.8 Approximate computer plot of wave field, horizontal plane

ray	amp.	az.	elev.	phase
1	1·0	45·0	15·0	0·0
2	0·9	50·0	15·0	0·0

3.8.2 Patterns in the ground plane

Figs. 3.5 to 3.12 are various examples of patterns in the plane $z = 0$ for two or three incident rays; the ray parameters are listed on the diagrams. In general, the ray azimuths have been given values that differ by several degrees, in order to show more than one set of kinks within the area of the plots. In practice, the kinks for two modes arriving from azimuths that differ by (say) 1° would be much further apart. The effect of a change of relative phase in a two-ray model is simply to translate the curves relative to the axes, without change of shape or orientation; the phase of both rays at the origin can therefore be set arbitrarily to zero, without any loss of generality. In each diagram the x scale is equal to the y scale; there is therefore no angular distortion.

The effects of ground reflection are ignored in Figs. 3.5 to 3.12. Strictly, the resultant amplitude is everywhere zero in the plane $z = 0$ with perfect ground reflection, but we can interpret these patterns as correct for some constant height just above the ground plane, e.g. at the height of the phase centres of short vertical monopoles. The ray parameters must also be interpreted as those of the effective rays obtained by combining each direct ray with its mirror image (see Section 2.9).

Fig. 3.9 Approximate computer plot of wave-field, horizontal plane

ray	amp.	az.	elev.	phase
1	1·0	45·0	15·0	0·0
2	0·9	45·0	50·0	0·0

Brief comments on each diagram will now be given.

Fig. 3.5 shows two rays arriving from the same elevation angle, and with azimuths of 45° and 55°. The CAs, as expected, are straight lines along azimuth 50°, i.e. they make an angle of 40° with the *x*-axis. The CPs appear to become approximately parallel to the *x*-axis near the *W* symbols.

Fig. 3.6 shows the effect of a change in the elevation angle of the weaker ray, compared with Fig. 3.5. The orientation of the CAs is now quite different, although they are still straight lines, as expected. This is the only set of lines in the horizontal plane with which both rays make the same angle in the same sense. The kinks in the CPs are now less severe, because the amplitudes resolved into the horizontal plane (i.e. 1·0 cos 15° and 0·9 cos 50°) differ more than the corresponding values in Fig. 3.5.

Fig. 3.7 shows what happens when the resolved amplitudes approach equality.

Fig. 3.8 shows the greater spacing between CAs (and kinks), as compared with Fig. 3.5, when the azimuthal separation of the rays is reduced.

Fig. 3.10 Approximate computer plot of wave-field, horizontal plane

ray	amp.	az.	elev.	phase
1	2·0	45·0	0·0	217·6
2	1·9	35·0	0·0	62·1
3	1·0	55·0	10·0	195·0

Fig. 3.9 shows the case in which the two rays arrive from the same azimuth. The CAs and CPs now form a set of parallel straight lines. Just as amplitude maxima in fading records tend to be broader than the minima, so the spacing of CPs near the *S* lines appears to be wider than near the *W* lines.

Fig. 3.10 shows, on a smaller scale, the more complicated structure of a three-ray field; the phases selected are quite arbitrary. The minimum resultant amplitude is zero (because the strongest ray is not greater than the sum of the other two amplitudes) and a 'tearing' of the CPs can therefore occur, as near the isolated φ symbol in the upper middle of the diagram; this symbol denotes an amplitude of 0·0 ± 0·05. The maximum possible amplitude for the ray parameters adopted is clearly 4·9, but this value is only likely to occur at isolated points. In order to show the shape of CAs for other amplitude values, the symbol 3 is printed for amplitudes of 3·0 ± 0·05. Some of these CAs approximate to straight lines, but they are not strictly parallel.

Fig. 3.11 shows a rather similar situation, but with rays more closely spaced in azimuth. Here the symbol 4 is used for an amplitude of 4·0 ± 0·05 and the symbol φ for an amplitude of 0·0 ± 0·05.

Fig. 3.12 shows what happens in a three-ray field when one ray is dominant. The CPs show little distortion as compared with a one-ray set, but the elliptical shape of the CAs, shown for amplitudes of 3·0 ± 0·05 and

1·0 ± 0·05, provides clear evidence of the presence of more than two rays. Presumably local maxima and minima of amplitude occur within these respective contours.

Fig. 3.11 Approximate computer plot of wave-field, horizontal plane

ray	amp.	az.	elev.	phase
1	2·0	45·0	15·0	0·0
2	1·4	40·0	15·0	123·4
3	1·0	50·0	15·0	234·5

3.8.3 Patterns in the vertical plane

Figs. 3.13 to 3.17 show wave-field patterns in a vertical plane. The vertical (z) co-ordinate is now height and the horizontal (x) co-ordinate is distance in the plane of zero azimuth, i.e. the N–S plane if azimuth angles are measured from due N. The line of dots on the z-axis can be interpreted as a vertical tower, 1λ in height. Elevation angle is, of course, measured from the x-axis, positive in the anticlockwise direction. The x and z scales are again made equal, in order to avoid distortion of angles.

For each mode of ionospheric propagation arriving at elevation angle Δ, we construct the wave-field for a pair of rays from elevation angles of Δ and $-\Delta$, i.e. we assume that the ground acts as a horizontal plane reflector. Thus a one-mode wave-field can be regarded as the field generated by one pair of rays, a two-mode field by two pairs, etc. The ray parameters are listed on the diagrams. In all except Fig. 3.16 the modes are assumed to arrive in the same azimuthal plane, i.e. the plane of the paper. In all except Fig. 3.14, the ground reflection is assumed to be perfect, i.e. the reflection coefficient is -1.

Fig. 3.12 Approximate computer plot of wave-field, horizontal plane

ray	amp.	az.	elev.	phase
1	2·0	45·0	15·0	0·0
2	0·7	40·0	30·0	123·4
3	0·5	50·0	23·0	234·5

Fig. 3.13 Approximate computer plot of wave-field, vertical plane

ray	amp.	az.	elev.	phase
1	1·0	0·0	30·0	0·0
2	1·0	0·0	−30·0	180·0

Fig. 3.13 shows the simple one-pair field. The reflected ray is given the sample amplitude as the direct ray and a phase shift of 180° on reflection. The resultant amplitude at the ground is therefore everywhere zero, as shown by the bottom row of φs (over-written in places by other symbols); phase symbols on the bottom row in this and other diagrams should be ignored. The CPs are vertical lines of alternately positive and negative segments (the small steps in some CPs are not real features) with the first amplitude maximum at a height of λ/2, the next at 3λ/2, etc. The CAs would be further apart for a lower elevation angle, closer for a higher angle.

Fig. 3.14 shows the effects of an imperfect ground reflection. The reflected ray has been given an amplitude less than that of the direct ray, and the phase change on reflection has been set quite arbitrarily to 70°. The CAs, shown for amplitudes of 8·0 ± 0·05 and 2·0 ± 0·05 are still horizontal; this is to be expected, because the CAs must bisect the angle between the direct and reflected rays. The kinked CPs are exactly as expected in a two-ray wave-field; each CP oscillates about a straight line perpendicular to the stronger ray. If the amplitude of the reflected ray were gradually increased towards equality, the CPs would tend towards a 'staircase' of alternately vertical and horizontal segments until the form shown in Fig. 3.13 is reached. Because the pattern is repetitive, it can be shown on a large scale.

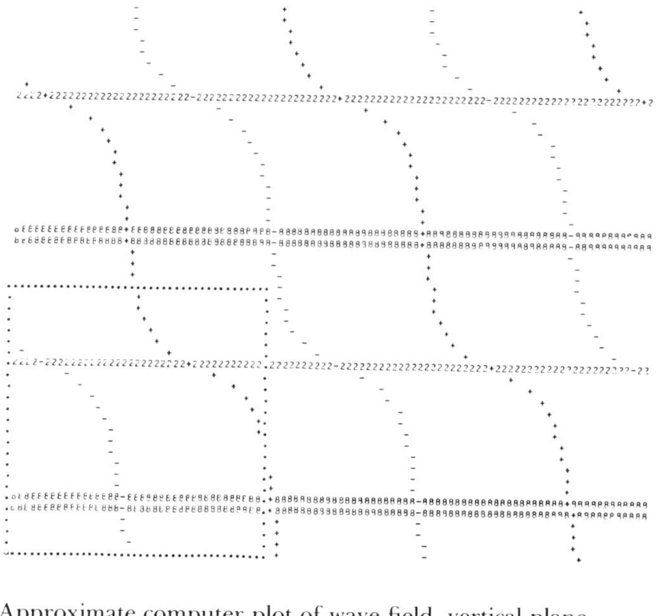

Fig. 3.14 Approximate computer plot of wave-field, vertical plane

ray	amp.	elev.	phase
1	5·0	30·0	0·0
2	3·0	−30·0	70·0

Fig. 3.15 shows a two-pair field with perfect reflection. The CAs, shown for 5·0 ± 0·05 and 1·0 ± 0·05, are no longer horizontal straight lines; they appear to be approximately elliptical. The two modes are in phase at the origin, and this condition may explain why the CPs become more nearly vertical near the left-hand margin of the diagram. The major kinks in the CPs clearly correspond to regions of low amplitude, as expected.

Fig. 3.16 is for the same ray parameters as Fig. 3.15 except that one pair arrives from an azimuth of −15° and the other from +15°, i.e. from opposite sides of the plane of the paper. There appears to be no major difference in character between the fields shown in Figs. 3.15 and 3.16.

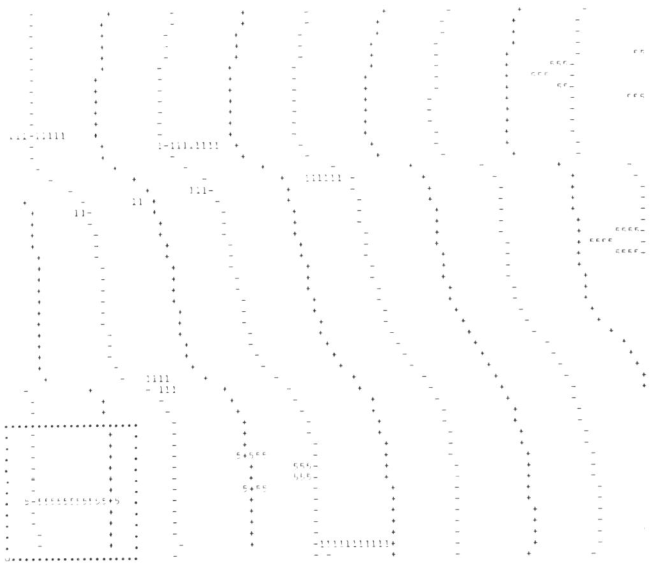

Fig. 3.15 Approximate computer plot of wave-field, vertical plane

ray	amp.	az.	elev.	phase
1	2·0	0·0	20·0	0·0
2	2·0	0·0	−20·0	180·0
3	1·0	0·0	30·0	0·0
4	1·0	0·0	−30·0	180·0

Fig. 3.17 shows a three-pair wave-field. Discontinuities in the CPs are now seen in regions where the resultant amplitude approaches zero. The amplitudes printed are 6·0 ± 0·2 and 1·0 ± 0·2. The occurrence in some CPs of near-vertical segments with lengths of the order of 1λ, particularly near the ground plane, is an interesting feature; note that the pattern *below* $z = 0$ can be pictured as a mirror reflection in $z = 0$ of the pattern shown when the ground reflection is perfect.

3.9 Summary

The results shown in the 13 examples presented in Section 3.8 seem to be entirely consistent with the theory developed in Sections 3.3 to 3.7. Although many other sets of ray parameters could be used to pursue the investigation further, it is hoped that these examples are sufficient to provide the reader with an appreciation of the principal features of wave-fields produced by a small number of rays. Ray separations in azimuth have been deliberately given comparatively large values in order to show the wave-field structure on a reasonable scale; it should be borne in mind that the kinks in a CP, such as those shown in Fig. 3.3, move further apart as the ray separation is reduced.

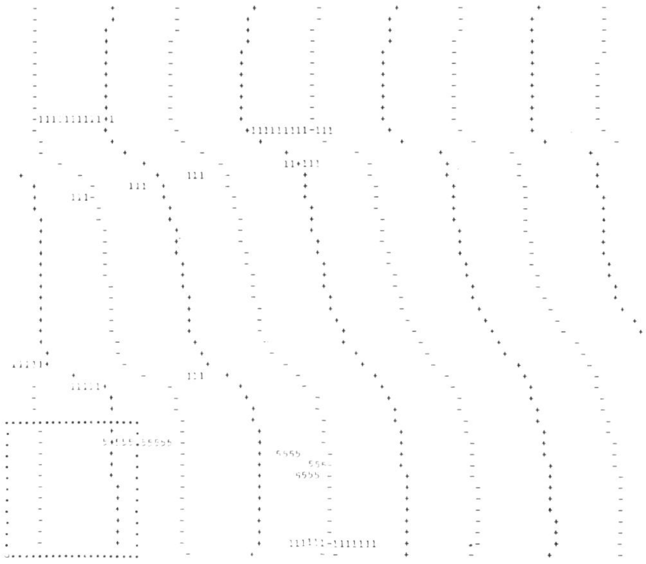

Fig. 3.16 Approximate computer plot of wave-field, vertical plane

ray	amp.	az.	elev.	phase
1	2.0	−15.0	20.0	0.0
2	2.0	−15.0	−20.0	180.0
3	1.0	15.0	30.0	0.0
4	1.0	15.0	−30.0	180.0

Although two-ray wave-fields are distinctly different from one-ray fields, and three-ray fields are different from two-ray fields, it seems unlikely that any major new phenomenon occurs when the number of rays exceeds three. This conclusion is confirmed to some extent by the nature of Figs. 3.15–3.17, which involve four and six rays.

It is important to note the restrictions that apply to each result derived above; for example, some theorems are true only if the number of rays does not exceed two, while others apply only if the rays arrive from the same elevation angle.

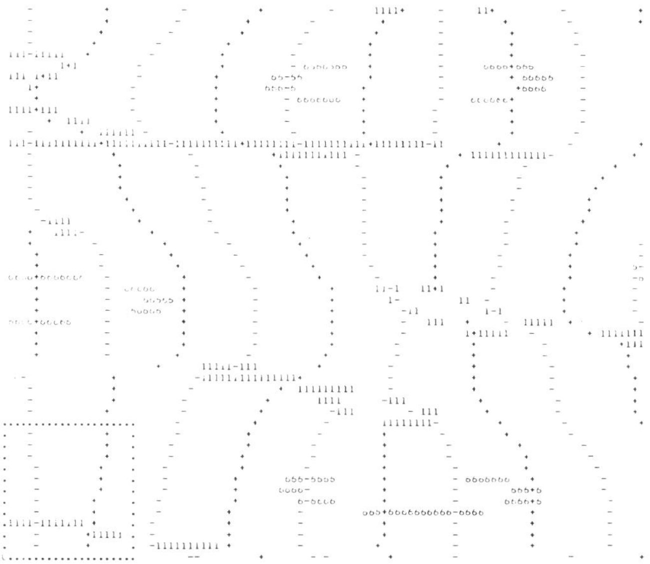

Fig. 3.17 Approximate computer plot of wave-field, vertical plane

ray	amp.	az.	elev.	phase
1	2.0	0.0	20.0	0.0
2	2.0	0.0	−20.0	180.0
3	1.0	0.0	30.0	0.0
4	1.0	0.0	−30.0	180.0
5	1.5	0.0	40.0	200.0
6	1.5	0.0	−40.0	20.0

References

AMES, J. (1964): 'Spatial properties of the amplitude fading of continuous HF radio waves', *Radio Sci.*, **68D,** pp. 1309–1318

BAIN, W.C. (1956): 'The theoretical design of direction-finding systems for high frequencies', *Proc. IEE.*, **103B,** pp. 113–119

DYSON, J.D. (1967): 'Determination of the phase centre and phase pattern of antennas', AGARD Conference Proceedings no. 15, Dusseldorf symposium July 1966, (Technovision Services), pp. 97–118

HAYDEN, E.C. (1961): 'Propagation studies using direction-finder techniques', *Radio Sci.* **65D,** pp. 197–212

MATHEWS, R.B., and SHERRILL, W.M. (1969): 'Ionospheric source simulation for the study of wave interference error in radio direction finders', Program and abstracts of URSI 1969 Spring meeting, Washington (URSI, Washington), p. 28

MIWA, S. (1969): 'Characteristics of wave-interference field and phase error', *Electron. Commun. Japan (USA)*, **52,** pp. 113–115

SWEENEY, L.E. Jr. (1970): 'Spatial properties of ionospheric radio propagation as determined with half degree azimuthal resolution', Thesis, Stanford University

WEBBER, R.V. (1979): 'Range estimation from wave-front curvature', Communications Research Centre, Ottawa, CRC Report 1322

Chapter 4

Zero-aperture bearings in two-ray wave-fields

4.1 Definition of zero-aperture bearing

We now regard a bearing as an azimuth angle measured by a direction finder at a particular moment. It is perhaps better to refer to this number as an indicated bearing, to distinguish it from the true great-circle bearing of the target transmitter. Even when the single-ray hypothesis underlying the design of most conventional direction finders is correct, the indicated bearing is not necessarily equal to the true bearing; ionospheric tilts and measurement errors can lead to small differences, typically a degree or so, between these numbers.

When two or more rays arrive, their relative fading produces fluctuations in the indicated bearing, as will be shown below. The direction finder attempts to interpret a distorted wavefront in terms of a single-ray model, i.e. a plane wavefront. The indicated bearing depends on the aperture of the DF array.

In this chapter we shall consider the 'bearing' defined by the normal to a curve of constant phase (CP) at a particular point in a two-ray wave-field. It must be emphasised that 'bearing' is thus being given a very special meaning; it is a geometrical property of the wave-field. In effect, we consider the bearing indicated by a phase-sensitive DF as its aperture shrinks to zero. Because this limiting case is an abstraction, we can continue to postpone any consideration of the behaviour of particular types of instrument; we are still dealing with the inherent properties of the field. Nevertheless, a number of general results can be derived that provide a useful insight into more practical situations.

It is convenient to measure azimuth angles relative to the x-axis, so that familiar mathematical results concerning the slopes of tangents and normals can be used without modification. The symbol θ will therefore be used for an azimuth angle measured anticlockwise from the x-axis.

We consider two waves:

$$A_1 \sin(\omega t + U)$$

and

$$A_2 \sin(\omega t + V)$$

where

$$U = \frac{2\pi}{\lambda}(x \cos\theta_1 \cos\Delta_1 + y \sin\theta_1 \cos\Delta_1 + z \sin\Delta_1) + \varphi_1$$

$$V = \frac{2\pi}{\lambda}(x \cos\theta_2 \cos\Delta_2 + y \sin\theta_2 \cos\Delta_2 + z \sin\Delta_2) + \varphi_2$$

and φ_1 and φ_2 are the ray phases at the origin.

4.2. Lines of constant phase

A surface of constant phase, $\tan\psi = t$, is given by

$$A_1 \sin U + A_2 \sin V = t(A_1 \cos U + A_2 \cos V) \tag{1}$$

To find the slope of the tangent to the curve defined by the intersection of this surface with the horizontal plane, we calculate (dy/dx) for $z = 0$, $t =$ constant. Differentiation of eqn. 1 gives

$$A_2 \cos U \frac{dU}{dx} + A_2 \cos V \frac{dV}{dx} = -t\left(A_1 \sin U \frac{dU}{dx} + A_2 \sin V \frac{dV}{dx}\right)$$

Substituting for t from eqn. 1, we obtain

$$A_1^2 \frac{dU}{dx} + A_1 A_2 \cos(U - V)\left(\frac{dU}{dx} + \frac{dV}{dx}\right) + A_2^2 \frac{dV}{dx} = 0$$

Also,

$$\frac{dU}{dx} = \frac{2\pi}{\lambda}\left(\cos\theta_1 + \frac{dy}{dx}\sin\theta_1\right)\cos\Delta_1$$

$$\frac{dV}{dx} = \frac{2\pi}{\lambda}\left(\cos\theta_2 + \frac{dy}{dx}\sin\theta_2\right)\cos\Delta_2$$

from which

$$\frac{dy}{dx} = -\frac{C}{D}$$

where

$$C = A_1^2 \cos\theta_1 \cos\Delta_1 + A_2^2 \cos\theta_2 \cos\Delta_2 + A_1A_2 \cos(U - V) \times \{\cos\theta_1 \cos\Delta_1 + \cos\theta_2 \cos\Delta_2\}$$

$$D = A_1^2 \sin\theta_1 \cos\Delta_1 + A_2^2 \sin\theta_2 \cos\Delta_2 + A_1A_2 \cos(U - V) \times \{\sin\theta_1 \cos\Delta_1 + \sin\theta_2 \cos\Delta_2\}$$

Using B to denote the bearing of the normal to the curve, we obtain

$$\tan B = \frac{D}{C}$$

$$= \frac{a + b \cos (U - V)}{c + d \cos (U - V)}$$

where

$$a = A_1^2 \sin \theta_1 \cos \Delta_1 + A_2^2 \sin \theta_2 \cos \Delta_2$$

$$b = A_1 A_2 (\sin \theta_1 \cos \Delta_1 + \sin \theta_2 \cos \Delta_2)$$

$$c = A_1^2 \cos \theta_1 \cos \Delta_1 + A_2^2 \cos \theta_2 \cos \Delta_2$$

$$d = A_1 A_2 (\cos \theta_1 \cos \Delta_1 + \cos \theta_2 \cos \Delta_2)$$

i.e.

$$\tan B = \frac{b}{d} + \frac{(ad - bc)}{d\{c + d \cos (U - V)\}} \qquad (2)$$

This is the basic equation from which the bearing can be calculated at any point in the *xy*-plane. Fig. 4.1 shows a plot of bearing B as a function of phase difference $(U - V)$ between the rays for the case $A_1 = 5{\cdot}0$, $\theta_1 = 44{\cdot}0°$, $\Delta_1 = 15{\cdot}0°$, $A_2 = 3{\cdot}0$, $\theta_2 = 46{\cdot}0°$, $\Delta_2 = 30{\cdot}0°$. Note that the indicated bearing can swing well outside the limit defined by the stronger ray, when the phase difference is near 180°. This condition corresponds to the kinks in the CPs.

4.3 Bearing swing

We now consider theoretically the fluctuation of the bearing as the phase difference between the two waves varies. We assume that $\cos (U - V)$ can

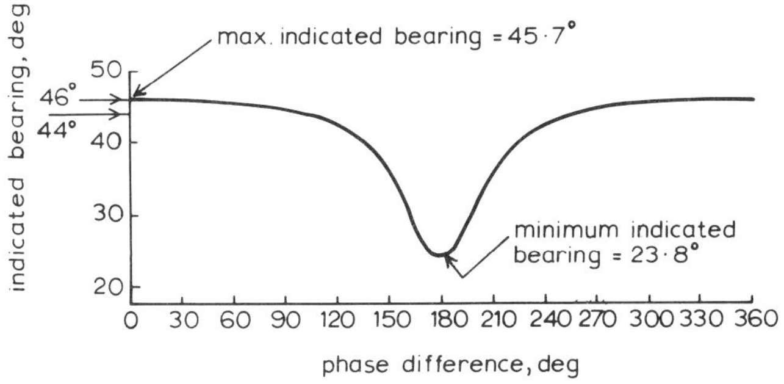

Fig. 4.1 Indicated bearing as a function of phase difference, two rays

fluctuate between -1 and $+1$ while A_1, A_2, θ_1, θ_2, Δ_1 and Δ_2 remain constant. We also assume for convenience that $\cos \theta_1$ and $\cos \theta_2$ are both positive. Then c and d are both positive and the limiting values of $\tan B$ for $\cos (U - V) = -1, +1$ are

$$(\tan B)_1 = \frac{b}{d} + \frac{(ad - bc)}{d(c - d)}$$

and

$$(\tan B)_2 = \frac{b}{d} + \frac{(ad - bc)}{d(c + d)}$$

If we define swing S as the difference between these values, we obtain

$$S = \frac{2(ad - bc)}{(c^2 - d^2)}$$

This expression simplifies slightly (without loss of generality) if we choose our axes such that $\theta_1 = -\theta_2$, i.e. so that the x-axis lies midway between the azimuths of the two waves. Writing θ for θ_1 and $-\theta$ for θ_2, we find

$$a = \sin\theta\,(A_1^2 \cos\Delta_1 - A_2^2 \cos\Delta_2)$$

$$b = A_1 A_2 \sin\theta\,(\cos\Delta_1 - \cos\Delta_2)$$

$$c = \cos\theta\,(A_1^2 \cos\Delta_1 + A_2^2 \cos\Delta_2)$$

$$d = A_1 A_2 \cos\theta\,(\cos\Delta_1 + \cos\Delta_2)$$

$$(ad - bc) = 2A_1A_2\,(A_1^2 - A_2^2) \sin\theta \cos\theta \cos\Delta_1 \cos\Delta_2$$

$$(c^2 - d^2) = (A_1^2 - A_2^2)\,(A_1^2 \cos^2\Delta_1 - A_2^2 \cos^2\Delta_2) \cos^2\theta$$

Then

$$S = \left(\frac{4A_1A_2 \cos\Delta_1 \cos\Delta_2}{A_1^2 \cos^2\Delta_1 - A_2^2 \cos^2\Delta_2}\right) \tan\theta \qquad (3)$$

As might be expected the swing is zero if either $A_1 = 0$ or $A_2 = 0$ (i.e. single wave conditions) or $\tan\theta = 0$ (two rays on same azimuth). For $\cos\Delta_1 = \cos\Delta_2$ and $A_1 \gg A_2$, the swing is approximately

$$S = 4\left(\frac{A_2}{A_1}\right)\tan\theta$$

For θ and B small, the swing in bearing is directly proportional to the azimuth separation, 2θ. Very large swings occur when the critical condition

$$A_1 \cos\Delta_1 = A_2 \cos\Delta_2$$

is approximately satisfied.

4.4 Mean bearing

We now consider the mean value $(\overline{\tan B})$ of tan B for a rectangular probability distribution of $(U - V)$, i.e. when all values of $(U - V)$ between 0 and 2π are equally likely.

From eqn. 2

$$(\overline{\tan B}) = \frac{b}{d} + \frac{(ab - bc)}{d} \cdot I$$

where

$$I = \frac{1}{2\pi} \int_0^{2\pi} \frac{dx}{c + d \cos x}$$

The value of the integral I is $\frac{1}{\sqrt{c^2 - d^2}}$ when $c^2 > d^2$, and is zero when $c^2 < d^2$. From the expression for $(c^2 - d^2)$ given in Section 4.3, it can be seen that $(c^2 - d^2)$ can only be negative if $(A_1^2 - A_2^2)$ and $(A_1^2 \cos^2 \Delta_1 - A_2^2 \cos^2 \Delta_2)$ are of opposite sign. In these conditions,

$$(\overline{\tan B}) = \tan \theta \left(\frac{\cos \Delta_1 - \cos \Delta_2}{\cos \Delta_1 + \cos \Delta_2} \right) \tag{4}$$

If $A_1^2 > A_2^2$ and $A_1^2 \cos^2 \Delta_1 > A_2^2 \cos^2 \Delta_2$ the appropriate expression is

$$(\overline{\tan B}) = \tan \theta \left\{ \left(\frac{\cos \Delta_1 - \cos \Delta_2}{\cos \Delta_1 + \cos \Delta_2} \right) + \frac{2 \cos \Delta_1 \cos \Delta_2}{(\cos \Delta_1 + \cos \Delta_2)} \sqrt{\left(\frac{A_1^2 - A_2^2}{A_1^2 \cos^2 \Delta_1 - A_2^2 \cos^2 \Delta_2} \right)} \right\} \tag{5}$$

When $\Delta_1 = \Delta_2$ this reduces to

$$(\overline{\tan B}) = \tan \theta \tag{6}$$

Eqn. 5 also holds when $A_2^2 > A_1^2$ and $A_2^2 \cos^2 \Delta_2 > A_1^2 \cos^2 \Delta_1$, provided that the negative square root is used (that this is the correct choice can be shown from eqn. 2). When $\Delta_1 = \Delta_2$ we obtain

$$(\overline{\tan B}) = -\tan \theta \tag{7}$$

The significance of eqns. 4, 6 and 7 can be seen by referring back to Fig. 3.4*a*. This shows lines of constant phase (say zero) for the two incident waves, intersecting at *A*, *B*, *C* etc. The curves of zero phase for the resultant must pass through these intersections, but we have to decide whether to link *A*, *B*, *C* etc. or *A*, *D*, *G* or *A*, *E*, *I*. We link *A*, *B*, *C* when $A_1 >> A_2$ and the mean bearing is close to tan θ. We link *A*, *D*, *G* when $A_2 >> A_1$ and the mean bearing is close to $-\tan\theta$. We link A, E, I in the intermediate condition when $A_1 > A_2$ and $A_2\cos\Delta_2 > A_1\cos\Delta_1$. The bearing of the straight line AE is given exactly by eqn. 4, as may be seen from Fig. 3.4*b*.

Eqn. 5 does not reduce to eqns. 6 and 7 when $\Delta_1 \neq \Delta_2$, but numerical calculations suggest that $(\overline{\tan B})$ is usually quite close to $\pm\tan\theta$. For example, with $A_1 = 2$, $A_2 = 1$, $\cos\Delta_1 = 0{\cdot}9$, $\cos\Delta_2 = 0{\cdot}7$, eqn. 5 gives

$$\overline{\tan B} = 0{\cdot}95\tan\theta$$

With $A_1 = 1$, $A_2 = 3$, $\cos\Delta_1 = 0{\cdot}9$, $\cos\Delta_2 = 0{\cdot}6$, eqn. 5 gives

$$\overline{\tan B} = -1{\cdot}11\tan\theta$$

4.5 Bearing as a function of phase

We know from eqn. 2 in Section 4.2 that the indicated bearing B is a function of the phase difference between the two rays at the DF. Suppose that this phase difference varies with time, but that all other ray parameters remain constant. The bearing will also vary with time, and the probability density function (p.d.f.) of B will depend on the p.d.f. of the phase difference.

It is important to distinguish between exploration of the wave-field in space and in time. A standing-wave pattern such as that shown in Fig. 3.3 can be regarded as a snapshot of the wave-field at a particular moment. Although the wavefronts (CPs) move with time, the position of the CAs and the kinks in the CPs do not change so long as the ray parameters remain constant. The bearing indicated by a static DF would then remain constant. The bearing variations indicated by a mobile DF depend on the path followed by the DF; if we move it along a CP, for example, we can be sure that we shall eventually sample all possible values of phase difference from 0 to 2π, whereas phase difference is constant along a CA.

In practice, we cannot 'freeze' the field and we expect the phase difference between two rays from the ionosphere to vary with time. We therefore assume that, over a reasonable sampling period, all values of phase between 0 and 2π radians are equally probable. It should be noted that this assumption could be unsafe for a comparatively short time interval, or for a static situation, e.g. with one ionospheric mode from a stable DOA and re-radiation of the ray from a fixed obstacle.

In the remainder of this chapter we shall deal with the rather restricted case of two rays arriving from the same elevation angle. Most of the theory to be presented here was first derived by Shirazi (1972).

Suppose that the two rays arrive from azimuths θ_1 and θ_2, that the mean of θ_1 and θ_2 is η and that the azimuthal separation is 2α. Thus

$$\theta_1 = \eta + \alpha$$

$$\theta_2 = \eta - \alpha$$

Let B be the indicated bearing from a DF of very small aperture. We can also define a normalised bearing

$$\xi = (B - \eta)/\alpha$$

On this scale the two rays arrive from azimuths of +1 and −1. The symbol R will be used for the ratio of ray amplitudes A_1/A_2 and δ for the phase difference between the rays. A rectangular probability function over the interval 0 to 2π radians will be adopted for δ.

Provided α is small, eqn. 2 in Section 4.2 can be written

$$B \simeq \eta + \frac{(R^2 - 1)\alpha}{R^2 + 1 + 2R\cos\delta} \tag{8}$$

This is an expression that occurs frequently in papers concerned with DF or with radar-glint theory. As R becomes very large, B approaches the bearing of the first ray, $(\eta + \alpha)$. If $R \ll 1$, B approaches the bearing of the second ray, $(\eta - \alpha)$. In both cases we obtain, as might be expected, the bearing of the dominant ray. It is convenient to assume from now on that R is greater than 1; this is simply an arbitrary convention and corresponding results apply when $R < 1$. Very large bearing errors are obtained when R is near 1 and cos δ is near −1; results of approximate theory should be distrusted in this situation.

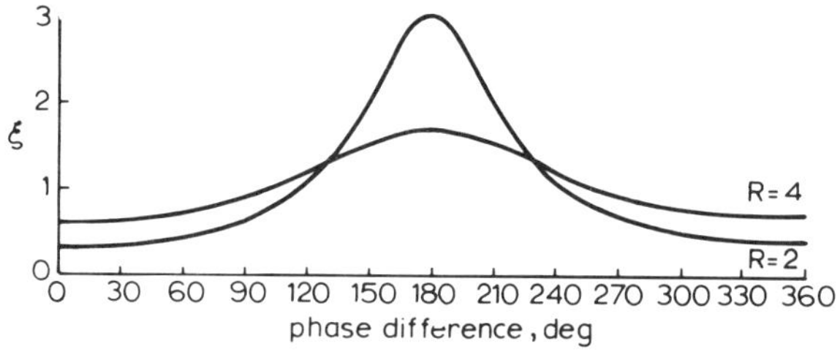

Fig. 4.2 Normalised bearing ξ as a function of phase difference, two rays of constant amplitude
stronger ray at $\xi = +1.0$
weaker ray at $\xi = -1.0$
R = amplitude ratio

Fig. 4.2 shows curves of normalised bearing ξ against δ for two values of R. The corresponding curves for $R = 0{\cdot}5$ and $R = 0{\cdot}25$ would be mirror reflections of the curves for $R = 2$ and $R = 4$ respectively in the line $\xi = 0$· These curves are very similar in character to Fig. 4.1 but the latter was calculated from an exact equation, and for two rays arriving from nonequal elevation angles. When R is large, ξ is always near +1. It is always positive for $R > 1$ and exceeds 1 for a range of δ values centred on 180°. The shape of these curves is, of course, directly related to the shape of the CPs; the large positive bearings correspond to the kinks in the CPs and to the '*W*' amplitudes in Figs. 3.5 to 3.9.

4.6 Statistical distributions with constant amplitudes

4.6.1. Probability density function

The p.d.f. of the indicated bearing for a constant value of R can be found from eqn. 8 as follows. For a rectangular-phase distribution, the probability that δ lies between δ and $(\delta + d\delta)$ is $d\delta/2\pi$. Hence the probability that ξ lies between ξ and $(\xi + d\xi)$ is

$$p(\xi) = \frac{1}{2\pi}\left|\frac{d\delta}{d\xi}\right|$$

$$= \frac{1}{2\pi}\frac{2R(R^2 - 1)\sin\delta}{(R^2 + 1 + 2R\cos\delta)^2}$$

From eqn. 8, we next obtain expressions for cos δ and sin δ to substitute in the above equation. After some manipulation the final result is

$$p(\xi) = \frac{1}{\pi|\xi|\{(\xi - \xi_1)(\xi_2 - \xi)\}^{1/2}} \tag{9}$$

where

$$\xi_1 = (R - 1)/(R + 1)$$
$$\xi_2 = (R + 1)/(R - 1)$$
$$\xi_1\xi_2 = 1$$

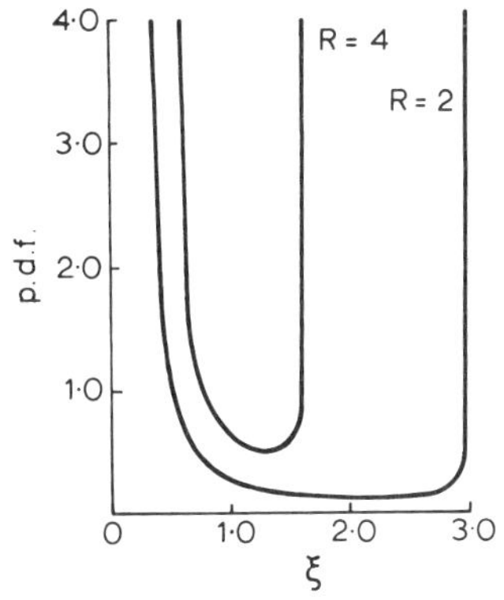

Fig. 4.3 P.d.f. of ξ, two rays of constant amplitude

Fig. 4.3 shows the U-shaped p.d.f.s for ξ calculated for the values $R = 2$ and $R = 4$.

From eqn. 9,

$$p(B) = \frac{\alpha}{\pi|B - \eta|\{(B - B_1)(B_2 - B)\}^{1/2}}$$

where

$$B_1 = \eta + \alpha(R - 1)/(R + 1)$$
$$B_2 = \eta + \alpha(R + 1)/(R - 1)$$

The mean value is

$$\bar{B} = \eta + \alpha \quad \text{when} \quad R > 1$$
$$= \eta - \alpha \quad \text{when} \quad R < 1$$

and the variance of the bearings about this mean is

$$\text{Var}\ (B) = 2\alpha^2 /(R^2 - 1)$$

However, the variance of a U-shaped distribution is not a particularly useful or meaningful parameter.

4.6.2 Cumulative distribution function

The c.d.f. for the normalised bearing ξ, i.e. the probability that ξ is less than or equal to a specified value t, can now be obtained as follows.

$$\text{c.d.f.}\ (\xi \leq t) = \int_{x=x_I}^{x=t} P(x)dx$$
$$= I\ (\text{say})$$

where

$$P(x) = \frac{1}{\pi|x|\{(x - x_1)(x_2 - x)\}^{1/2}}$$

$$x_1 = \xi_1$$

and

$$x_1 x_2 = 1$$

Put
$$x = \xi_1 \cos^2\theta + \xi_2 \sin^2\theta$$

with
$$\theta = 0 \text{ at } x = \xi_1$$
$$\theta = \frac{\pi}{2} \text{ at } x = \xi_2$$

Then
$$x - x_1 = (\xi_2 - \xi_1)\sin^2\theta$$
$$x_2 - x_1 = (\xi_2 - \xi_1)\cos^2\theta$$

and
$$I = \frac{4}{\pi}\int_{\theta=0}^{\theta} \frac{d\theta}{(\xi_1 + \xi_2) + (\xi_1 - \xi_2)\cos 2\theta}$$

which is a standard integral:

$$I = \frac{2}{\pi}\tan^{-1}\left(\frac{1}{\xi_1}\tan\theta\right)$$

$$= \frac{2}{\pi}\tan^{-1}\left\{\frac{1}{\xi_1}\sqrt{\frac{(x-\xi_1)}{(\xi_2-x)}}\right\}$$

When $x = \xi_2$, $I = 1$

We have thus checked that the total cumulative probability is 1, as expected.

The c.d.f. integral can therefore be written in the following form:

$$\text{c.d.f. } (\xi \leqslant t) = \frac{2}{\pi}(\tan^{-1}\gamma + \tan^{-1}\xi_1)$$

where

$$\gamma = D + \frac{ST}{D + \sqrt{D^2 - T^2}}$$

$$D = (\xi_2 - \xi_1)/2$$

$$S = (\xi_2 + \xi_1)/2$$

and

$$T = t - S$$

The shape of this distribution, for $R = 2$ and $R = 4$, is shown in Fig. 4.4. When R is large, the c.d.f. tends towards a nearly-vertical line near $t = 1$, i.e. near the normalised bearing of the stronger ray.

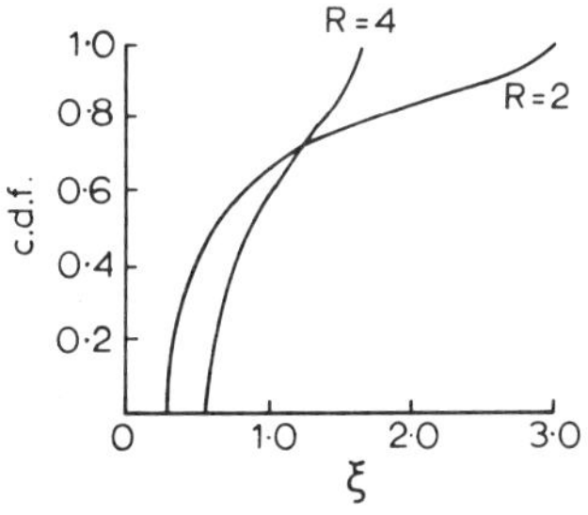

Fig. 4.4 C.d.f. of ξ, two rays of constant amplitude

4.6.3. Determination of the ray parameters

We note here the academic possibility of determining some of the ray parameters from an observed p.d.f. For example, if we know $\bar{B}$, B_1 and B_2 we can certainly calculate η, α and R. From the equations in Section 4.6.1 we obtain

$$\eta = (B_1 B_2 - \bar{B}^2)/B_3$$
$$\alpha = (B_2 - \bar{B})(\bar{B} - B_1)/B_3$$
$$R = (B_2 - B_1)/B_3$$

where

$$B_3 = B_1 + B_2 - 2\bar{B}$$

and is a measure of the asymmetry of the distribution. B_1 and B_2 are limits reached for particular phase values (0 and π radians respectively) and do not depend on the phase distribution provided these values are attained at some time in the sampling interval. The value of $\bar{B}$ does, however, depend on the assumption of a rectangular phase distribution.

If amplitude measurements can be made at closely-spaced points in the wave-field, η can also be determined from the orientation of the CAs, since these bisect the azimuth angle between the rays. This method requires a spatial exploration of the field and does not depend on the p.d.f. of δ.

However, these results all depend on rather artificial assumptions: that the two rays arrive from the same elevation angle and have constant amplitudes throughout the sampling period. In effect, amplitude fading is neglected over a sampling period sufficiently long to give a near-rectangular phase distribution. It would be more satisfactory to adopt (say) a Rayleigh fading law; we now proceed to a derivation of the corresponding bearing distribution.

4.7 Statistical distributions with Rayleigh fading

4.7.1 P.d.f. of the amplitude ratio

In order to represent two rays with Rayleigh-distributed amplitudes a, b, but with different r.m.s. values a_r, b_r, we write the amplitude p.d.f.s as

$$p_1(a) = \frac{2a}{a_r^2} \exp\,(-a^2/a_r^2)$$

$$p_2(b) = \frac{2b}{b_r^2} \exp\,(-b^2/b_r^2)$$

and put

$$R = b/a,$$

and

$$S = a$$

The joint probability distribution of R and S is

$$p_3(R, S) = p_1(a)p_2(b)/|J|$$

where the Jacobian J is

$$J = \begin{vmatrix} \dfrac{\partial R}{\partial a} & \dfrac{\partial S}{\partial a} \\ \dfrac{\partial R}{\partial b} & \dfrac{\partial S}{\partial b} \end{vmatrix} = -\frac{1}{a}$$

The required p.d.f. of R is obtained by integrating over all values of S, i.e

$$p_4(R) = \int_{S=0}^{S=\infty} \frac{4RS^3}{a_r^2\, b_r^2} \exp\left\{ \frac{-S^2}{a_r^2} - \frac{R^2 S^2}{b_r^2} \right\} dS$$

$$= \frac{2r^2 R}{(r^2 + R^2)^2}$$

where

$$r^2 = b_r^2/a_r^2$$

This is the p.d.f. of the amplitude ratio in terms of the ratio, r^2, of the mean powers in the two rays. This result is required in the next sub-section.

4.7.2 P.d.f. of the bearing

Eqn. 9 can be written

$$p(\xi|R) = \frac{1}{\pi|\xi|} \sqrt{\frac{R^2 - 1}{(\xi + 1)^2 - R^2\,(\xi - 1)^2}}$$

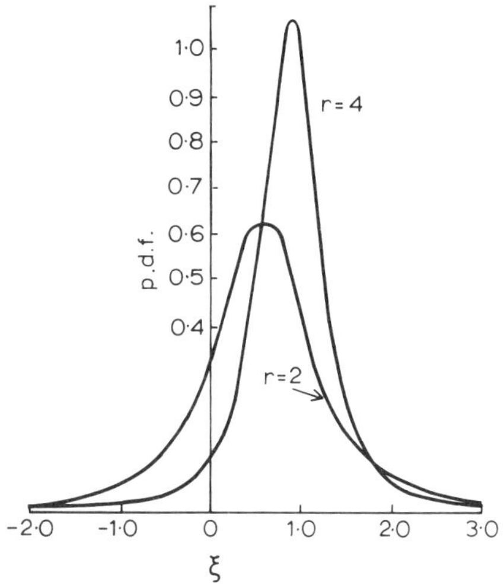

Fig. 4.5 P.d.f. of ξ, two rays with Rayleigh fading

The p.d.f. of ξ independent of R may therefore be written

$$p_5(\xi) = \int_{R=0}^{R=\infty} p(\xi|R)p_4(R)dR$$

$$= \frac{2r^2}{\pi|\xi|} \int_{R=0}^{R=\infty} \frac{RdR}{(r^2 + R^2)^2} \sqrt{\frac{R^2 - 1}{(\xi + 1)^2 - R^2 (\xi - 1)^2}}$$

Integration over valid areas of the R, ξ plane, i.e. areas where the expression under the square root sign is positive, yields eventually

$$p_5(\xi) = \frac{2r^2}{(r^2 + 1)^2} \left\{ \left(\xi - \frac{r^2 - 1}{r^2 + 1} \right)^2 + \left(\frac{2r}{r^2 + 1} \right)^2 \right\}^{-3/2} \tag{10}$$

for all values of ξ. Details of the derivation of this expression are given in Appendix 9.5.3. of Shirazi's thesis (1972). The shape of this distribution, for $r = 2$ and $r = 4$, is shown in Fig. 4.5

The p.d.f. is symmetrical about the value

$$\bar{\xi} = (r^2 - 1)/(r^2 + 1)$$

which is therefore the mean.

The maximum of the p.d.f. curve occurs at this point and is

$$p_5(\bar{\xi}) = (r^2 + 1)/4r$$

The p.d.f. of the indicated bearing B may therefore be written in terms of the mean bearing $\bar{B}$ and peak probability P as

$$p_6(B) = P\{1 + 4P^2(B - \bar{B})^2\}^{-3/2}$$

A simple check on the value of $\bar{\xi}$ can be obtained as follows. The probability that the amplitude b exceeds amplitude a, integrated over all possible values of a, is

$$p_7(b>a) = \int_{a=0}^{a=\infty} \exp\left(\frac{-a^2}{b_r^2}\right) \frac{2a}{a_r^2} \exp\left(\frac{-a^2}{a_r^2}\right) da$$

$$= \frac{b_r^2}{a_r^2 + b_r^2}$$

Similarly, amplitude a exceeds amplitude b during a fraction

$$a_r^2/(a_r^2 + b_r^2)$$

of the sampling period. If we sort the indicated bearings into two populations, according to whether $b > a$ (i.e. $R > 1$) or $a > b$ (i.e. $R < 1$), we expect a mean bearing of $+1$ for the first population and -1 for the second.

The meaning bearing is therefore

$$\bar{\xi} = \frac{b_r^2}{a_r^2 + b_r^2} - \frac{a_r^2}{a_r^2 + b_r^2}$$

$$= \frac{(r^2 - 1)}{(r^2 + 1)}$$

in agreement with the mean of the p.d.f. The p.d.f. appears to have no finite moments other than the first (the mean).

4.7.3 C.d.f. of the bearing

By means of the substitution

$$\tan x = (1 + r^2)(\xi - \bar{\xi})/2r$$

into eqn. 10, we obtain

$$p_5(\xi)d\xi = ½ \cos x dx$$

Hence,

$$\text{c.d.f.}(\xi \leqslant \gamma) = \tfrac{1}{2}\int_{x=-\pi/2}^{x=x_0} \cos x dx$$

where

$$x_0 = \tan^{-1}\{(1 + r^2)(\gamma - \bar{\xi})/2r\}$$

from which

$$\text{c.d.f.}(\xi \leqslant \gamma) = \tfrac{1}{2}\{1 + \sin x_0\}$$

and

$$\text{c.d.f.}\{B \leqslant (\bar{B} + \alpha t)\} = \tfrac{1}{2}\{1 + 2tP\alpha(1 + 4t^2P^2\alpha^2)^{-1/2}\}$$

where t is the normalised bearing. Curves of this function for $r = 2$ and $r = 4$ are shown in Fig 4.6.

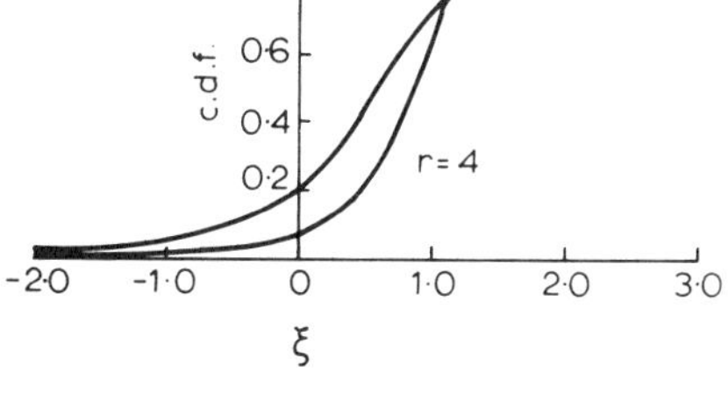

Fig. 4.6 C.d.f. of ξ, two rays with Rayleigh fading

4.7.4. Determination of the parameters of fading rays

Values of η, α and r^2 can be estimated from the above expressions if the p.d.f. and the c.d.f. have been observed and a Rayleigh fading law is assumed, but there is a fundamental ambiguity: a mirror reflection of the ray azimuths in the mean bearing gives the same p.d.f. as the true azimuths.

4.8 Conclusions

The results presented in this chapter refer entirely to the geometry of surfaces of constant phase in a two-ray wave-field. The curves of constant phase defined by the intersections of these surfaces with the horizontal plane have been considered. The 'resolved' mode amplitudes $A_1 \cos \Delta_1$ and $A_2 \cos \Delta_2$ appear in the theoretical expressions because we have chosen to plot curves in the horizontal plane.

The indicated bearing varies as a function of phase difference and other parameters; its behaviour when ray amplitudes are assumed constant can be summarised as follows:

(*a*) The indicated bearing fluctuates about the bearing of the stronger signal.
(*b*) The swing of the indicated bearing is approximately proportional to the amplitude of the weaker ray relative to that of the stronger and to the ray separation in azimuth (for small angular separations).
(*c*) The mean bearing over all phase conditions is rigorously equal to that of the stronger ray when $\Delta_1 = \Delta_2$.
(*d*) The mean bearing is close to that of the stronger ray, even when $\Delta_1 \neq \Delta_2$.
(*e*) Very large bearing swings occur when the critical condition $A_1 \cos \Delta_1 = A_2 \cos \Delta_2$ is approximately satisfied.

These conclusions must be read in conjunction with the special meaning given to 'bearing' in Section 4.1. The action of a practical DF has not yet been discussed, but numerical simulations to be described later suggest that the tendency to 'capture' the bearing of the stronger signal still applies, provided 'stronger' is replaced by 'apparently stronger': the elevation pattern of a single element and the phasing of a set of elements are among the factors that affect apparent strength. On physical grounds, it would be expected that the bearing swing measured on a DF of finite aperture would be less than that given by eqn. 3 and that the swing would decrease with increasing aperture.

Probability distributions have been obtained for constant ray amplitudes and Rayleigh-fading rays. In the former case there is a clear connection between the *U*-shape of the distribution, i.e. a tendency for errors to assume one of two values, as in Fig. 4.3, the two limits of the indicated bearing in Fig. 4.2 and the alternate long and short segments, both nearly straight, of the kinked CPs in, for example, Fig. 3.3. With fading rays the mean-indicated bearing, averaged over all phase conditions, is a weighted mean of the two ray azimuths, each weight being proportional to the fraction of the sampling period for which the corresponding ray was dominant.

It must be stressed that the bearing distributions obtained in Sections 4.6 and 4.7 apply only to the model of two rays with the same elevation angle. The theory has been based on eqn. 8, which is a valid approximation to eqn. 2 when bearing variations are comparatively small, i.e. when α is small and R is not near 1. With fading rays, large bearing swings can occur when the effective amplitudes pass through equality. Thus the p.d.f. and the c.d.f. for this model are based on a logical inconsistency: the effect on the distribution of using the full expression, eqn. 2, might be quite minor, but the 'tails' of the distributions obtained here should be treated with some caution. The infinitely large variance associated with the p.d.f. is an unsatisfactory feature.

In spite of these reservations it is gratifying to have a set of self-consistent p.d.f.s and c.d.f.s for the cases of two steady rays and two Rayleigh-fading rays.

Reference

SHIRAZI, S.M.S.A. (1972): 'Signal processing for high frequency direction finding'. PhD Thesis, Electronic and Electrical Engineering Department, University of Birmingham.

Chapter 5

Directive array patterns

5.1 Radiation patterns and spatial angular spectra

The aim in this chapter is to summarise those aspects of antenna array theory that are relevant to the design of DF systems. The reader is assumed to be familiar with the more basic standard properties of directive arrays, as described in textbooks such as Kraus (1988) and Smith (1988).

In the orthodox DF instrument, some or all of the antenna elements are connected together to form an array, the radiation pattern of which possesses desirable directional properties; in general terms, directivity is clearly an essential property of an array to be used for directional measurements. We shall see in later chapters that there is an alternative to physically combining the elements into an array: the addition of phasors can instead be carried out in a computer. Even with this processing scheme, we can talk of synthetic arrays formed by the computer. In either case, therefore, familiarity with array theory can provide useful insight into the physical interpretation of mathematical results.

The radiation pattern of a transmitting antenna or array is a function of the current distribution across its aperture. A receiving antenna or array is often regarded as exhibiting the same pattern, although strictly it produces a single output determined by the distribution of signal sources. Various relationships between current distributions, radiation patterns and the spatial harmonics of the far field can be most easily understood if a circular array is used for illustrative purposes. The following treatment is based largely on a paper by Longstaff, Chow and Davies (1967).

Let $D(\theta)$ be the directional pattern of a single-ring circular array in the plane containing the circle, where θ is the azimuth angle in this plane. Let the distribution of signal sources in the far field be represented by $H(\theta)$. If the main beam of the array points to azimuth α at a particular moment, the output $V(\alpha)$ of the array will be

$$V(\alpha) = \int_{-\pi}^{\pi} H(\theta)D(\alpha - \theta)d\theta$$

where the RF terms have been suppressed in $H(\theta)$ and $V(\alpha)$. V is the convolution of H and D; the Fourier transform (FT) of V is the product of the Fourier transforms of H and D, i.e.

$$\{V(\theta)\} = \{H(\theta)\}\{D(\theta)\}$$

where the curly brackets denote FTs with respect to spatial frequency, a variable with dimensions of cycles per 360° of θ. For example, a current distribution round the ring that can be represented by one cycle of a sine wave produces a spatial frequency of one cycle in the radiation pattern. If the directional pattern of the array is continuously rotated, $V(\theta)$ and $D(\theta)$ become repetitive functions of time and their spatial spectra can be interpreted as frequency spectra. A receiving array acts as a filter to the far-field source distribution.

This concept led Bracewell and Roberts (1954) to suggest that an antenna behaves like a lowpass filter and that spatial harmonics above a cutoff value are completely lost; they suggested a cutoff value for an aperture of width W of W/λ for linear processing. However, this conclusion has been challenged by several later authors on the grounds that higher harmonics are present in the radiation pattern, even though those above the cutoff value are relatively weak. Lo (1961), for example, showed that, in the absence of noise, the exact incident field distribution could be determined: but Chisholm (1963) showed that noise imposes a *practical* limit. The criterion suggested by Bracewell and Roberts provides a useful guide to the significant harmonics in the radiation pattern; beyond this value, super-directivity is likely to be involved (see Section 5.3 below).

The pattern $D(\theta)$ of a circular (transmitting) array, for a specified excitation in each of n elements, can be calculated by summing n corresponding phasors at suitable angular intervals in the far field. When the number of elements becomes very large we can replace the discrete current distribution at points round the ring by a circular current sheet (the current flow should still be pictured as vertical) and the far-field pattern is then the Fourier transform of the current distribution $I(\theta)$. Suppose

$$I(\theta) = I_s \cos s\theta$$

The discrete sum is now replaced by a continuous integral:

$$D(\alpha) = \int_0^{2\pi} I_s \cos s\theta \exp\left\{\frac{2\pi jR}{\lambda} \cos(\alpha - \theta)\right\} d\theta$$

where R is the radius of the array, λ is the wavelength and $j = \sqrt{-1}$. Hence,

$$D(\alpha) = I_s \cos s\alpha (j)^S J_s\left(\frac{2\pi R}{\lambda}\right) \quad (1)$$

The pattern of a circular ring with a constant current distribution is therefore circular (as might be expected from symmetry considerations), with an amplitude proportional to $J_0(2\pi R/\lambda)$.

5.2 Deconvolution of a continuous distribution

As an illustration of the concepts outlined above, we consider the deconvolution of a continuous distribution of external noise, $H(\theta)$, with the aid of a rotatable array with a radiation pattern of known form, $D(\theta)$. We express each of these functions as a Fourier series:

$$\begin{aligned} H(\theta) &= A_0 + A_1 \cos\theta + A_2 \cos 2\theta + \ldots \\ &\quad + B_1 \sin\theta + B_2 \sin 2\theta + \ldots \\ D(\theta) &= a_0 + a_1 \cos\theta + a_2 \cos 2\theta + \ldots \\ &\quad + b_1 \sin\theta + b_2 \sin 2\theta + \ldots \end{aligned}$$

For a symmetrical pattern all the bs are zero, but they are included here for the sake of generality. The output signal, when the array maximum points to angle α, is

$$\begin{aligned} V(\alpha) &= C_0 + C_1 \cos\alpha + C_2 \cos 2\alpha + \ldots \\ &\quad + S_1 \sin\alpha + S_2 \sin 2\alpha + \ldots \end{aligned}$$

It may be shown from the convolution integral that

$$\begin{aligned} C_0 &= 2\pi A_0 a_0 \\ C_r &= \pi(A_r a_r + B_r b_r) \quad \text{for} \quad r \geqslant 1 \\ S_r &= \pi(B_r a_r - A_r b_r) \quad \text{for} \quad r \geqslant 1 \end{aligned}$$

In practice, the Cs and Ss can be determined from the measured output and the pattern coefficients are known. Then

$$\begin{aligned} A_0 &= C_0/2\pi a_0 \\ A_r &= (a_r C_r - b_r S_r)/\pi(a_r^2 + b_r^2) \\ B_r &= (a_r S_r + b_r C_r)/\pi(a_r^2 + b_r^2) \end{aligned}$$

Note that the rth harmonic of the output depends only on the rth harmonic of the radiation pattern and the rth harmonic of the input distribution, for all values of r.

The azimuthal power polar diagram of an arc of a circular array can be closely approximated by an empirical expression of the form

$$D(\theta) = \left(\frac{\sin N\theta}{N\theta}\right)^p$$

Suppose, for example, $N = 45{\cdot}0$ and $p = 1{\cdot}5$, and that an arbitrary noise-power distribution is defined by

$$H(\theta) = 5{\cdot}0 + 2{\cdot}0 \cos 3\theta + 3{\cdot}0 \sin 3\theta$$

The Fourier coefficients for $D(\theta)$, $V(\alpha)$ and the deduced coefficients of $H(\theta)$ are shown in Table 5.1.

Table 5.1 Fourier coefficients

	Antenna pattern		Output		Deduced input	
Harmonic	Cosine	Sine	Cosine	Sine	Cosine	Sine
0	0·0149	–	0·4683	–	5·00	–
1	0·0283	0·0	0·0000	−0·0002	0·00	0·00
2	0·0270	0·0	0·0000	0·0000	0·00	0·00
3	0·0261	0·0	0·1641	0·2461	2·00	3·00
4	0·0253	0·0	0·0000	0·0000	0·00	0·00
5	0·0246	0·0	0·0000	0·0000	0·00	0·00
		etc		etc		

The coefficients of the noise distribution are seen to be recovered correctly. Thus any noise distribution that can be closely approximated by the sum of a reasonable number of Fourier terms can be reconstructed with this technique. Two simplifying assumptions have been made. Firstly, by working in terms of power distributions and patterns, we have avoided the problem of the determination of the phase of the distant field distribution, which requires the use of complex coefficients. Secondly, we have ignored the possible variations of $D(\theta)$ and $H(\theta)$ with elevation angle; in general, such variations cannot be explored with a pattern rotated in azimuth only. Both assumptions are reasonable if, for example, we wish to determine the azimuths of major nearby sources of man-made noise, which are likely to lie in or near the horizontal plane. In such a situation we are not interested in the noise phase; the purpose of deconvolution is to 'sharpen up' the noise picture, which has been blurred by the finite beamwidth and sidelobes of the measuring instrument.

5.3 Pattern synthesis

The radiation pattern of an array of n antenna elements depends partly on the geometry of the array and partly on the way in which the received signals are combined in a beam-forming network, i.e. a network, usually additive, with n inputs and one output to which the receiver is attached. The radiation pattern can be greatly modified by altering the amplitude and phase taper applied in the BFN.

In principle, we can make the pattern of an array of fixed aperture 'sharper' by giving increased weight to the higher-order harmonics in the aperture current distribution. The higher harmonics correspond to rapid reversals in the sign of the current for small movements along the aperture. Eqn. 1 provides a mathematical basis for pattern synthesis with a circular array. To synthesise a specified far-field pattern, it is only necessary to analyse it into its spatial harmonic components and to excite each of these components in the correct amplitude and phase superimposed around the circumference of the array. If the rth harmonic of the directional pattern is required to have a relative amplitude a_r, the rth harmonic of the excitation or taper function must be

$$I_r(\theta) = \frac{a_r \cos r\theta}{(j)^r J_r\left(\dfrac{2\pi R}{\lambda}\right)}$$

In principle there is no limit to the maximum directivity that can be achieved, and enhanced directivity would seem, almost by definition, to be a desirable quality for a DF. In the HF band, where external noise dominates over receiver noise, there is no objection to sacrificing some absolute gain if necessary. In practice, however, it is difficult to obtain superdirectivity, which is defined in an IEEE standard (1983) to be the condition that occurs when the ratio of the maximum directivity to standard directivity obtained with a uniform amplitude equiphase current distribution significantly exceeds unity. Superdirectivity can only be obtained at the expense of a large increase in the ratio of average stored energy to energy radiated per cycle. The radius of a circular array and the number of elements, n, both impose constraints on the highest harmonic that can be easily produced in practice. If the current distribution changes in phase by more than 2π radians in a distance λ round the circumference, the resulting field is highly reactive, corresponding to superdirectivity. The minimum size of array to produce a Fourier component of order k and to avoid the very severe tolerances on superdirective current distributions has a circumference $k\lambda$, and hence a radius $k\lambda/2\pi$. If the pattern of the array is to be rotated, $\cos\{(n/2 - 1)\theta\}$ is the highest harmonic that can be generated with n elements; see Longstaff, Chow and Davies (1967).

The performance of a circular array can be analysed in terms of phase modes $a_r \cos r\theta$ that are identical, apart from a constant of proportionality, with the harmonic terms described above, such that the zero mode is a constant term, the first mode has a periodicity of 2π, the second $2\pi/2$, the third $2\pi/3$ and so on. The arrival of a single ray at a circular array generates these modes, which can be separated in a Butler matrix; see, for example, a paper by Karavassilis and Davies (1985) describing the action of a four-element array of active loops. The phase of the mth phase mode is equal to m times the bearing of the incident ray; hence the bearing can be determined from the difference of phase of any two successive modes.

The concept of phase modes provides considerable insight into the properties of circular arrays and has led to new ideas in direction finding and adaptive array design. However, there are difficulties to be overcome in any practical system: for example, in the phase-measuring process used by Karavassilis and Davies amplitude modulation interfered with envelope modulation. The arrival of more than one ray spoils the simple relationship between bearing and mode phases, although the array can be used to point a null at an interfering signal and then find the azimuth of arrival of a wanted co-channel signal, as shown by Cvetkovic (1989).

There are a number of ways of exploiting phase modes, as shown by Travers and Green (1972). For example, the lower-order modes can be used for ambiguity resolution only, with the bearing determined from the higher-order modes. The effective increase in directivity obtained in this way appears to lead to some reduction in the errors arising from site re-radiators.

Applications of the concept of spatial harmonics are not confined to circular arrays, or to linear processing of the element signals. It provides the background for much of the theory that has been developed for multiplicative/correlation arrays (Mathur, 1969), minimum-redundancy arrays (Moffet, 1968) and aperture-synthesis techniques (Ryle and Hewish, 1960). For example, Shearman (1963a) has shown how multi-element receiving patterns can be synthesised from the outputs of two physical elements only. He has also shown (1963b) that, for an array in which the element outputs are multiplied together in pairs and the smoothed products added, the directional pattern is independent of whether or not the various pairs are collinear.

5.4 Beam-forming networks

5.4.1. Components

A knowledge of the action of the beam-forming network is required before the radiation pattern of an array can be calculated. In the receiving case, the BFN determines how the signals in various elements are combined; in the transmitting case, how the output from the transmitter is divided among the elements.

If we regard the BFN for a receiving array of n elements as a device with n inputs and one output to the receiver, it is clear that all frequency-changing circuitry is likely to be incorporated in the receiver. (Alternatively, we do not need a BFN if we have an n-channel receiver). The phasor representing the signal in a given element on the frequency to which the receiver is tuned is completely defined by two numbers, amplitude and phase. These are the only characteristics that can be modified in the BFN before the phasors are combined.

For a DF array intended for use over a wide frequency band, it is convenient to replace a phase shift with an equivalent time delay, since the time delays required to bring signals from a given direction into phase coincidence (i.e. to form a cophasal beam on the selected direction) depend on the array geometry but not on frequency. The basic components usually used in BFNs are therefore amplitude taper pads (attenuators) and delay lines; the latter may be of either lumped-constant or distributed form. A length a of coaxial cable is equivalent to a distance a/k in free space, where k is the velocity ratio for the cable; k is usually in the range 0·65 to 0·85. The time delay imposed on a pulse signal is then a/kc, where c is the velocity of light in free space, *provided* the cable is correctly matched at one or both ends. That is to say, the cable must be terminated in its characteristic impedance or driven by a source of this impedance if unwanted reflections are to be avoided and the line is to give the correct delay. An equivalent phase shift is produced in a CW signal.

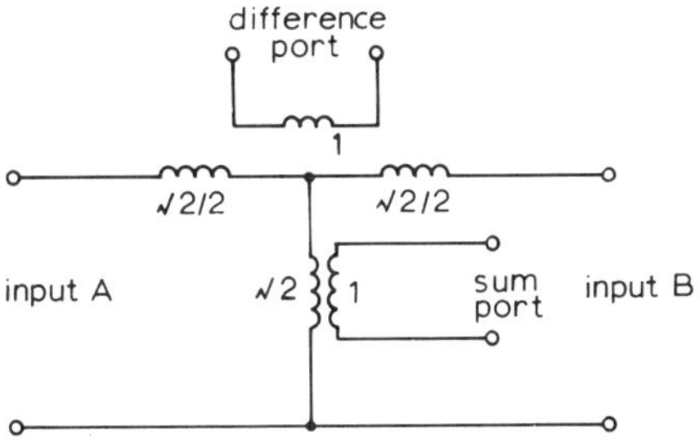

Fig. 5.1 Four-port sum and difference hybrid

Signals are usually combined, after suitable amplitude and phase changes, in adding units consisting either of hybrids or resistive networks. A standard sum-and-difference hybrid, designed for a 75Ω system, is shown in Fig. 5.1. With inputs A and B this unit gives outputs proportional to $(A + B)$ and $(A - B)$ when correctly terminated. For example, the sum port might be connected to a 75Ω receiver by 75Ω cable and the difference port might be terminated with a 75Ω resistor; the apparent impedance at each input port would then be 75Ω. Hybrids connected in cascade arrangements can be used as signal splitters (one input, many outputs) or combiners (many inputs, one output).

We defer until Section 5.8 a discussion of the possibility of taking the product rather than the sum of two signals.

5.4.2. Doublet

As a simple example of the use of a beam-forming network, Fig. 5.2 shows how the signals from two receiving vertical monopoles A and B can be combined in order to form a cardioid-shaped radiation pattern in the horizontal plane. The signal from B is delayed by an amount equivalent to the spacing d between elements and is then subtracted from the signal from A. For a single plane wavefront arriving from azimuth α measured from the line BA and elevation Δ, the signals S_A and S_B in A and B respectively can be written

$$S_A = a \sin \omega t$$

$$S_B = a \sin\left(\omega t - \frac{2\pi d}{\lambda}\cos\alpha\cos\Delta\right)$$

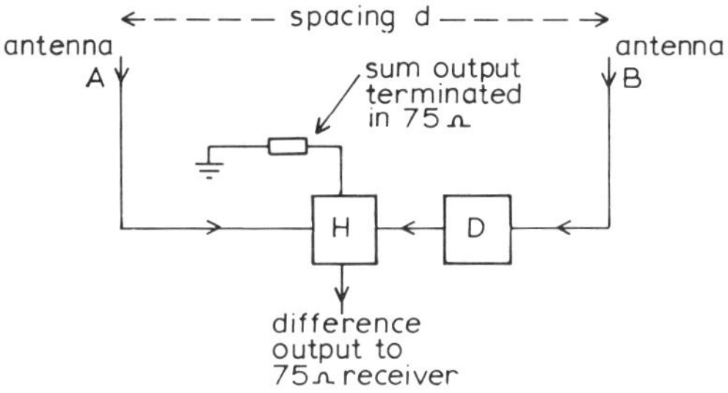

Fig. 5.2 Doublet circuit diagram
H = hybrid
D = time delay equivalent to spacing *d*

where a is proportional to the amplitude of the wave. The signal from B is now delayed by an amount equivalent to a distance d in free space and becomes

$$S'_B = a \sin\left\{\omega t - \frac{2\pi d}{\lambda}(1 + \cos\alpha\cos\Delta)\right\}$$

Hence

$$S_A - S'_B = 2a \sin\left\{\frac{\pi d}{\lambda}(1 + \cos\alpha\cos\Delta)\right\} \times$$

$$\cos\left\{\omega t - \frac{\pi d}{\lambda}(1 + \cos\alpha\cos\Delta)\right\}$$

The 'pattern term', $2 \sin\left\{\frac{\pi d}{\lambda}(1 + \cos\alpha\cos\Delta)\right\}$ is zero when $\alpha = \pi$, $\Delta = 0$ and the pattern therefore contains, as expected, a null in this direction.

5.4.3. Goniometer

The rotating goniometer used with many types of circular WADF is a particular example of a BFN. A full description of the design and construction of goniometers is beyond the scope of this book, but

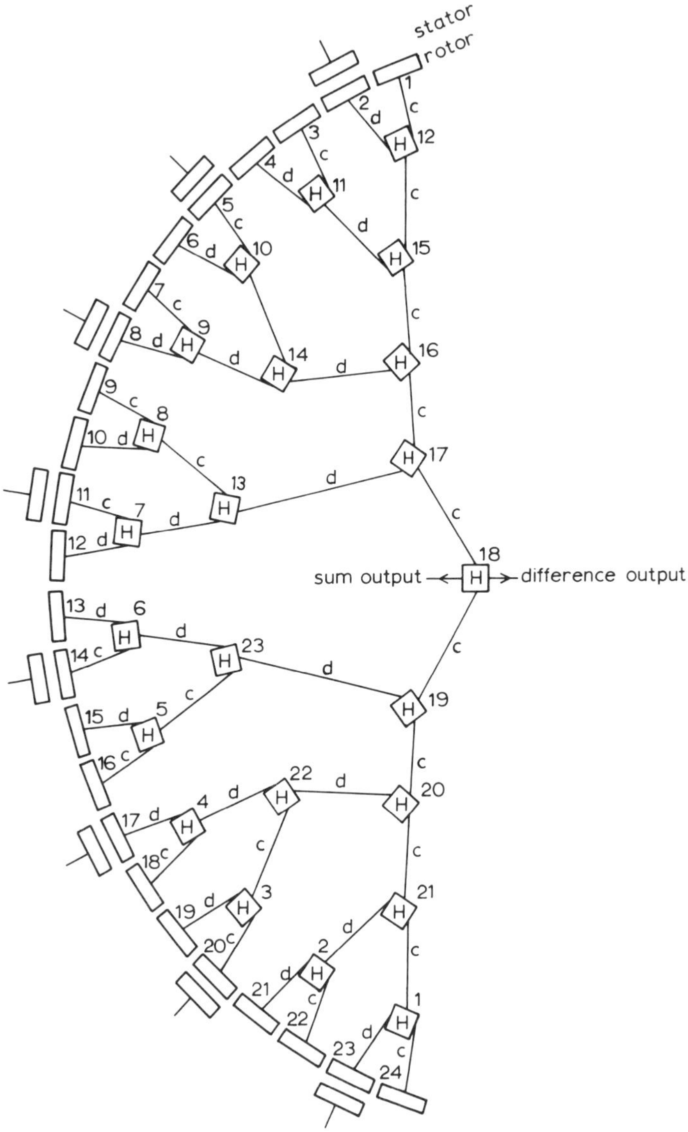

Fig. 5.3 Block diagram of hybrid goniometer
d delay cables
H hybrid transformers
c connection cables

information can be found in papers by Wundt (1968) and Hockley (1973). Hockley described a goniometer designed for use with a 24-element circular array, in which the signals from an arc of eight elements are used at any one moment; see Fig. 5.3. Let A represent the sum of the signals from elements 1 to 4 and B the sum from elements 5 to 8, both sets of phasors being cophased for a direction defined by the axis of symmetry midway between elements 4 and 5. The two outputs from the goniometer are a sum pattern $(A + B)$ and a difference, or 'rabbits-ears' pattern, $(A - B)$. The effect of goniometer rotation is to rotate the arc of elements in use, i.e. 1–8, then 2–9, then 3–10 etc. Because coupling to the elements is by means of capacitive combs, the transitions can be made smoothly, with little change of pattern shape.

The design of a tapped delay-line network for use over a wide frequency band poses considerable problems. This is perhaps most clearly seen if we think of the goniometer temporarily as a device for connecting the output of a transmitter to an arc of elements. In the above example, there is no problem in splitting the output power equally between the two arcs, 1–4 and 5–8. Of the power going to (1–4), we first split off one quarter for element 1. Then, after a suitable delay, one third of the remainder goes to element 2, and so on. To maintain the correct fractions and the correct delays over (say) a 4:1 frequency band, in which the impedance of each antenna is likely to vary widely, is obviously difficult. Hockley therefore adopted a simpler arrangement, based on a network of hybrid combiners (which could be regarded as splitters with two outputs in the transmitting case).

Various forms of electronic goniometer have been suggested, in order to obtain rapid beam scanning and to avoid mechanical problems. Bain (1960) and Davies and McCartney (1965) described methods of using all the elements in a circular array. If the number of elements is large, the beam cophasal pattern in azimuth is then proportional to

$$J_0\left(\frac{4\pi R}{\lambda}\sin\frac{\theta}{2}\right)$$

where R is the radius of the array and θ is the angle relative to the beam maximum at $\theta = 0$. The disadvantage of using all elements is that the vertical pattern is likely to break up into multiple lobes.

5.5 Phase centre

Returning to the theory of the doublet given in 5.4.2, we now consider the 'RF term',

$$\cos\left\{\omega t - \frac{\pi d}{\lambda}(1 + \cos\alpha\cos\Delta)\right\}.$$

It is instructive to compare this term with the signal S_c that would be found in a monopole identical with those used in the doublet, but placed at the midpoint C of the line AB:

$$S_c = a \sin \left\{ \omega t - \frac{\pi d}{\lambda} \cos \alpha \cos \Delta \right\}$$

The phase angle in both expressions depends on α and Δ, but the difference of phase between the doublet output and S_c is $(\pi d/\lambda + \pi/2)$, which is a constant for a given frequency and spacing. It is therefore legitimate to regard C as the phase centre of the doublet; for example, the phase difference between the signals from two identical doublets is the same as the phase difference between two monopoles at the mid-points.

Dyson (1967) has shown that many types of antenna do not possess a unique, fixed, phase centre. For the special case of a linear receiving array it is convenient to define the phase centre as follows (Gething, 1973). Suppose such an array is receiving plane waves from a single point source in the far field. The phase centre is one of the points on the line of the array at which an identical element would give an output of the same phase as the output of the BFN. For a signal arriving from angle θ relative to the broadside direction, these points are spaced at intervals of $\lambda/\sin\theta$; the phase centre is the point which remains within a finite distance of each element as θ approaches $0°$. In this way, it is possible to define a unique point for any value of θ, but the position of the point may vary with θ.

5.6 Mutual coupling

The radiation pattern of a practical DF array will depart from the pattern calculated by simple phasor addition methods if mutual coupling effects between antenna elements are significant. Ideally, we wish to know the practical radiation pattern, since the action of a DF array depends on its directional properties. As with any other form of measuring instrument, our aims must be to reduce all forms of instrumental error as far as possible and then to apply suitable calibration corrections, if possible, to what remains. Physically, it seems sensible to disturb the incident wave-field as little as possible; DF antenna elements should therefore be designed as short, high-impedance voltage probes rather than as devices for extracting as much power as possible from the field. For example, in an interferometer system described by Burtnyk, McLeish and Wolfe (1965), short (3·5 m) monopoles were used and were connected through cathode followers to the transmission lines.

Some information about the magnitude of mutual coupling effects can be obtained by making impedance measurements at the antenna terminals. Suppose we measure the current I_1 for an applied voltage V_1 at the terminals of one antenna in the presence of a second antenna with voltage V_2 and current I_2 at the terminals.

We expect linear relationships of the form

$$V_1 = z_{11}I_1 + z_{12}I_2$$
$$V_2 = z_{21}I_1 + z_{22}I_2$$

and, for identical elements, $z_{12} = z_{21}$, $z_{11} = z_{22}$. The apparent impedance, V_1/I_1, of the first element when the second is open-circuit ($I_2 = 0$) is

$$z_{0c} = z_{11}$$

and the apparent impedance when the second is short circuited at the terminals ($V_2 = 0$) is

$$z_{sc} = z_{11} - z_{12}^2/z_{11}$$

Measurements with an impedance bridge in these two conditions therefore enable us to determine z_{11} and z_{12}. Note that z_{11} is strictly not the self-impedance of element 1 alone, but of element 1 in the presence of the open-circuit element 2. An alternative technique described by Altshuler (1960) is to drive the two elements first in phase and then in antiphase. z_{12}/z_{11} can be regarded as a measure of the magnitude of mutual coupling and depends on several physical factors: the frequency, the separation of elements, the nature of the elements and the shape of the element in the region of the feedpoint. In practice, it is usually necessary to measure the impedances rather than calculate them.

More generally, for an n-element array we can write in matrix notation

$$\{V\} = [Z]\{I\}$$

where z_{rs} is an element of the impedance matrix for elements r and s. The impedance matrix can be inverted into an admittance matrix with elements y_{rs}

$$\{I\} = [Y]\{V\}$$

If (say) the admittance matrix is required for a particular type of calculation, it may be possible to devise experiments to measure admittances directly, rather than measure impedances and then invert.

As a first approximation, terms such as z_{13}, z_{14}, z_{24} . . . are sometimes neglected and only the coupling between neighbouring elements is taken into account through z_{12}. This approximation is not strictly the same as neglecting y_{13}, y_{14} etc in comparison with y_{12}, although physically there may be nothing to choose between these assumptions. Analytic expressions can be found for y_{rs} for a uniformly spaced circular or linear array when $z_{13} = z_{14}$ etc $= 0$ and, since these expressions can provide a useful check on numerical inversions in a computer, they are quoted here. For a circular array of n elements, the author has shown (1965) that

$$y_{rs} = \frac{-1}{z_{12}}\left\{\frac{w^{p+1} + w^{n+1-p}}{(1 - w^n)(1 - w^2)}\right\}$$

where $p = |r - s|$ and w is either root of the complex quadratic

$$w^2 z_{12} + w z_{11} + z_{12} = 0$$

For a linear array the corresponding solution (Gething, 1966) is

$$y_{rs} = \frac{-1}{z_{12}} \left\{ \frac{w^{p+1} + w^{2n+3-p} - w^{r+s+1} - w^{2n+3-r-s}}{(1 - w^{2n+2})(1 - w^2)} \right\}$$

These solutions are exact and do not depend on any assumption about the relative magnitudes of z_{11} and z_{12}.

Impedance measurements relate to currents at the terminals of the elements, whereas radiation pattern calculations depend on the current distribution over the whole of each element. When mutual coupling is significant, it may no longer be legitimate to make the usual assumption that the current distributions on identical elements are of identical form. The common experience seems to be that the effects of mutual coupling on radiation patterns must be investigated by means of measurements on radiation patterns; neither theory nor impedance measurements alone seem to be sufficient. Nevertheless, theoretical calculations have been made for various DF systems by, for example, Bain (1956), Travers (1957) and Harrison (1961) and give some insight into the magnitude of the effects to be expected. A paper by Byatt (1967) illustrates the effects on doublet patterns of changes in the terminating impedances of nearby elements.

The infinitely-deep nulls often present in theoretical patterns represent a perfect cancellation, in a particular direction, of both the in-phase and quadrature components of n phasors, where n is the number of elements in the array. This perfect cancellation is unlikely to be preserved when coupling is significant; the field intensity patterns for the phase and quadrature components may differ slightly and their combination, in the power radiation pattern, will produce some filling of the nulls, which may now be (say) 20 or 30 dB deep. Some examples of this effect may be seen in a paper by Mellors, Davies and Withers (1970).

The position of a minimum in a radiation pattern can also be 'pulled' slightly by mutual coupling. However, the position of the principal null, for a DF array that is symmetrical in both lay-out and current distribution, should not be affected. Because it is not easy to measure, calculate or control mutual coupling effects, it is important to preserve symmetry in any array for which coupling is likely to be significant.

As an example of the value of symmetry, let us consider the four-element Adcock, with elements placed on the *N–S* and *E–W* diagonals of a square. In the receiving mode, the Adcock azimuth indication is obtained from the ratio of the difference of signals in the *E* and *W* elements to the difference of signals in the *N* and *S* elements. Since impedance measurements are made by connecting a signal source to an element, it is more convenient to discuss the case of transmission in this context. If voltages V_N, V_E etc are applied to the *N*, *E* etc elements, the currents at the terminals will be linear functions of these voltages:

$$V_N = z_{NN}I_N + z_{NE}I_E + z_{NW}I_W + z_{NS}I_S$$

$$V_E = z_{EN}I_N + z_{EE}I_E + z_{EW}I_W + z_{ES}I_S$$

etc where the z's represent impedances in an obvious notation. On the reasonable assumptions that

$$z_{NN} = z_{EE} = z_{SS} = z_{WW} = z_{11} \quad \text{(say)}$$

$$z_{NE} = z_{NW} = z_{EN} = z_{ES} \text{ etc} = z_{12} \quad \text{(say)}$$

$$z_{NS} = z_{EW} = z_{13} \quad \text{(say)}$$

we obtain

$$V_N - V_S = z_{11}(I_N - I_S) - z_{13}(I_N - I_S)$$
$$V_E - V_W = z_{11}(I_E - I_W) - z_{13}(I_E - I_W)$$

Hence

$$\frac{(V_E - V_W)}{(V_N - V_S)} = \frac{(I_E - I_W)}{(I_N - I_S)}$$

which is exactly the ratio obtained in the absence of mutual coupling. This simple result, when taken in conjunction with the principle of reciprocity, does suggest that azimuth measurements made with a four-element Adcock are not affected by mutual coupling. The same useful result does not appear to apply to a 4-element array when the bearing is determined from the phases of the phase modes.

Perrott (1985) considered the effects of mutual coupling on the relative phases of antenna currents in an interferometer array under single-ray conditions. The ray azimuth angle was calculated to a first approximation by ignoring the mutual effects. The angle was then used to find the perturbation to the phase measurements caused by mutual coupling, so that the effect could be largely removed. It was shown that this process, continued iteratively, leads to substantially-reduced bearing errors.

5.7 Radiation patterns: additive processing

5.7.1. Circular WADF

Some of the properties of the directional radiation patterns obtained with different types of DF array can be conveniently discussed at this stage. We first describe some of the features of the horizontal polar diagrams of a circular wide-aperture DF with a diameter of about a thousand feet. The array contains 96 equispaced doublets, of which 32 are used at any one moment to form the sum and difference beams. Further details of the array dimensions and of the simplifying assumptions used in calculating the patterns are given in Appendix 1. Figs. 5.4*a* and 5.4*b* show the sum and

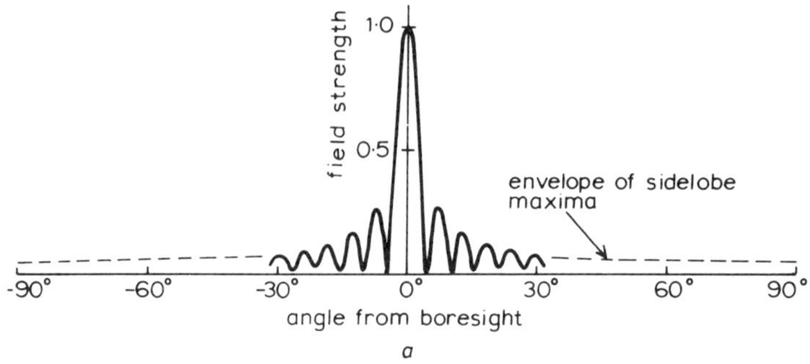

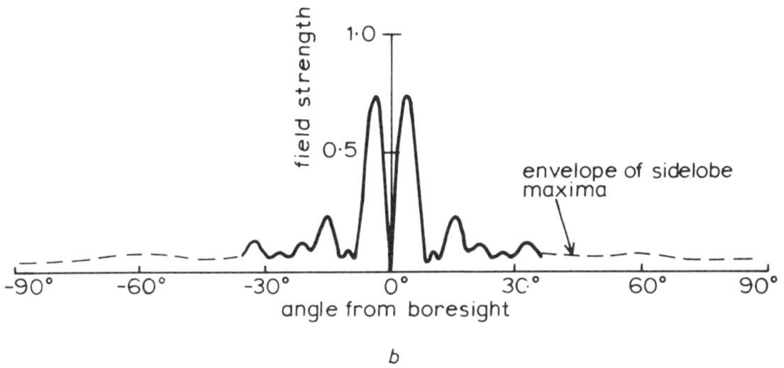

Fig. 5.4 Sum and difference radiation patterns
(a) Sum pattern, circular array of doublets, 12 MHz
(b) Difference pattern, circular array of doublets, 12 MHz (same vertical scale as Fig. 5.4a)

difference patterns respectively at a frequency of 12 MHz. The difference pattern would normally be used for DF purposes; the bearing of the (single) target transmitter is indicated by the null between the two main lobes. Strictly, the pattern shows the varying output obtained with a fixed goniometer, as the target is moved in azimuth, but it is a close approximation to the pattern obtained with a spinning goniometer and a fixed target. In the latter case, we can be confident that the null plane is vertical when it coincides with one of the 192 directions of symmetry, but between these directions the plane may waver slightly from strict verticality and small variations in pattern shape can also occur. The resulting small error in the indicated bearing is sometimes referred to as cogging error.

The azimuthal radiation pattern of a cophasal arc can be represented in terms of an infinite Fourier–Bessel series. This expression is based on the Jacobi–Anger relation

$$\exp(jx \cos \alpha) = \sum_{m=0}^{\infty} e_m j^m J_m(x) \cos m\alpha$$

where

$$e_m = 1 \quad \text{for} \quad m = 0$$

and

$$e_m = 2 \quad \text{for} \quad m = 1, 2, \ldots$$

In practice, some six or eight terms of the series might give a good approximation to the pattern, but it is generally simpler to calculate the pattern numerically, by phasor addition.

5.7.2 Linear array

The sum and difference patterns shown in Figs. 5.4*a* and 5.4*b* are very similar to those that would be obtained near the broadside direction from a linear array of 32 doublets with about the same total aperture. The first sidelobe level for a linear array of n monopoles approaches −13·6 dB as n becomes large. A theoretical sidelobe level of about −10 dB, and a practical level of −8 or −9 dB, are commonly found for a circular arc of monopoles.

In comparing the outputs of sum and difference patterns in a monopulse phased-array DF system, Lo and Vu (1988) obtained improved performance by applying a form of Chebyshev weighting to both, in order to reduce sidelobe levels.

5.8 Multiplicative processing

The following discussion of multiplicative processing is based largely on a paper by Shaw and Davies (1964). For illustrative purposes we consider a linear array of eight elements with uniform spacing d between elements. The response obtained by adding the outputs of all eight elements is

$$D_a(p) = \frac{\sin 8p}{8 \sin p}$$

where

$$p = \frac{\pi d}{\lambda} \sin \theta$$

and θ is an angle measured from the broadside direction. If the array is divided into two halves and the resultant signals are multiplied together, the resultant directional response can be shown to be

$$D_m\ (p) = \left(\frac{\sin 4p}{4 \sin p}\right)^2 \cos 8p$$

The multiplicative pattern is found to have half the beamwidth of the rectified additive array. However, because the process of multiplication produces a square-law characteristic, a fairer comparison is with the square-law-rectified additive array. The multiplicative beamwidth is about 70% of the beamwidth of the square-law-rectified additive pattern, as may be seen from the comparison in Fig. 5.5.

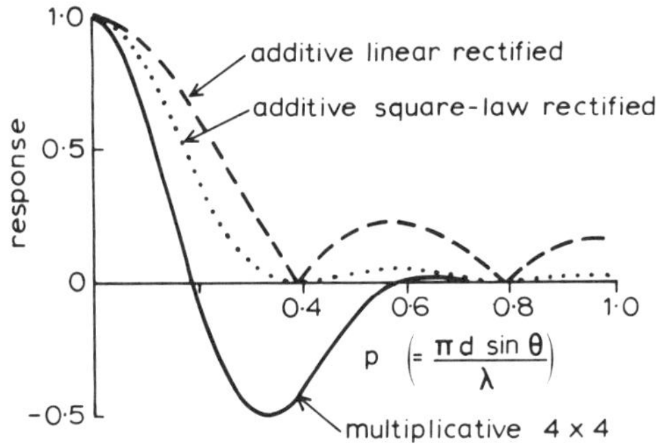

Fig. 5.5 Directional responses for eight-element array

These results might suggest that multiplicative arrays would be preferred for DF applications. However, the conventional polar diagram shows the response of an array to a *single* test transmitter. Since resolution, i.e. the ability to separate two or more arriving rays, is an important property of a DF array, a comparison of the behaviour of additive and multiplicative arrays in the presence of two or more targets is necessary before an informed choice can be made. Unfortunately, the calculation of the two-target response of a multiplicative array is not as straightforward as the corresponding calculation for an additive array, because the principle of superposition does not apply. A derivation of the appropriate expression is given in Appendix 2, for the pattern of an electronically-scanned linear array in which the two subarrays may be of unequal length; the case of mechanical scanning, by rotation of the whole array, is also considered.

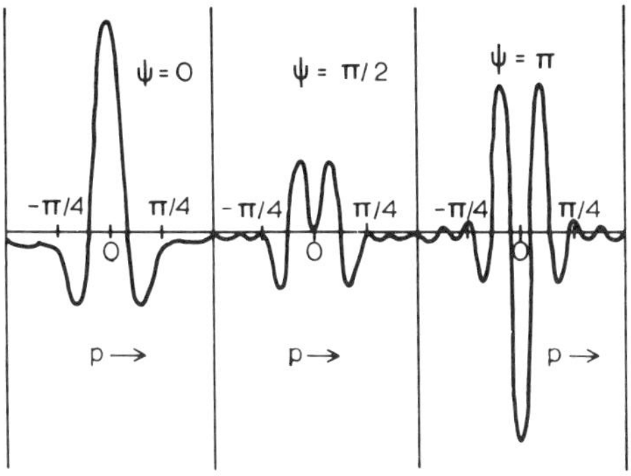

Fig. 5.6 Calculated responses for electronic scanning, 4 × 4 array, interelement spacing = $\lambda/2$, targets at $p_x = \pi/16$, $p_z = +\pi/16$ radians, ψ = phase difference at origin

The response of an array consisting of two four-element subarrays, as it is scanned electronically across two targets, is shown in Fig. 5.6. The response depends on the phase difference between the signals from the two targets, and curves are shown for three values of this parameter. The scan angle θ is replaced on the x axis by the parameter p defined above. On this scale, the targets are at $p = \pi/16$ and $p = -\pi/16$.

The patterns are calculated on the assumption that certain amplitude product terms, such as A_sB_c and A_cB_s in Appendix 2, are removed by integration over many RF cycles. Other terms such as e_2 and e_3 also depend on the product of target amplitudes and become zero if averaged over all values of ψ from 0 to 2π, i.e. over all values of the phase difference between the two targets. Terms of both types are frequently encountered in theoretical treatments of various types of correlation arrays and of aperture-synthesis schemes (Ryle and Hewish, 1960). Clarke (1970) emphasised the importance of the question of whether these unwanted terms are suppressed in particular processing schemes. If they are, the benefits of time-sampling as well as space-sampling are being exploited effectively to resolve incoherent sources.

Shaw and Davies (1964) concluded that, for two targets of equal strength, the multiplicative system considered in Appendix 2 always gives superior resolution to the additive array of the same aperture by a significant amount, although the improvement is less than the corresponding change in beamwidth. However, when examples with two targets of unequal amplitude are examined, it is found that the multiplicative system tends to suppress evidence of the weaker signal when the amplitude ratio differs substantially from unity. Shaw and Davies concluded that the system cannot maintain reasonable resolution over a wide range of amplitude ratios.

Multiplicative processing is closely related to the aperture-synthesis technique used extensively in radioastronomy. Because, in this application, sampling rates of the order of 10^6 per second can be used in bandwidths of the order of 1 MHz, product terms rapidly become insignificant compared with the wanted terms. It is also used in sonar, sometimes combined with hard-limiting (Nairn, 1968).

Whereas radioastronomy sources can be loosely described as 'very incoherent', this description does not necessarily apply in the HF band, where the frequency difference between two modes from the same transmitter may be 0·1 Hz or less when movements in the ionosphere are comparatively small. Integration times of the order of minutes may be required to reduce product terms to insignificant levels and, over such a time interval, DOAs may change or the transmitter may even cease to transmit.

Because the principle of superposition does not apply, it is difficult to visualise the response of a multiplicative system to several rays and the complexity of any mathematical analysis increases with the number of rays. Presumably the tendency to suppress the weaker ray in two-ray problems applies also in multiray fields. Because of these various disadvantages, multiplicative processing is not normally used in HF DF. Shearman et al. (1973) showed that a form of aperture synthesis can be used in HF backscatter experiments; note that the length of transmission and the modulation are then under the control of the experimenter. The relative advantages of additive and multiplicative processing when fading targets can be observed over a period of time perhaps deserve further study.

References

ALTSHULER, E. (1960): 'The measurement of self and mutual impedances', *IRE Trans.*, **AP-8,** pp. 526–527

BAIN, W.C. (1956): 'Possible errors of a particular wide-aperture direction finder', *Proc. IEE.*, **103C,** pp. 313–324

BAIN, W.C. (1960): 'Improvements in or relating to high frequency direction finding systems', British Patent 853662.

BRACEWELL, R.N., and ROBERTS J.A. (1954): 'Aerial smoothing in radio astronomy', *Aust. J. Phys.*, **7,** pp. 615–640

BURTNYK, N., McLEISH, C.W., and WOLFE, J. (1965): 'Performance of an interferometer direction finder for the hf band', *Proc. IEE.*, **112,** pp. 2055–2059

BYATT, D.W.G. (1967): 'Wullenweber arrays using doublet aerials', *Marconi Rev.*, **30,** pp. 1–16

CHISHOLM, R.M. (1963): 'Transfer functions and the resolving power of radio telescopes', in 'EM theory and antennas' (McMillan, New York), pp. 1281–1284

CLARKE, J. (1970): 'Crossproduct terms in synthetic-aperture receivers', *Proc. IEE.*, **117,** pp. 1627–1632

CVETKOVIC, M. (1989): 'A four element circular array direction finding and null steering system', *ICAP 89, IEE. Conf. Publn.* **301, 1,** pp. 168–172

DAVIES, D.E.N., and McCARTNEY, B.S. (1965): 'Cylindrical arrays with electronic beam scanning', *Proc. IEE.*, **112,** pp. 497–505

DYSON, J.D. (1967): 'Determination of the phase centre and phase pattern of antennas', AGARD Conference Proceedings no. 15, Dusseldorf symposium, July 1966, (Technivision Services), pp. 97–118

GETHING, P.J.D. (1965): 'Inversion of an impedance matrix', *IEEE. Trans.*, **AP-13,** pp. 830–831

GETHING, P.J.D. (1966): 'Mutual-impedance effects for a linear aerial array', *Electron. Lett.*, **2,** pp. 387–388
GETHING, P.J.D. (1973): 'Concept of phase centre of an array applied to elevation angle measurements', *Proc. IEE.*, **120,** pp. 541–543
HARRISON, C.W. (1961): 'Antenna coupling error in direction finders', *J. Res. Natl. Bur. Stand.* **65D,** pp. 363–369
HOCKLEY, J. (1973): 'A goniometer for use with high-frequency circularly disposed aerial arrays', *Radio & Electron. Eng.*, **43,** pp. 475–485
IEEE (1983): 'Standard definitions of terms for antennas', *IEEE Standard* 145 (New York, IEEE.)
KARAVASSILIS, N., and DAVIES, D.E.N. (1985): 'An adaptive circular array for HF direction finding and null steering', *ICAP 85, IEE. Conf. Publn.* **248,** pp. 295–299
KRAUS, J.D. (1988): 'Antennas', (McGraw–Hill), second edition
LO, K.W. and VU, T.B. (1988): 'Improving performance of monopulse phased array in direction finding', *Proc. IEE.*, **135H,** pp. 391–394
LO, Y.T. (1961): 'On the theoretical limitation of a radio telescope in determining the sky temperature distribution', *J. Appl. Phys.*, **32,** pp. 2052–2054
LONGSTAFF, I.D., CHOW, P.E.K., and DAVIES, D.E.N. (1967): 'Directional properties of circular arrays', *Proc. IEE.*, **114,** pp. 713–718
MATHUR, N.C. (1969): 'A pseudodynamic programming technique for the design of correlator supersynthesis arrays', *Radio Sci.*, **4,** pp. 235–244
MELLORS, M., DAVIES, D.E.N., and WITHERS, M.J. (1970): 'Zero steering in the directional pattern of a linear array in the presence of mutual coupling', *Proc. IEE.*, **117,** pp. 35–40
MOFFET, A.T. (1968): 'Minimum-redundancy linear arrays', *IEEE. Trans.*, **AP-16,** pp. 172–175
NAIRN, D. (1968): 'Clipped-digital technique for the sequential processing of sonar signals', *J. Acoust. Soc. Am.*, **44,** pp. 1267–1277
PERROT, R.A. (1985): 'The effect of mutual coupling on interferometer arrays', *ICAP 85, IEE. Conf. Publn.*, **248,** pp. 292–294
RYLE, M., and HEWISH, A. (1960): 'The synthesis of large radio telescopes', Monthly Notices RAS., **120,** pp. 220–230
SHAW, E., and DAVIES, D.E.N. (1964): 'Theoretical and experimental studies of the resolution performance of multiplicative and additive arrays', *Radio & Electron. Eng.*, **28,** pp. 279–291
SHEARMAN, E.D.R. (1963a): 'Synthesis of multi-element directional patterns using a two-element single-frequency receiving array', *J. Brit. IRE.*, **26,** pp. 59–66
SHEARMAN, E.D.R. (1963b): 'Non-collinear and cylindrical multiplicative arrays', *J. Brit. IRE.*, **26**, pp. 481–484
SHEARMAN, E.D.R., BICKERSTAFF, P., and FOTIADES, L. (1973): 'Synthetic-aperture sky-wave radar: techniques and first results', Int. conf. on radar – present and future, London 23–25 Oct 1973 (IEE London, 1973), pp. 414–421
SMITH, M.S. (1988): 'Introduction to antennas', (Macmillan Education, Basingstoke and London)
TRAVERS, D.N. (1957): 'The effect of mutual impedance on the spacing error of an eight-element Adcock', *IRE. Trans.*, **AP-5,** pp. 36–39
TRAVERS, D.N., and GREEN, T.C. (1972): 'Wider aperture sine-cosine direction finders', *Radio Sci.*, **7,** pp. 1025–1032
WUNDT, R.M. (1968): 'Wullenweber arrays', Chapter 5 in 'Signal processing arrays', editor BLACKBAND, W.T. (Technivision, Wokingham).

Chapter 6

Instrumental and site errors

6.1 Introduction

Although this book is not primarily concerned with DF hardware, it is convenient at this point to summarise the physical sources of measurement errors arising from instrumental and site effects. Re-radiating objects in the vicinity of the receiving site are of particular interest here, because the rays from such objects add to the complexity of the wave-field to be resolved.

When the direction finder is not defined more specifically, it may be helpful to visualise it as circular, with a diameter of (say) 300 m, and erected on a site that is 'good' in some sense to be defined below. The DF might be used to make measurements on a variety of sky-wave signals in the HF band (2–30 MHz) for ionospheric research purposes, e.g. to study the structure and movements of travelling ionospheric disturbances. The main types of wide-aperture systems (WADF) in current use can be grouped under three broad headings:

(*a*) The circular WADF with a spinning goniometer and single-channel receiver,
(*b*) The crossed-arm interferometer, with a twin-channel receiver used as a phase meter,
(*c*) An array of *n* elements used with an *n*-channel receiver.

Measurements of elevation angles of arrival, as well as azimuth angles, can be made with types (b) and (c); the latter includes vertical arrays of elements.

This list is not exhaustive; it does not include, for example, various forms of commutated antenna DF (Earp and Cooper-Jones 1958) and Doppler systems (Anderson, 1958) or amplitude-comparison (monopulse) systems, as discussed by Cooper and Wyndham (1973). Further details of hardware and instrumental techniques can be found in a number of textbooks such as those by Keen (1947), Kukes and Starik (1962) and Watson and Wright (1971).

The sources of error in DF can be broadly divided into four categories:

(*a*) Instrumental errors, which can cause incorrect measurements in even the simplest situation, namely a single plane wavefront arriving at the DF array,
(*b*) Site errors, i.e. distortions and deviations of the arriving wave-fronts caused by features in the neighbourhood of the receiving site such as re-radiating conductors, ground contours and gradients of ground conductivity,
(*c*) Wave-interference errors, when the arrival of two or more rays produces swings in the indicated bearing.
(*d*) Propagation errors, i.e. changes in ray direction produced by ionospheric tilts or magneto-ionic effects (see Chapters 14 and 15).

A DF system should be regarded as an instrument for the determination of the direction of arrival (DOA) of one or more rays at the receiving site; that is to say, it cannot be expected to remove errors arising on the propagation path. For a good WADF, i.e. a well designed and well maintained system on a well-chosen site, the total effect of instrumental and site errors might be of the order of 0·1° and is therefore small in comparison with wave-interference and propagation errors, which can easily exceed 1°.

6.2 Instrumental errors

A detailed discussion of the instrumental errors that can arise in various forms of DF system will not be attempted here. It is sufficient to note that the designer of any new instrument should be familiar with the meaning of terms such as spacing error, cogging error, polarisation error and mutual-coupling effects; he must also consider errors arising from a lack of exact channel equality in systems employing two or more receiving channels. A guide to relevant literature up to 1965 can be found in a review paper by Gething (1966a).

Some instrumental errors arise from approximations introduced in the analogue processing of the antenna signals; octantal error in the four-element Adcock provides a convenient example. The indicated bearing α' and the true azimuth of arrival α are related by the equation

$$\tan \alpha' = \frac{\sin\left(\dfrac{\pi d}{\lambda}\cos\Delta\sin\alpha\right)}{\sin\left(\dfrac{\pi d}{\lambda}\cos\Delta\cos\alpha\right)}$$

where d is the diagonal spacing. Writing γ for $(\pi d\cos\Delta)/\lambda$, we obtain

$$\alpha' \simeq \alpha + \frac{\gamma^2}{24}\sin 4\alpha$$

when γ is small and the angles are measured in radians. Since the error $(\alpha' - \alpha)$ depends on $\cos^2\Delta$, it cannot be removed by a ground-wave calibration. This difficulty would be entirely avoided if true phase-difference measurements were made between the *N* and *S* antennas and between the *E* and *W* antennas, or if phase modes were used to determine the angle α (see Section 5.3). An alternative approach would be to combine the measurement of α' with some independent measurement involving Δ and α in some different combination: the author suggested (1966b) that the additional measurement might consist of a comparison of the sum signals, $(E + W)$ and $(N + S)$.

Various forms of error affecting multi-element Adcock systems and Doppler systems have been analysed in a paper by Hammerle (1989); the improvement factor *F* obtained by using a Doppler system rather than an Adcock in a two-ray field is

$$F = \xi / J_1\,(2\xi)$$

where $$\xi = \frac{\pi D}{\lambda} \sin\left(\frac{\delta}{2}\right)$$

$$\delta = \text{azimuth difference}$$

$$D = \text{diameter of the array}$$

$$J_1 = \text{Bessel function of the first kind, order 1}$$

The improvement factor reduces to unity when the aperture is very small, because

$$J_1\,(2\xi) \rightarrow \xi$$

and hence $F \rightarrow 1$

In a modern WADF, instrumental errors can generally be reduced to very small limits by

(*a*) careful attention to detail at the design and construction stage,
(*b*) a high standard of maintenance,
(*c*) regular calibration checks.

So far as (a) above is concerned, it is often easier to adopt the best available standards, e.g. in cutting cables to electrical equality or in surveying the positions for antenna elements, than to spend time and effort on a theoretical justification of lower standards. Cost savings become more important, however, if many instruments are to be produced to the same design.

6.3 Calibration procedures

Calibration signals can be injected at various points in a receiving chain, e.g. to check channel balance in a twin-channel receiver used as a phase meter. When these checks are satisfactory, the antenna signals corresponding to a single plane wavefront from some selection directed can be simulated with the aid of a signal generator and a network of delay lines and hybrids, often referred to as a wavefront generator; this can be used for laboratory-type tests of the DF under simplified conditions.

Calibration signals can then be radiated from a test transmitter at a known position, or a series of known positions, in the neighbourhood of the DF site, e.g. at a distance of about 100 m from a NADF or (say) 1 km from the centre of a WADF. Such tests should be regarded as providing information about instrumental errors with the real antenna system rather than site errors and it should be borne in mind that the transmitter is likely to be within the near-field region of the array.

A mobile-calibration transmitter can also be taken round the array at a distance near the limit at which adequate signals can be heard by ground-wave propagation; this limit might be of the order of 50 km, depending on transmitter power, frequency and terrain. Such a test provides some information about the magnitude of site errors for signals *arriving in the horizontal plane*. In order to obtain a calibration at nonzero elevation angles, but without the propagation errors associated with ionospheric paths, the transmitter must be carried aloft in an aircraft or balloon and must be tracked accurately by optical or radar means.

When a new DF system is being commissioned, it is logical to carry out the various kinds of calibration checks in the above order. Because the aircraft calibration is the most difficult and expensive, it is not usually attempted until other tests have given satisfactory results.

6.4 Effects of re-radiators on measurements of azimuth angle

6.4.1 Introduction

The ideal DF site would be flat, of uniformly high conductivity and entirely free of any building, tree or other structure out to a radius of several miles. In practice the area of ground under the direct control of the DF authority is usually very limited and the presence of artefacts such as lamp-posts, vertical metallic drain-pipes and metallic buildings such as hangars or agricultural barns in the neighbourhood of the site cannot always be avoided.

A paper by Knight (1967) shows how the re-radiated field from a vertical mast varies with frequency, distance etc; although Knight refers mainly to VHF and UHF problems, some of his theoretical results should also apply to the HF band; see also Hill (1964). Even brick buildings are known to produce reflections and 'shadowing' effects at VHF and are clearly undesirable site features in any band. Also at VHF, Bagley (1986) demonstrated that large errors can arise if the desired direct ray from the transmitter is not received at full strength, because of terrain screening, and the indicated bearing is determined by scattered components; unwanted components of the orthogonal polarisation can also be picked up in linearly-polarised elements as a result of one or more stages of polarisation conversion at scattering objects.

Propagation from a distant target transmitter to the DF site is by one or more ionospheric modes. Even in nominally single-mode conditions, the direct signal can be re-radiated by conductors near the DF site, thus producing a multiray field. Any error in the measured DOA arising from such effects, usually referred to for brevity as a site error, is clearly relevant to the main theme of this book, namely the action of DF systems in multiray wave-fields. The main conclusions reached by a number of investigators will therefore be summarised below.

6.4.2 Calculation of susceptibility to site errors

The most common type of calculation found in published literature is of susceptibility to site error. The term susceptibility is used to describe a statistical comparison of the errors calculated for different DF systems in a model situation; no use is made of any information about the real site. The following model was adopted by Hopkins and Horner (1949) in an analysis of several types of DF system.

Consider a direct plane wave of unit amplitude, incident on the DF from elevation angle Δ, and an interfering plane wave horizontally incident, of amplitude ρ at the DF, both waves being of the polarisation for which the instrument is designed. The error due to the interfering wave will depend on its azimuth θ and phase γ relative to the direct ray. By adjusting γ, we can bring the magnitude of the error to a maxmimum value e_m, which can be interpreted as a measure of the 'nuisance value' of the re-radiated ray.

We now find an average value for e_m^2 over all possible azimuths:

$$e_\sigma^2 = \frac{1}{2\pi} \int_0^{2\pi} e_m^2 d\theta$$

An arbitrary constant value of $\rho = 0.1$ is used in the evaluation of the integral; at each azimuth, the value of γ is adjusted to produce maximum error. The quantity e_σ then measures susceptibility to site errors; it is basically a property of the array radiation pattern, not of the site, although the ground constants do have an effect on the vertical radiation pattern.

It must be emphasised that there is no direct connection between this type of calculation, involving the *average* maximised effect of a single obstacle, and the *total* effect of a number of obstacles near a real site. Agreement between a susceptibility calculation and an observed r.m.s. error for various positions of a distant calibration transmitter can be forced by adopting an appropriate value of ρ, but such a procedure is rather meaningless. However, it does seem reasonable to suppose that the order of merit of different DF systems obtained from susceptibility calculations for a single re-radiator would also apply to models involving many re-radiators.

It is usual to assume in susceptibility calculations that the re-radiator is in the far field of the array; wavefronts are assumed to be plane and to give rise to signals of the same amplitude in all the antenna elements. In order to provide a simple comparison, the direct ray is often assumed to arrive from zero elevation angle. In summary, susceptibility calculations are based on a rather artificial model that cannot be directly equated with real situations.

We now consider three specific DF systems; it will be assumed that the direct ray arrives from zero elevation and that the reflection coefficient of the site is zero.

Adcock system

For a small 4-element Adcock, Bain (1953) showed that the appropriate expression for e_m is

$$e_m = \rho \sin \alpha$$

Hence

$$e_\sigma^2 = \frac{\rho^2}{2}$$

$$e_\sigma = \frac{\rho}{\sqrt{2}} \text{ radians}$$

$$= 4{\cdot}05^\circ \quad \text{with} \quad \rho = 0{\cdot}1$$

Two-element interferometer

For two elements at a wide spacing S, with some extraneous means for resolving bearing ambiguities, the appropriate expressions are

$$e_m = \frac{\rho}{a} \sin (a \sin \theta)$$

where

$$a = \pi S/\lambda,$$

and

$$e_\sigma = \frac{4{\cdot}05^\circ}{a} \{1 - J_0(2a)\}^{1/2}$$

for

$$\rho = 0{\cdot}1.$$

This expression gives

$$e_\sigma = 0{\cdot}13^\circ \quad \text{when} \quad S = 10\lambda$$

Single-lobe beam system

As a first approximation to the radiation pattern of a WADF, consider a radiation pattern consisting of a single beam of width B radians to half-power points, *with no side lobes*. Hopkins and Horner (1949) give the expressions

$$e_m = \frac{2\rho B}{\pi} \sin\left(\frac{\pi\theta}{2B}\right) \text{ radians within the beam}$$

$$= 0 \text{ outside the limits of the beam}$$

$$e_\sigma = 1{\cdot}45B^{1.5} \qquad \text{with } \rho = 0{\cdot}1$$

$$= 0{\cdot}04^\circ \qquad \text{for } 5^\circ \text{ beamwidth.}$$

Sum beam, WADF system

The calculation of e_m and e_σ for a more realistic radiation pattern with sidelobes is best carried out by numerical integration, with the re-radiator moved round the array in small steps of azimuth. As a typical example, we consider a ring of monopoles of diameter 327 m in front of a reflecting screen, with the beam formed from a sector of one-third of the elements. At 8 MHz (full diameter ≡ 8.7λ) the value of e_σ for $\rho = 0.1$ is found to be

$$e_\sigma = 0{\cdot}06^\circ$$

From the table given by Hopkins and Horner, it appears that the corresponding value for a difference pattern might be slightly smaller.

6.4.3 Maximum DF error

It is clear that the error produced by a re-radiator varies with azimuth as well as phase. For example, in the single-lobe beam system considered above, the error is zero when the weaker ray coincides with the stronger ray and when the weaker ray is outside the main beam. It is of interest to calculate the azimuth at which the maximum value of e_m occurs.

The weaker ray has maximum effect, on a given azimuth, when it is in phase or antiphase with the stronger ray. We consider these two cases separately. Suppose that the shape of the beam in the horizontal plane, when centred on azimuth 0°, is given by

$$F = \cos(\pi\alpha/2B)$$

where B is the (full) beamwidth in radians to 3 dB points. If the stronger ray is on azimuth α and the weaker on azimuth $(\theta - \alpha)$, the total signal for the in-phase condition is

$$F = \cos\left(\frac{\pi\alpha}{2B}\right) + \rho\cos\left\{\frac{\pi(\theta - \alpha)}{2B}\right\}$$

To find the 'indicated bearing' for the sum pattern, we find the value of α giving the maximum output.

For
$$\frac{dF}{d\alpha} = 0$$

$$\tan(k\alpha) = \frac{\rho \sin(k\theta)}{1 + \rho\cos(k\theta)} \tag{1}$$

where $k = \pi/2B$. The correction α to the indicated bearing (0°) has a positive maximum α_m when θ is given by

$$\cos(k\theta) = -\rho \tag{2}$$

for which

$$\alpha_m = \frac{1}{k}\tan^{-1}\left\{\frac{\rho}{\sqrt{1-\rho^2}}\right\}$$

The total signal for the antiphase condition is

$$F = \cos\left(\frac{\pi\alpha}{2B}\right) - \rho\cos\left\{\frac{\pi(\theta-\alpha)}{2B}\right\}$$

from which it is found that the maximum negative correction α'_m occurs when

$$\cos(k\theta) = \rho \tag{3}$$

and that

$$\alpha'_m = \frac{-1}{k}\tan^{-1}\left\{\frac{\rho}{\sqrt{1-\rho^2}}\right\}$$

Thus α_m and α'_m, which are of equal magnitude, are proportional to beamwidth B.

If ρ is very small, eqn. 2 leads to the approximate condition

$$k\theta \simeq -\frac{\pi}{2} \tag{4}$$

$$\theta \simeq -B$$

and, as might be expected, the weaker ray must be placed at a negative azimuth to produce a negative swing in the indicated bearing. The values of α_m and α'_m are directly proportional to B. Eqn. 4 shows that we cannot assume that $k\theta$ in eqn. 1 is a small angle. Note that k can be comparatively large for a narrow-beam system; for example, $k = 18$ if $B = 5°$. But when α, θ and ρ are all very small, it is legitimate to write eqn. 1 as

$$\alpha \simeq \rho\, \theta$$

In these circumstances, the indicated bearing is 'pulled' by an amount proportional to ρ and θ but approximately independent of beamwidth. Physically, this result is clearly sensible: when the stronger and the weaker rays are very close in azimuth, the indicated bearing is a weighted mean and the limited beamwidth of the array provides little protection against re-radiators close to the axis of the main beam.

6.4.4 An alternative model

An alternative method of estimating the effects of site errors on a two-element interferometer, which may be slightly more realistic, is to assume that the DF is situated in a wave-field consisting of a steady wanted signal plus an angular-noise distribution, uniform in azimuth, of randomly-phased scattered radiation. The statistics of the signals received by a pair of spaced antennas in this field can be calculated in terms of the ratio B of signal and noise powers in one element; Section 2.6, on phase measurements with specular and diffracted components, is relevant. The mean-square phase error for a spacing s is

$$\varphi_0^2 = (1 - R)/B$$

where

$$R = \frac{1}{2\pi}\int_0^{2\pi} \cos(2a \cos\alpha)d\alpha$$

$$= J_0(2a)$$

and $a = \pi s/\lambda$ as before.
The variance σ^2 of the indicated bearing is therefore

$$\sigma^2 = \left(\frac{\varphi_0}{2a}\right)^2$$

$$= \{1 - J_0(2a)\}/(4a^2B)$$

For a narrow aperture system, i.e. for small a, this reduces to

$$\sigma_0^2 \simeq 1/(4B)$$

Thus the error reduction factor obtained by increasing the aperture is measured by

$$\frac{\sigma^2}{\sigma_0^2} = \frac{\{1 - J_0(2a)\}}{a^2}$$

which tends to $1/a^2$ for very wide apertures. The r.m.s. error should then be approximately inversely proportional to the aperture s.

6.4.5. Summary

The calculations in Section 6.4.2 provide a useful comparison of the susceptibility to site errors of three types of WADF with about the same aperture. The multilobed interferometer, as might be expected, is more susceptible to site errors than a single-lobed system with no sidelobes or a single-lobed system with low sidelobes; even so, its susceptibility is only about 1/30th of that of a NADF. The analysis in Sections 6.4.2 and 6.4.4 shows that in certain circumstances e_σ is approximately proportional to $1/a$; the result for the single-lobe beam system, however, suggests that e_σ is approximately proportional to $1/a^{1.5}$. The contribution to the variance of the indicated bearing attributable to site errors is therefore proportional to something between $1/a^2$ and $1/a^3$.

Analytical expressions for e_σ for a number of other DF systems can be found in the Appendix of the paper by Hopkins and Horner (1949). A general conclusion from numerous calculations and experiments is that an aperture of about 1λ is sufficient to reduce the site-error contribution to variance, for a good site, to a negligible level.

We have, throughout this Section, assumed that the DF system provides a single indication of azimuth at any one moment. If the system is capable of resolving the two rays postulated and of returning a separate azimuth measurement on each there is, in principle, no site error and the above calculations are not applicable.

6.5 Effects of the site on the measurement of elevation angles

6.5.1 Introduction

A phase-difference measurement between the signals in two elements with identical radiation patterns in the vertical plane, such as two parallel horizontal dipoles mounted at the same height above flat ground of uniform properties, is not affected by the presence of the ground. Under these assumptions, vertical angle measurements made with a horizontally-disposed array such as a two-arm interferometer are independent of the reflection coefficient of the ground. Errors can, however, arise if the ground is bumpy or if the ground constants vary significantly over the area of interest.

Techniques of vertical-angle measurement that involve the use of a vertically-disposed array generally require some knowledge of the vertical-radiation patterns of the elements and are affected by the nature of the ground. To illustrate the magnitudes of the effects that can occur, we consider here the case of measurements made with a vertical stack of eight horizontal loop antennas, uniformly spaced in height. The dimensions of the array, which will be used in several numerical examples in later chapters, are given in Appendix 3. We visualise a single direct ray of strength E (in units of μV/m, say) arriving from an elevation Δ. The vertical pattern of an individual loop in free space is assumed to be proportional to $\cos \Delta$ and the signal in a loop at height h above flat ground with a reflection coefficient of -1 (see Fig. 6.1) is then

$$S = 2kE \cos \Delta \sin \left(\frac{2\pi h}{\lambda} \sin \Delta \right)$$

where k is a constant with dimensions of length (m) and S is in μV. The factor of two could be absorbed into k, but is retained as a reminder that the signal represents the vector addition of two rays of equal amplitude, the direct ray and the reflected ray. The value of Δ could be deduced from measured signals, S_1 to S_8 in the eight loops at heights h_1 to h_8 either by finding the sine curve that best fits the observations of S against h or by using the Wilkins and Minnis technique (1956) of comparing loop amplitudes in pairs.

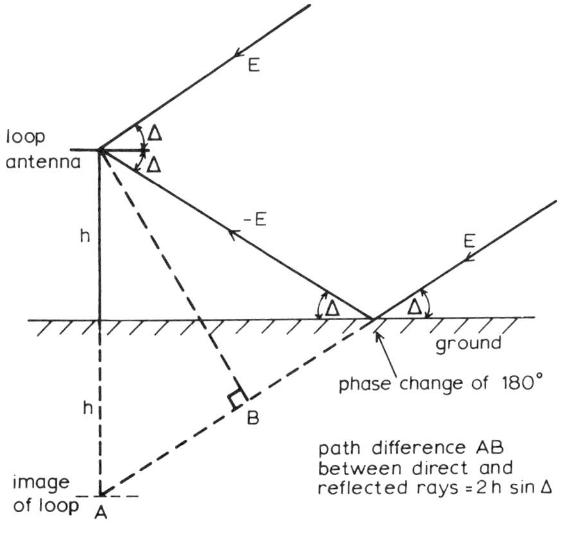

Fig. 6.1 Direct and reflected waves reaching loop

6.5.2 Size of Fresnel zones

The principal contribution to the ground-reflected ray originates from an area of ground defined by the boundary of the first Fresnel zone. This is the locus of points for which the total path length from transmitter to receiver is greater by $\lambda/2$ than the minimum path length corresponding to specular reflection and is approximately elliptical. For plane-earth geometry and an antenna at height h, the radial limits of the zone (which is now truly elliptical) for elevation angle Δ are at distances x_1 and x_2 from a point vertically below the element, where x_1 and x_2 are the roots of the equation

$$x^2 \sin^2 \Delta - 2x \cos \Delta \left(h \sin \Delta + \frac{\lambda}{2}\right) + h^2 \cos^2 \Delta - h\lambda \sin \Delta - \frac{\lambda^2}{4} = 0$$

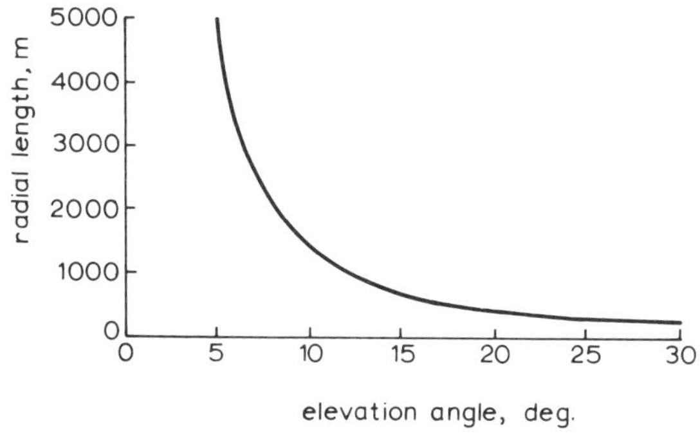

Fig. 6.2 Radial length of Fresnel zone for element at 38 m above flat earth frequency = 10 MHz

In Fig. 6.2 the outer limit is shown as a function of elevation angle for a frequency of 10 MHz and an element height of 38 m. Fig. 6.3 shows the outer limit as a function of frequency for a fixed elevation angle of 10°, again for an element height of 38 m.

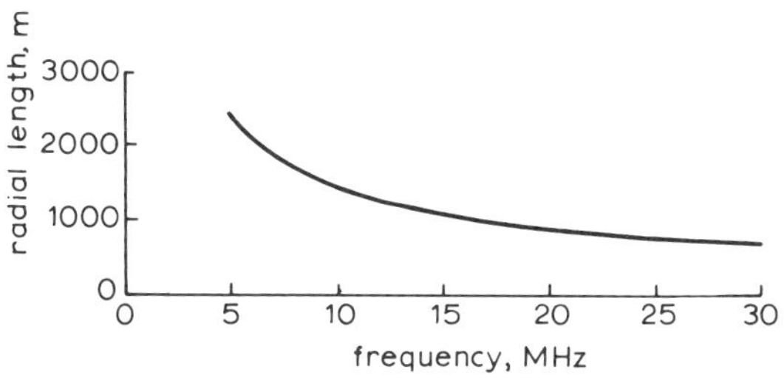

Fig. 6.3 Radial length of Fresnel zone for element at 38 m above flat earth elevation angle = 10°

The ellipses are long and thin. The lengths are reduced when earth curvature is taken into account and can be obtained by multiplying the roots of the above equation by a divergence factor $D(<1)$ which is also a function of elevation angle. The appropriate value of D can be read from a nomogram given by Beckmann and Spizzichino (1963).

Figs. 6.2 and 6.3 are sufficient to show that the reflection coefficient of the ground depends on the average properties over a comparatively large area and that minor bumps etc. are of little significance at HF. The effect of site roughness is approximately equivalent to a reduction in the magnitude of the reflection coefficient (Clarke, 1963).

6.5.3 Uniform ground slope

For a uniform ground slope of angle t down from the tower on which the loops are mounted and for a ground reflection coefficient of -1, it may be shown that each loop signal is of the form

$$S = 2kE \cos \Delta \sin \left[\frac{\pi h}{\lambda} \{\sin \Delta + \sin (\Delta + 2t)\} \right]$$

For small values of t this simplifies to

$$S \simeq 2kE \cos \Delta \sin \left[\frac{2\pi h}{\lambda} \sin (\Delta + t) \right]$$

The error in a measured value of Δ if no account is taken of ground slope is therefore t, the average tilt over a Fresnel zone.

6.5.4. Imperfect reflection

Effect on loop amplitudes

We again suppose that a single direct ray of unit amplitude is incident on the array of loops from an elevation angle Δ. The phase of the mode at the base of the tower can be taken to be zero without loss of generality. The phase of the direct ray at height h is then α, where

$$\alpha = \frac{2\pi h \sin \Delta}{\lambda}$$

Suppose that the ground is level, but is an imperfect reflector. The reflected ray is of amplitude R, where $R \neq 1$ and the phase change on reflection is $(\pi - \beta)$, where $\beta \neq 0$. The resultant signal S is given by

$$\begin{aligned} S^2 &= \cos^2 \Delta \{\sin \alpha + R \sin (\alpha + \beta)\}^2 \\ &\quad + \cos^2 \Delta \{\cos \alpha - R \cos (\alpha + \beta)\}^2 \qquad (5) \\ &= \cos^2 \Delta \{1 + R^2 - 2R \cos (2\alpha + \beta)\} \end{aligned}$$

This cosine formula is exact. We now introduce approximations for the case of horizontally-polarised waves, for which R is close to 1, by writing

$$R = 1 - \epsilon$$

and neglecting ϵ^2. Then

$$S^2 = \cos^2 \Delta \{2 - 2\epsilon - 2(1 - \epsilon) \cos(2\alpha + \beta)\}$$

$$S = 2 \cos \Delta \sqrt{1 - \epsilon} \sin \left(\alpha + \frac{\beta}{2}\right)$$

The factor $\sqrt{1 - \epsilon}$ merely reduces the apparent mode amplitude by the same factor in each loop. The effect of β not being equal to zero is a change in the variation of amplitude with height and is therefore of more interest. It can be shown that β is given by

$$\tan \beta = \frac{-2k \sin \Delta \sin \gamma}{(k^2 - \sin^2 \Delta)}$$

where k and γ are defined by the complex equation

$$k^2 (\cos \gamma + j \sin \gamma)^2 = \kappa - 1800\, \sigma/f - \cos^2 \Delta$$

and κ is the relative permittivity of the ground, σ is the conductivity in S/m and f is the frequency. For large k and small Δ, β is given approximately by

$$\beta \simeq \varphi \sin \Delta$$

where $\varphi = -2 \sin \gamma/k$ and is a function of frequency and ground constants; the approximate proportionality of β to $\sin \Delta$ was noted in Section 2.9.1.

The amplitudes S_r for heights h_r therefore lie on a sine curve of the form

$$S_r \propto \sin \left\{ \frac{2\pi(h_r + \delta h)}{\lambda} \sin \Delta \right\}$$

where

$$\delta h = \frac{\varphi \lambda}{4\pi}$$

Up to this point, the question of sign conventions has not been discussed. However, we can now be sure that δh is positive, corresponding to penetration of the waves into an imperfect ground. For ground of high conductivity, δh is typically a few tenths of a metre. For example, δh for 'wet, rich soil' with a conductivity of 0.03 S/m and a permittivity of 25 is found to be 0.29m at 12 MHz.

The situation is shown diagrammatically in Fig. 6.4; the solid curve shows the expected amplitude/height relationship for a perfect reflector and the dashed curve shows that expected for an imperfect reflector. The difference has been exaggerated for clarity. In practice, the fitting of real observations to either type of curve would probably give an excellent approximation to Δ; the error in Δ resulting from the neglect of δh is unlikely to exceed a few tenths of a degree.

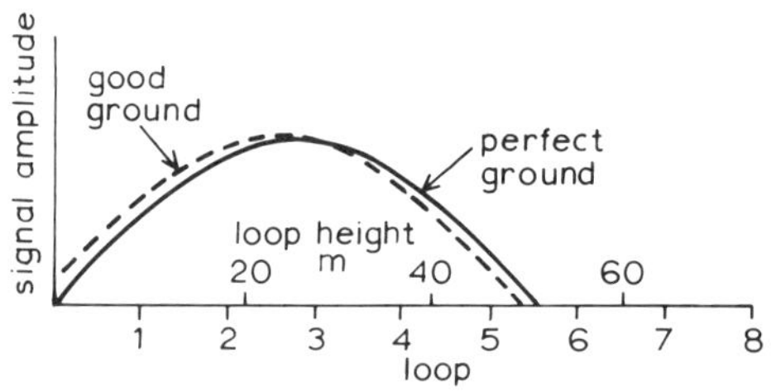

Fig. 6.4 Effect of imperfect reflection on loop amplitudes; separation between curves exaggerated for clarity

If real data from (say) an aircraft calibration are sufficiently good to give a reliable determination of δh, this seems to be an attractive method of determining ground constants; observations made at various elevation angles (but at the same frequency) can be combined in a least-squares solution for δh. Alternatively, δh could be calculated from estimates of the ground constants obtained in some other way, e.g. by the wave-tilt method used by Gething *et al.* (1965).

Effect on loop phases

Eqn. 5 above gives S^2 in terms of two orthogonal components. The tangent of the angle γ by which the loop phase is changed by imperfect reflection is given by

$$\tan \gamma = \frac{\cos \alpha - R \cos (\alpha + \beta)}{\sin \alpha + R \sin (\alpha + \beta)}$$

For small β,

$$\tan \gamma \simeq \frac{\epsilon \cos \alpha + \beta \sin \alpha}{(2 - \epsilon) \sin \alpha + \beta \cos \alpha}$$

and, provided $(2 - \epsilon) \sin \alpha >> \beta \cos \alpha$

$$\tan \gamma \simeq \frac{\epsilon \cot \alpha + \beta}{(2 - \epsilon)}$$

Near $\alpha = 0°$, $180°$ etc, the phase correction can become very large, as is physically obvious from the phasor diagrams; the corresponding amplitude is comparatively small. In general, the phases appear to be complicated functions of the other parameters. An exact equation is

$$\gamma = \frac{\beta}{2} + \tan^{-1}\left\{\left(\frac{1-R}{1+R}\right)\cot\left(\alpha + \frac{\beta}{2}\right)\right\} \tag{6}$$

On transferring the phase reference to $-\delta h$ we obtain

$$\tan\gamma = \left(\frac{1-R}{1+R}\right)\cot\alpha \tag{7}$$

In order to demonstrate the phase spreads that might occur in practice, a numerical example will now be given for a single incident mode at an elevation angle of 22·6° on a frequency of 12 MHz, for various sets of ground constants. The constants adopted for various types of soil are shown in Table 6.1. They are approximately consistent with the empirical relationship between conductivity σ in mS/m and permittivity κ, suggested by Eaton (1974)

$$\kappa = 20(1 - e^{-0.23\sigma})$$

Table 6.1 Ground constants

Ground	Conductivity, S/m	Permittivity
Perfect	(Reflection coefficient −1)	
Excellent	0·03	25·0
Good	0·01	20·0
Poor	0·001	10·0

The ground descriptions used in the Table ('excellent' etc) refer to reflection properties and are adopted purely for ease of reference. Note that 'excellent' corresponds to the 'wet, rich soil' referred to above.

Loop amplitudes, normalised to 100 units in loop 1, and loop phases relative to loop 1, are shown in Table 6.2. Both are rounded off to integer values. The phase variations are particularly large in loop 7 because of the low amplitude in this loop. The phase spread deduced from the figures shown in Table 6.2, i.e. the departure from the strict phase/antiphase condition expected with perfect ground, is shown in Table 6.3.

If the phase/antiphase criterion is to be used to distinguish between single-mode conditions and multimode conditions, it is clear that

(a) the phases of the signals with low amplitudes (say less than 10% of the maximum signal) should be ignored,

(b) some suitable tolerance must be used in examining the phase spread in the remaining loops, of the order of 30° for good ground.

Table 6.2 Calculated amplitudes and phases

	Perfect		Excellent		Good		Poor	
Loop	Amp	Phase deg.	Amp	Phase deg.	Amp	Phase deg.	Amp	Phase deg.
1	100	0	100	0	100	0	100	0
2	126	0	122	3	122	4	124	7
3	58	0	54	8	55	12	58	20
4	53	180	55	176	55	174	55	170
5	124	180	122	181	122	182	123	184
6	104	180	100	184	100	186	102	191
7	6	180	6	246	10	255	17	258
8	96	0	96	0	96	0	96	359

Table 6.3 Calculated Phase Spreads

Ground	Phase spread	
	All loops	Omitting loop 7
Perfect	0°	0°
Excellent	70°	12°
Good	81°	18°
Poor	88°	30°

6.5.5 Effects on elevation angles of nulls

Suppose that the signals from the eight loops are added together in a beam-forming network that incorporates amplitude taper, but not phase taper. The resulting array, located above a perfect ground, will, provided the frequency is high enough, possess nulls in the vertical radiation pattern. If the pattern is now re-calculated for an imperfect ground reflection it is found (Gething, 1973) that some nulls remain at exactly the same position while others move by (typically) a few tenths of a degree and become minima rather than infinitely-deep nulls.

The first family corresponds to the nulls of the array in free space. These are symmetrical about the broadside direction even if the amplitude taper is asymmetrical. Thus, a null at $+\Delta$ is matched by a null at $-\Delta$ and the array above level ground must exhibit a null at $+\Delta$, whatever the ground reflection coefficient. The other family corresponds to ground-interference minima; the array and its image can be regarded as a two-element interferometer, the elements being separated by twice the height h_c of the phase centre of the array above the ground. With imperfect ground, h_c must be replaced by $(h_c + \delta h)$, where δh is as defined in Section 6.5.4. The elements of the interferometer are now further apart and the minima tend to be lower (for excellent and good ground) than corresponding nulls.

An example of this effect, and of the distinction between the two families of nulls and minima, is shown in Table 6.4 for a uniformly weighted array. The ground constants are as in Table 6.1.

Table 6.4 Elevation angles of nulls and minima at 15 MHz for various types of ground reflection

Perfect (Reflection −1)	Excellent		Good		Poor		Free space (no reflection)
Nulls	Nulls	Minima	Nulls	Minima	Nulls	Minima	Nulls
13·9		13·9		13·9		14·0	
15·7	15·7		15·7		15·7		15·7
28·8		28·7		28·7		28·9	
32·8	32·8		32·8		32·8		32·8
46·3		46·0		46·1		46·5	
54·4	54·4		54·4		54·4		54·4
74·6		73·8		74·4		–	

The array null at 0°, and the element null at 90°, have been omitted for simplicity.

It will be shown in later chapters that some methods of vertical-angle measurement depend on synthesising nulls at the mode-arrival angles. If the reflection coefficient of the real ground is assumed to be −1, the errors in the measured angles will be zero if free-space nulls are used; from Table 6.4 it is seen that the errors are of the order of a few tenths of a degree if ground-interference minima are used.

6.5.6. Calibration with airborne transmitter

It is difficult to predict theoretically the errors in elevation angle measurements that might be caused by the height variations and by the imperfect reflection coefficient at a real site. The above calculations all suggest that, given a site of comparatively high conductivity that is level to within a few metres out to a distance of (say) 1000 m, the angle errors are not likely to exceed a few tenths of a degree. However, it is desirable to check this conclusion, and to test the correct functioning of the measurement equipment, by means of a calibration procedure with an airborne transmitter; it is usually only possible to sample a few azimuths, elevation angles and frequencies.

The fact that the aircraft carrying the transmitter is at a finite distance from the tower introduces wavefront curvature effects, which will now be considered. For a loop at height h and the aircraft A at distance r from the base of the tower (Fig. 6.5) we assume that high powers of h/r can be neglected. This approximation would not, of course, be valid if the aircraft approached within a hundred metres.

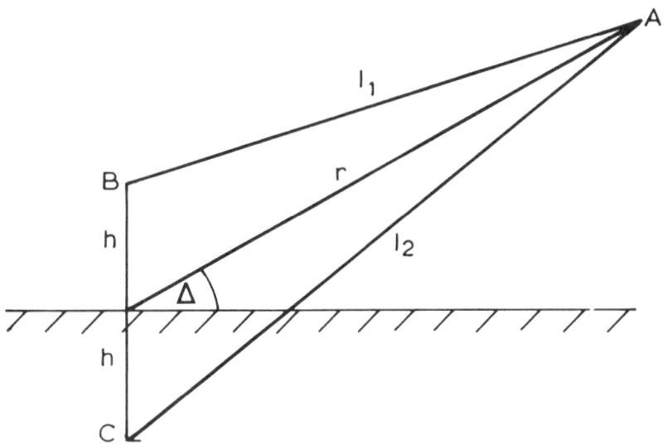

Fig. 6.5 Path lengths from aircraft to loop and image

Following the usual practice, we neglect amplitude effects arising from the finite distance of the aircraft. To calculate phases of the direct signal to B and of the signal to the loop image at C we note that

$$l_1^2 = r^2 + h^2 - 2rh \cos\left(\frac{\pi}{2} - \Delta\right)$$

$$l_2^2 = r^2 + h^2 - 2rh \cos\left(\frac{\pi}{2} + \Delta\right)$$

Hence

$$l_1 \simeq r \left\{1 - \frac{h}{r} \sin \Delta + \frac{h^2 \cos^2 \Delta}{2r^2} + O\left(\frac{h^3}{r^3}\right)\right\}$$

$$l_2 \simeq r \left\{1 + \frac{h}{r} \sin \Delta + \frac{h^2 \cos^2 \Delta}{2r^2} + O\left(\frac{h^3}{r^3}\right)\right\}$$

where $O(x)$ denotes 'terms of the order of x'. The amplitude of the loop signal depends on $(l_2 - l_1)$. To the order of approximation used here,

$$(l_2 - l_1) = 2h \sin \Delta$$

This is the same expression as that obtained for an aircraft at infinite distance; note that there is no dependence on r. The phase of the loop signal is closely related to $\frac{1}{2}(l_1 + l_2)$: the correction term is $(h^2 \cos^2 \Delta)/(2r)$ and the measured phases should be altered by $(\pi h_s^2 \cos^2 \Delta)/(r\lambda)$ radians, where h_s is the height of the sth loop. If the sign convention is chosen correctly, the normal phase/antiphase criterion should be more closely satisfied by the corrected phases than by the uncorrected phases. To a first approximation, therefore, loop amplitudes are unaffected by wavefront curvature but small corrections to the loop phases should be made.

6.6 Conclusions

Broadly, we have been concerned in Chapters 5 and 6 with the determination of the DOA of a single ray, bearing in mind that re-radiating objects near the site can give rise to additional rays. The basic tool required is a directive array with a known radiation pattern. Various factors that can lead to small departures of the real pattern from the pattern calculated from simple theory have been considered.

The calculations presented should be regarded as illustrative rather than exhaustive but it seems that, provided suitable precautions are taken, errors arising from site effects such as slopes, imperfect ground reflection and re-radiators are likely to be very small for arrays with apertures of a few wavelengths or more. The effects of mutual coupling are more problematic and every effort should be made in the design of an array to reduce mutual impedances to the lowest possible level relative to self impedances.

Calculations of site effects and of susceptibility to site errors are inevitably based on incomplete and idealised mathematical models of the real site. Susceptibility calculations must in any case be regarded as comparative rather than absolute. It is therefore valuable to test the real radiation pattern by means of some form of calibration exercise, using a low-power mobile transmitter; measurements made under carefully-controlled conditions for even a limited sample of directions and frequencies are worthwhile.

The next few chapters will be devoted to the action of a direction finder in multiple ray situations and it will be convenient to assume, in the simulated problems used to illustrate the theory, that instrumental and site errors can be neglected. The results presented in this chapter provide some justification for this assumption.

References

ANDERSON, R.E. (1958): 'Bearing memory improves direction finder', *Electronics*, **31,** 31 Jan. 1958, pp. 44–48

BAGLEY, G.C. (1986): 'The effects of transmitting site pollution on tactical VHF DF', Colloquium on direction finding systems in their operating environment, *IEE. Digest No. 1986*/**10**, pp. 2/1–5

BAIN, W.C. (1953): 'Calculation of wave interference errors on direction finder employing cyclical differential measurements of phase', *Proc. IEE*, **100,** pp. 253–260

BECKMANN, P., and SPIZZICHINO, A. (1963): 'The scattering of electromagnetic waves from rough surfaces', (Pergamon), p. 235

CLARKE, R.H. (1963): 'Theoretical characteristics of radiation reflected obliquely from a rough conducting surface', *Proc. IEE.*, **110**, pp. 91–100

COOPER, D.C., and WYNDHAM, B.A. (1973): 'A comparison of monopulse techniques for direction finding in the presence of interference', International Conference on Radar – Present and Future, London 23–25 October 1973 (IEE), pp. 154–159

EARP, C.W., and COOPER-JONES, D.L. (1958): 'Practical evolution of commutated-aerial direction-finding system', *Proc. IEE.*, **105B,** Supp. 9, pp. 317–332

EATON, J.L. (1974): 'MF aerials: the determination of ground-wave CMF and mean ground conductivity from radial field measurements by an optimisation technique', BBC Engineering, No. 99, pp. 41–44

GETHING, P.J.D. (1966a): 'High-frequency direction finding', *Proc. IEE.*, **113,** pp. 49–61

GETHING, P.J.D. (1966b): 'Determination of octantal error on a 4-element Adcock direction finder', *Electron. Lett.* **2,** pp. 388–389

GETHING, P.J.D. (1973): 'Concept of phase centre of an array, applied to elevation-angle measurements', *Proc. IEE.*, **120,** pp. 541–543

GETHING, P.J.D., MORGAN, A.D., SHEPHERD, E.G., and TIBBLE, D.V. (1965): 'Wave-tilt measurements of ground properties', *Electron. Lett.* **1**, pp. 106–107

HAMMERLE, R. (1989): 'Factors limiting the accuracy of Doppler and Adcock direction finding systems', Passive DF Colloquium, *IEE.* Digest No. 1989/**16,** pp. 3/1–13

HILL, P.C.J. (1964): 'The measurement of re-radiation from lattice masts at VHF', *Proc. IEE.*, **111,** pp. 1957–1968

HOPKINS, H.G., and HORNER, F. (1949): 'Direction-finding site errors at very high frequencies', *Proc. IEE.*, **96,** pp. 321–332

KEEN, R. (1947): 'Wireless direction finding', (Iliffe) 4th edition

KNIGHT, P. (1967): 'Re-radiation from masts and similar obstacles at radio frequencies', *Proc. IEE.*, **114,** pp. 30–42

KUKES, I.S., and STARIK, M. Ye. (1962): 'Osnovy Radio-pelengatsii (Principles of radio direction finding), (Soviet Radio Publishing House, Moscow).

WATSON, D.W., and WRIGHT, H.E. (1971): 'Radio direction finding', (Van Nostrand Reinhold)

WILKINS, A.F., and MINNIS, C.M. (1956): 'Arrival angle of hf waves', *Wireless Engr.*, **33,** pp. 47–53

Chapter 7

An introduction to resolution techniques

7.1 The resolution goal

We shall assume from now on that a DF system is available that is capable of measuring correctly, in single-ray conditions, at least the azimuth angle of arrival of the ray. In effect, instrumental errors and site errors will be regarded as negligible, although it is not essential to assume an absence of re-radiating objects near the DF; some of the discrete rays in multi-ray problems could arrive from such objects.

As an example of the order of resolution required from the DF system, suppose we wish to determine, to an accuracy of 0·1°, the separate azimuths of arrival of two modes separated by 1° in azimuth. Broadly, narrow-aperture DFs (NADF) with apertures of a few tenths of a wavelength (λ) at the frequencies of interest are incapable of providing the required resolution in any reasonable time interval for any realistic signal/noise ratio. Very large linear arrays designed primarily for ionospheric research, such as the 2·5 km array at Stanford University described by Sweeney (1970) and the 1·1 km array used by the Canadian Communications Research Centre (Rice, 1971), are capable of meeting the goal for signals arriving from near-broadside directions, but would be very expensive to implement in a circular form to provide the same resolution in any azimuth sector.

Between these two extremes of NADF and very wide aperture DF (VWADF) lie WADFs with array dimensions of the order of a few λ; the physical dimensions fall generally in the range 50 to 500 m. It is with such systems that we shall be concerned, and the underlying question is whether their classical resolving power can be enhanced by intelligent analysis of the time-varying wave-field produced by the complex fading of two or more rays.

Ability to resolve a multicomponent wave-field depends on several physical factors in addition to array aperture: it depends on the number of rays, their separations in azimuth and elevation, the signal/noise ratio and the length of the observation period in relation to fading rates. It also depends on the nature of the signal processing scheme or analysis method used to calculate the DOA's of the rays; we therefore begin with a general survey of possible processing schemes. The advent of multichannel receivers and the availability of digital computers have opened up new possibilities which will be examined in greater detail in subsequent chapters.

7.2 A simple example

Let us consider a specific example of the general resolution problem, in order to provide a picture of the magnitudes of some of the physical parameters involved. Suppose that two rays are received, corresponding to 1*E* and 1*F*2 modes from a transmitter at a distance of about 1000 km, that the *E*-mode arrives from the true bearing and that the *F*-mode arrives from about 1° off the true bearing as a result of a tilt in the F-layer. The elevation angles might be about 9° and 26° for the *E*- and *F*-modes, respectively. The travel time of the *F*-mode is about 0·44 ms longer than the time over the E-layer path. Finally, the *F*-mode might exhibit a Doppler shift of up to about 1 Hz relative to the *E*-mode, because of a motion of the F-layer affecting the reflection point.

These characteristics immediately suggest four possible methods of separating the two modes:

(*a*) In azimuth,
(*b*) In elevation,
(*c*) In time,
(*d*) In frequency.

All of these methods are potentially valuable.

The conventional DF, used for azimuth measurements only, does not usually have a large enough aperture to resolve modes directly; if the Rayleigh criterion for the resolving power of a telescope objective of diameter D is adopted, the resolving power is $1{\cdot}22\lambda/D$ radians and a diameter of 70λ is required to separate modes 1° apart, as in this example. However, the Rayleigh criterion is pragmatic (Hausz, 1964), and takes no account of time averaging. It should also be possible to exploit the much greater separation of the modes in elevation angle; in a crossed-arm interferometer, for example, the direction cosines of the arrival angles relative to the arms are determined and this facilitates mode separation.

Separation of modes in the time domain is, of course, easy with a transmitter radiating short pulses. In the case of longer unresolved pulses, e.g. on/off keyed signals, the first and possibly the last arrival can be isolated by appropriate triggering and gating. So called 'dot-lock' techniques to select the first mode have often been tried in the past (Hatch, 1947), with rather limited success in terms of improved accuracy; the rejection of a large proportion of the arriving energy appears to be an undesirable feature. Time separation is not available, at least in any simple form, on

CW signals. The autocorrelation function of signal amplitude against time may, in favourable circumstances, display humps which reveal the number of modes and their time separations, but does not provide separate DOA measurements on the various modes.

Techniques have been developed that enable the relative Doppler shifts of two or more modes to be recorded and displayed; see Davies and Baker (1966). The techniques work best on transmitters with carriers that are very stable in frequency. From amplitude/frequency displays obtained with two antenna beams pointing in slightly different directions it is possible, in principle, to determine an amplitude ratio for each mode and hence to decuce its bearing. However, apart from the need for frequency stability in the transmitter, there is the added difficulty that the ionosphere does not always produce a significant difference of Doppler shifts between the modes. The technique is particularly suitable for the study of travelling ionospheric disturbances, and considerable progress has been made by combining bearing measurements and Doppler measurements, as in experiments described by Jones and Reynolds (1975). Others who have used the Doppler technique for mode separation include Clarke and Tibble (1978) and Afraimovich and Panchenko (1979).

It would be rash to rule out the use of any combination of these methods in future generations of equipment. Here we shall assume that mode separation in the angle coordinates is the most promising line of research in the near future; this method is likely to be the most straightforward to implement in a wide variety of situations because it does not impose special requirements on transmitter modulation, frequency stability or ionospheric instability.

7.3 More complicated situations

In practice, the wave-field may be a good deal more complicated than in the simple example just described. Several rays may be incident on the DF, some from the target transmitter, some from sources of interference and some from re-radiating obstacles near the site. Each ray may be surrounded by a cone of diffracted energy, and is likely to exhibit some 'jitter' in DOA as a function of time. If this jitter becomes very large (several degrees, say) it is questionable whether DF measurements are of any value.

If the number of incident rays of comparable strengths is very large, and if a continuous distribution of 'sky noise' is present in addition to the discrete arrivals from unwanted sources, it is unlikely that the wave-field can be resolved by simple analysis methods. It then becomes necessary to resort to the deconvolution methods of radioastronomy. The radio astronomer scans the incident field with a narrow-beam array and then uses mathematical techniques known as deconvolution or restoration to 'sharpen up' his map, i.e. to allow for the effects of the known beamwidth. Because more than one field distribution could have produced the map, there is no unique answer to this process, but the 'principal solution' discussed by Bracewell (1958) is usually sought by Fourier transform methods. Even so, there are theoretical difficulties introduced by the finite-grid size used in digitising the information, which lead to debates about the best *practical* method of restoration (Newman, 1973).

7.4 The value of elevation-angle measurements

As noted above, it may be possible to detect the presence of more than one mode more easily from elevation-angle measurements than from azimuth measurements. This argument alone provides a good reason for attempting to measure both angles on each arriving ray. In addition, elevation-angle measurements combined with some knowledge of the state of the ionosphere produce a number of other potential benefits, the value of which depends on the exact application.

Mode identification has already been mentioned as a pre-requisite for the intelligent analysis of bearing errors and the possible application of corrections for ionospheric tilts. Given a suitable ray-tracing program the arriving rays can, in effect, be launched back from the receiving site towards the transmitter and the ray trajectories studied. Ideally, we should find that all the rays meet (in some cases after multiple reflections) at the true position of the transmitter. Any measured relative time delays between modes should also be correctly explained, in the sense that wave packets should trace back to a common time origin as well as a common space origin.

In practice, knowledge about the state of the ionosphere at a particular moment is likely to be very imperfect, and the use of a sophisticated ray-tracing program is rarely justified in real problems, although it may be very valuable for theoretical studies. However, quite approximate ionospheric information may be sufficient for the determination of path geometry. The first step is to decide whether a ray re-launched at the measured elevation angle is likely to be reflected at the E-, F1-, or F2-layers. Since the properties of the normal E- and F1-layers are very predictable as functions of solar zenith angle and season, the only real uncertainty is the possible presence of *E*s ionisation. Secondly, a knowledge of the true range of the transmitter, if available, can be used to decide the probable number of hops.

Bailey and McClurg (1963) and Treharne (1967) pointed out that vertical triangulation can be used in certain circumstances to deduce the range of a transmitter of unknown position. Interest in techniques of single station location, in which bearing and range are determined from the same site, continues; see, for example Treharne (1981), Barabashov, Vertogradov and Khondu (1987), McNamara (1988), Baker and Lambert (1989) and Warrington and Jones (1991). The possibility of estimating the range of the target transmitter provides an additional reason for making elevation-angle measurements.

Finally, when a DF is used to track TIDs by means of measurements on one or more transmitters of known position, it seems intuitively obvious that a study of correlated variations in elevation and azimuth angles (plus

Doppler measurements if possible), will yield much more information than a study of either alone. A two-dimensional array of elements distributed in the horizontal plane can be used, in principle, to measure both angles. Some vertical aperture is desirable, however, if low-elevation angles are to be measured.

7.5 Terminology and notation

Because we shall be so much concerned in the next few chapters with the measurement of azimuth angles and elevation angles of arriving rays, it is now time to make use of the abbreviation HAM for horizontal angle measurement and VAM for vertical angle measurement. As noted in the Introduction, HAM may refer either to a technique for measuring ray azimuths or to the action of taking an azimuth observation, and similarly for VAM; the exact meaning should be clear from the context.

In multiray conditions there is no reason to suppose that the indicated bearing (as defined in Chapter 4) is equal to the azimuth of any one ray. It is important that the distinction between true bearing (one number), indicated bearing (one number) and the azimuths of arrival $\alpha_1, \alpha_2, \ldots, \alpha_n$ for n rays (n numbers) should be constantly borne in mind. The symbol Δ will be used for elevation angles. It is sometimes convenient to refer to rays from the target transmitter as modes or 'wanted' rays, in order to distinguish them from 'unwanted' rays from interfering transmitters within the receiver passband. Full resolution requires the correct identification of all wanted rays, and the correct determination of all their parameters: DOA, relative amplitude and relative phase. A more practical objective is to determine the DOAs of the two or three strongest rays; if these angles are correctly determined within (say) 0·1° we can reasonably claim that the structure of the wave-field has been explained, even if the existence of much weaker rays (say 20–30 dB down on the strongest) passes undetected. Once the DOAs have been determined, it is usually not too difficult to infer something about relative average amplitudes and relative phases; the latter are usually of little interest. As an absolute minimum, we wish to be able to determine the correct azimuth, within some suitably small tolerance, of the strongest ray. Unfortunately the strongest ray, or even the strongest few rays, are not always from the target transmitter.

Fluctuations in indicated bearing, produced by relative fading between rays, are often described in published literature as arising from wave interference (WI). We shall see that a study of WI fluctuations alone can provide some information about the structure of the wave-field.

When two rays, A and B, are incident on the DF, the surfaces of constant phase are of the kinked form described in Chapter 3. At moments when ray B fades to some insignificant amplitude level compared with A, the surfaces become plane and the indicated bearing is equal to the azimuth of ray A. It is convenient to refer to such a situation as quasiunimoded propagation (QUMP); the prefix quasi indicates that the condition is transitory. Clearly, if moments of QUMP can be identified in some way, separate HAM can be obtained on A (when B is very weak) and on B (when

A is very weak). The subject of wavefront testing, to detect moments of QUMP automatically, will therefore receive some attention. In principle, moments of QUMP could occur in (say) a three-ray field if two rays faded simultaneously to a low level, but such periods are likely to be much rarer than in a two-ray field.

Wavefront testing is one example of criteria that can be incorporated in what will be referred to as the DF processor. In digital processing, the processor may consist of nothing more than a computer algorithm. In general, however, the term has a wider significance and includes the way in which the DF array is to be used and the way in which the output is to be interpreted. Consider, for example, a circularly-disposed antenna array (CDAA) of the Wullenweber type; the operator might be instructed, or an automatic device might be programmed, to use the low-angle difference output of the spinning goniometer, to set the cursor on the principal minimum of the display, to take 10 observations at intervals of 10 seconds and to average the results. Any change in these instructions, such as using the maximum of a sum beam, and rejecting observations when the display is markedly asymmetric, or of low amplitude, would generally produce a different indicated bearing. Another important aspect of the processor is the way in which any ambiguities are to be resolved, e.g. on an interferometer system: the order in which observations are taken at narrow and wide spacings on one or more arms and the way in which observations on different arms are related can have an effect on performance in multiray conditions. The aim is always to adopt a processor that works perfectly in single-ray conditions and in an optimum or near-optimum fashion in multiray conditions.

7.6 Restriction on the type of direction finder

It is possible to suggest a wide variety of array configurations and, for each configuration, many different processors. Each of these combinations can be tested, either against a range of toy problems or in real situations. In order to keep the topics for discussion within bounds, we shall largely restrict our attention to four main types of DF that seem most suitable for the measurement of DOAs in the HF band. These are

(*a*) The CDAA, with a spinning goniometer and single-channel receiver (HAM only),
(*b*) The crossed-arm interferometer, with a twin-channel receiver used as a phase meter (HAM and VAM),
(*c*) A vertically-stacked array of horizontally-polarised elements (VAM only),
(*d*) Any horizontally-disposed set of n elements used with an n-channel receiver (HAM and VAM).

The dimensions adopted in Chapter 5 for a typical CDAA of 96 doublet elements will again be adopted here (see Appendix 1). Interferometers, which may be defined briefly as systems with multilobed radiation patterns, can be of many forms (Sherrill, 1971); elements are usually arranged in two or more straight arms, with the arms forming a cross, triangle, or *L*-shape. Because we shall be mainly concerned (in Chapter 9) with angle measurements relative to one arm, it is not necessary to restrict discussion to a particular arrangement of arms.

In order to illustrate some of the problems of VAM, the array of eight equally-spaced loops on a vertical tower considered in earlier chapters will again be used; the dimensions are defined in Appendix 3. It will be assumed that an eight-channel receiver is available for use with the array, the channels being carefully matched in amplitude and phase characteristics.

The horizontally-disposed set of *n* elements used with an *n*-channel receiver might be arranged in a circle, as in (*a*) above, or in crossed-arm form as in (*b*), or in a more irregular shape. Whatever its form it differs in principle from (*a*) because the outputs of the elements are *not* combined in a beam-forming network (BFN) to produce a directional radiation pattern. It also differs from (*b*), in which elements are usually selected in pairs. Instead, the amplitudes and phases of the element voltages are recorded digitally for subsequent (or, less commonly, real-time) computer analysis; the same procedure is generally used with the vertical array (*c*).

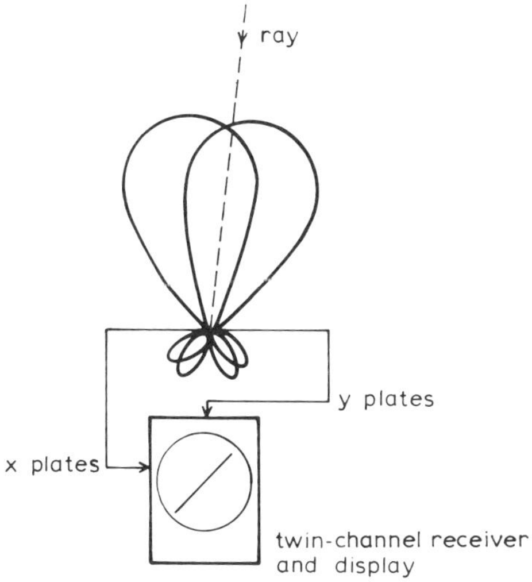

Fig. 7.1 Amplitude-comparison DF system

7.7 The processor: general considerations

7.7.1 Amplitude and phase systems

Some DF processors are based on amplitude comparisons, as in radar monopulse systems, and some on phase difference measurements, as in interferometers. Fig. 7.1 shows diagrammatically a very simple form of amplitude comparison system; the amplitudes from two antenna beams are compared by means of a twin-channel receiver and oscilloscope display. A straight line of inclination 45°, for example, shows equality between the signals and might imply a single ray arrival on the axis of symmetry, as shown.

In the simplest form of such a system, the antennas used have coincident phase centres and identical radiation patterns. A number of general points are worth making. Firstly, there may be several intersections between the two patterns, suggesting that any given angle on the visual display can correspond to more than one DOA. Secondly, the amplitude ratio (for a single ray) is a function of elevation angle and azimuth angle; without some additional measurement, it is not possible to determine these quantities separately. On the boresight direction, the amplitude ratio (unity) is not affected by changes of elevation angle. Thus the system provides no way of discriminating between one mode and several modes in or very near this plane. This comparative sensitivity to azimuth changes and insensitivity to elevation changes, over a narrow sector near boresight, is sometimes an advantage when only HAM is required but must be counted as a disadvantage when the objective is ray resolution.

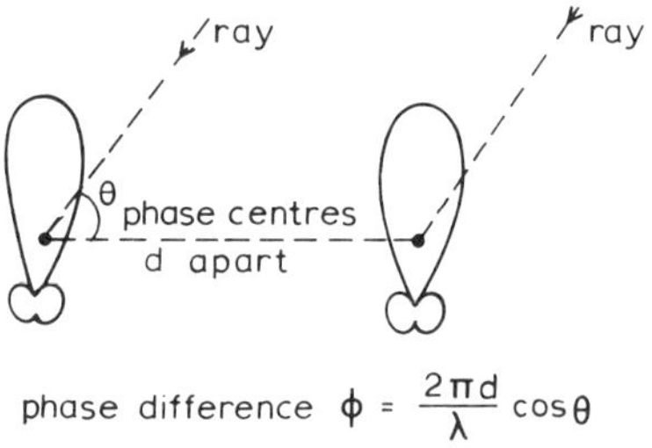

Fig. 7.2 Phase-difference DF system

Fig. 7.2 shows an alternative arrangement, in which the phase centres of the elements or directive arrays are separated by some distance *d*. If directive patterns are used, it is now desirable to have them identical and pointing in the same direction. Suppose that the output phasors for a single ray are *a* and *b*; these phasors will be of equal amplitude but will differ in phase by φ, where

$$\varphi = \frac{2\pi d}{\lambda} \cos\theta \qquad (1)$$

and θ is the angle between the ray and the line joining the phase centres. The twin-channel receiver and oscilloscope can now be used as a phase-meter by displaying $(a + b)$ against $(a - b)$ changed in phase by $\pi/2$. It is easily shown (see Fig. 7.3) that the angle φ can then be found from the relation

$$\varphi = 2\gamma$$

where γ is the inclination of the line display on the oscilloscope CRT. When a and b are not equal in amplitude the line becomes an ellipse, as noted by Ross, Bramley and Ashwell (1951); the above relationship still holds, with γ as the inclination of the major axis of the ellipse.

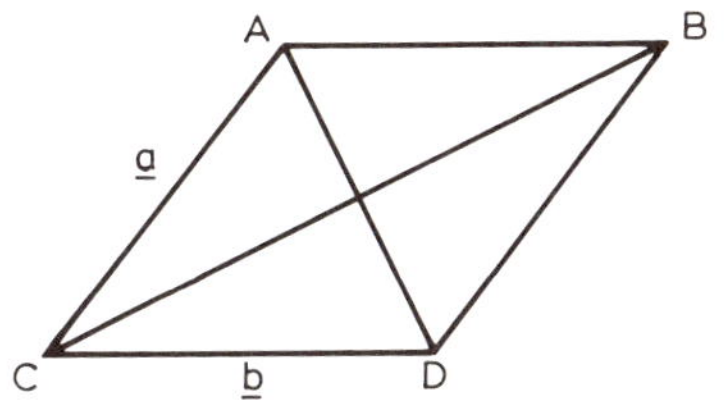

Fig. 7.3 Sum-and-difference phase meter
a and b are of equal magnitude and $A\hat{C}D = \varphi$
$CB \perp AD$
sum $CB = 2CD \cos (\varphi/2)$
difference $AD = 2CD \sin (\varphi/2)$

$$\therefore \quad \frac{AD}{CB} = \tan \frac{\varphi}{2}$$

We again encounter a lack of sensitivity to elevation angle near the boresight (i.e. broadside) direction, a dependence in other directions of θ on both azimuth and elevation, and the possibility of two or more rays producing what appears to be a single-ray display, if they happen to be on the surface of a cone whose axis passes through the line of phase centres. That is to say, various combinations of azimuth α (relative to the axis) and elevation Δ can give the same value of θ, because

$$\cos \theta = \cos \alpha \cos \Delta$$

For the moment we regard a two-element phase system as providing a capability to measure the angle θ.

Thus a distinction can be drawn between the amplitude-comparison method, based on squinted beams with coincident phase centres, and the phase difference method, using parallel beams and separated phase centres; in the former, directive patterns of known shape are essential, whereas directivity is optional in the latter. It is, of course, possible to suggest more sophisticated systems involving some combination of amplitude and phase measurements.

7.7.2 The value of aperture

We now consider the phase-measuring system a little further, still under single-ray conditions. It is convenient to refer to the angle θ in eqn. 1 as the cone angle; it is the angle between the ray and the line joining the two antennas, which might form one arm of an interferometer. The error in the determination of θ resulting from an error in the measurement of φ can be inferred from the equation

$$\frac{d\varphi}{d\theta} = -\frac{2\pi d}{\lambda}\sin\theta$$

For maximum instrumental accuracy we seek to make $d\varphi/d\theta$ large; hence we need d large and θ near $\pi/2$. Thus, instrumental accuracy for a given aperture is a maximum when $\theta = \pi/2$, and drops to zero when $\theta = 0$ or π. It follows that, given a selection of possible arms in some larger array, it is best to choose an arm approximately perpendicular to the ray. In the usual terminology applied to linear arrays, we prefer a broadside array to an end-fire array.

Strictly, we have considered the accuracy of the determination of θ for one ray, rather than resolving power, the ability to separate two or more rays. However, similar geometrical considerations are likely to apply: since the end-fire arrangement is insensitive to small changes in the ray DOA, it is hardly likely to detect small separations between two rays.

For accurate measurements of α, the preferred geometry is therefore a long horizontal arm approximately perpendicular to the ray direction; for the measurement of *small* elevation angles a vertical array is desirable, but above 45° a horizontal end-fire array might be superior. In summary, we aim to make the projected aperture, on a line defined by the intersection of the plane normal to the ray direction and the plane containing the angle to be measured, as large as possible.

It is well known that ambiguities in the determination of θ begin to appear when $d > \lambda$. Consider, for example, angles θ_1 and θ_2 such that

$$\cos\theta_1 = \frac{\lambda\ \varphi_1}{2\pi d}$$

$$\cos\theta_2 = \frac{\lambda\ (\varphi_1 + 2\pi)}{2\pi d}$$

Because the whole number of cycles $2\pi n$ in a phase difference cannot be determined, φ_1 is indistinguishable from $(\varphi_1 + 2\pi)$, $(\varphi_1 + 4\pi)$ . . . $(\varphi_1 + 2\pi n)$. The number of possible solutions therefore increases as d increases. The standard method of overcoming this difficulty (Sherrill, 1969) is to use a comparatively narrow spacing ($d < \lambda$) for an unambiguous but approximate determination of ray direction, followed by a measurement at a wider spacing to improve the accuracy. These measurements are often referred to as coarse and fine respectively. In some large interferometers, it is necessary to take successively coarse, medium and fine measurements.

So far we have assumed that the accuracy with which phase can be measured is independent of aperture. This is a reasonable assumption if we adopt an idealised model of a single, 'clean' ray and suppose that the sources of measurement error are entirely instrumental. We could then write

$$\sigma_\varphi = k$$

where σ_φ is the standard deviation of a phase measurement and k is some constant. If, on the other hand, the standard deviation is entirely due to small fluctuations in θ, we expect a variation with aperture of the form

$$\sigma_\varphi \propto d$$

and the variance of phase measurements would then increase with d^2.

In practice, a single mode may consist of a specular component plus a diffracted cone of energy. It is therefore not realistic to visualise a mode as a clean, steady, single ray in a noiseless environment. Because of the decrease in the correlation function R, as described in Section 2.6.1, with increasing aperture, the accuracy advantage obtained with a large aperture is somewhat less than the theoretical limit. The same remark applies, for rather similar reasons, when two or more rays are present; the departures of the kinked lines of constant phase from the straight lines for a single mode tend to increase (for moderate apertures) with increasing aperture. The aperture advantage is further reduced, when the accuracy of (say) azimuth measurements is tested against true great circle bearing rather than the (unknown) true azimuth of arrival, because systematic errors such as those arising from ionospheric tilts affect systems of any aperture. The proportion of incorrectly resolved ambiguities, usually small, may also increase slightly when the fine aperture is increased.

Table 7.1 Variation of standard deviation with aperture

Source of error	s.d. proportional to
Phase-measurement or other instrumental errors	$\frac{1}{d}$
WI within and between modes	$\frac{1}{\sqrt{d}}$ approx.
Systemastic errors (tilts)	constant
Site error susceptibility	$\frac{1}{d}$ or $\frac{1}{d^{1\cdot 5}}$

The expected variation of standard deviation (s.d.) with aperture for various types of error is therefore as shown in Table 7.1. The entry for susceptibility to site errors is based on the results presented in Section 6.4. The overall improvement for a particular system depends on which of these errors is dominant. When a simple rule of thumb is required, it is both convenient and reasonable to assume that *variance* is inversely proportional to aperture.

Results obtained by Burtnyk, McLeish and Wolfe (1965) provide some support for this rule. They made phase difference measurements on a variety of sky-wave signals at antenna spacings of 7·5 m and 122·5 m. The

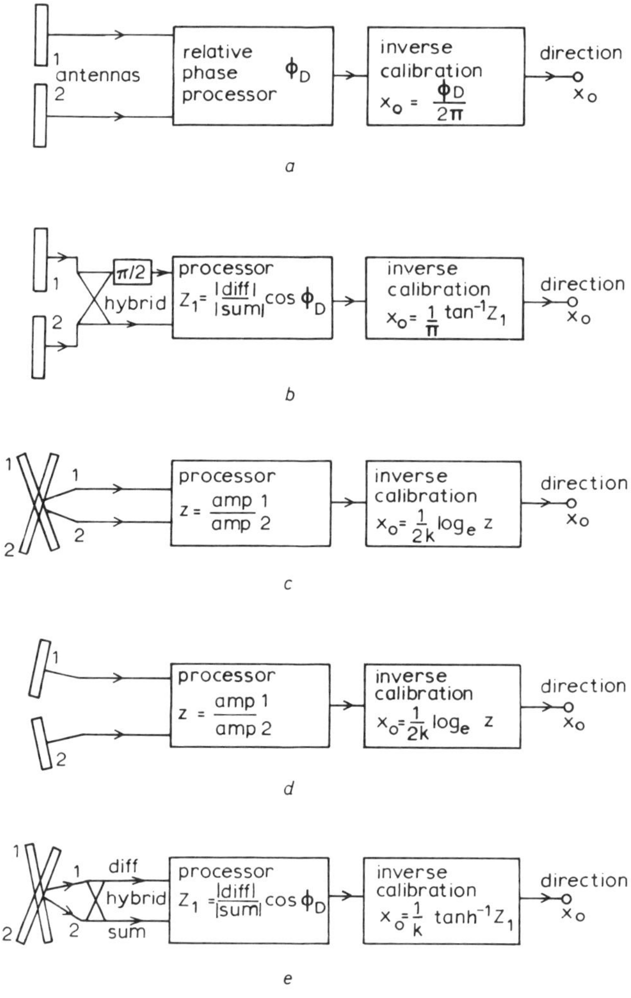

Fig. 7.4 DF processors examined by Cooper and Wyndham
(*a*) 'Direct phase' system
(*b*) 'Phase sum and difference' system
(*c*) 'Video' system with coincident phase centres
(*d*) 'Video' system with separated antennas
(*e*) 'Amplitude sum and difference' system

s.d.s of the phase measurements at the wider spacing were greater by an average factor of about eight than the s.d.s for the narrower spacing. When the ratio of apertures, 16·3, is taken into account, the bearing error reduction factor (BERF) resulting from the use of the bigger aperture is of the order of 2. The authors obtained a mean BERF of 2·4 from their analysis; when the estimated effects of lateral deviation were removed, a value of about 4 was obtained, close to the expected value of (aperture ratio)$^{1/2}$.

7.7.3 Comparison of processors in 2-ray fields

Sections 7.7.1 and 7.7.2 are largely concerned with the action of a processor in a single-ray wave-field. Cooper and Wyndham (1973) have considered the action of various processors in a two-ray field and have examined the accuracy of the determined direction of the stronger ray. Fig. 7.4 is based on diagrams in their paper and serves to define the processing schemes examined. For present purposes it is convenient to regard the antennas as linear arrays, with the rays arriving from some near-broadside direction in the plane of the paper; x is the normalised direction, i.e. the bearing in beamwidths. The aim is to determine the value of x for the wanted (stronger) ray as accurately as possible. The beam shape, which must be known for the amplitude-comparison systems, is represented by the expression $\exp(-kx^2)$, where $k = \log_e 4$; this expression is used, for example, in the 'video' system, Fig. 7.4*c*. The 'inverse calibration' shown in the diagrams can be regarded as part of the total processor.

Cooper and Wyndham's results show that, for the same overall aperture, systems *a* and *e* are least upset by interference and are superior to the other schemes in this respect. This conclusion serves to demonstrate that the processor is one of the factors affecting performance. The next four chapters are devoted to a more detailed study of specific processing schemes.

References

AFRAIMOVICH, E.L., and PANCHENKO, V.A. (1979): 'Doppler filtering in SW direction finding: analog-digital processing of modulated signals', Radiotekhnika, **34,** no. 9, pp. 3–5. English translation in Telecom. Radio Eng. Part 2, **34,** no. 9. pp. 54–57

BAILEY, A.D., and McCLURG, W.C. (1963): 'A sum-and-difference interferometer system for HF radio direction finding', *IEEE. Trans.*, **ANE-10,** pp. 65–72

BAKER, D.C., and LAMBERT, S. (1989): 'Range estimation for SSL HFDF systems by means of a multi-quasiparabolic ionospheric model', *Proc. IEE.*, **Pt.H, 136,** pp. 120–125

BARABASHOV, B.G., VERTOGRADOV, G.G., and KHONDU, A.A. (1987): 'Assessment of vertical sounding utilization efficiency in decameter-wave angle-and-range measuring systems', Radiotekhnika, **42,** no. 9, pp. 3–7. English translation in Telecom. Radio Eng., **42,** no. 10, pp. 27–31

BRACEWELL, R.N. (1958): 'Restoration in the presence of errors', *Proc. IRE.*, **46,** pp. 106–111

BURTNYK, N., McLEISH, C.W., and WOLFE, J. (1965): 'Performance of an interferometer direction finder for the hf band', *Proc. IEE.*, **112,** pp. 2055–2059

CLARKE, R.H., and TIBBLE, D.V. (1978): 'Measurement of the elevation angles of arrival of multicomponent HF skywaves', *Proc. IEE.*, **125,** pp. 17–24

COOPER, D.C., and WYNDHAM, B.A. (1973): 'A comparison of monopulse techniques for direction finding in the presence of interference', International Conference on Radar – Present and Future (IEE), pp. 154–159

DAVIES, K., and BAKER, D.M. (1966): 'On frequency variations of ionospherically propagated HF radio signals', *Radio Sci.*, **1,** pp. 545–556

HATCH, J.F. (1947): 'Developments in HF direction-finder shore station using Adcock aerials', *J. IEE.*, **94,** pp. 683–692

HAUSZ, W. (1964): 'Angular location, monopulse and resolution', *Microwave J.*, **7,** pp. 60–65

JONES, T.B., and REYNOLDS, J.S.B. (1975): 'Ionospheric perturbations and their effect on the accuracy of h.f. direction finders', *Radio and Electron. Eng.*, **45,** pp. 63–73

McNAMARA, L.F. (1988): 'Ionospheric modelling in support of single station location of long range transmitters', *J. Atmos. & Terr. Phys.*, **50,** pp. 781–795

NEWMAN, W.I. (1973): 'A new method of restoration', *Astrophys. J.*, **184,** pp. 453–459

RICE, D.W. (1971): 'The CRC high frequency direction finding research facility', Conference Digest, 1971 Int. electrical and electronics conference, Toronto. (IEEE, New York)

ROSS, W., BRAMLEY, E.N., and ASHWELL, G.E. (1951): 'A phase-comparison method of measuring the direction of arrival of ionospheric radio waves', *Proc. IEE.*, **98,** pp. 294–302

SHERRILL, W.M. (1969): 'Bearing ambiguity and resolution in interference direction finders', *IEEE. Trans.*, **AES-5,** pp. 959–966

SHERRILL, W.M. (1971): 'A survey of HF interferometry for ionospheric propagation research', *Radio Sci.*, **6,** pp. 549–566

SWEENEY, L.E. Jr. (1970): 'Spatial properties of ionospheric radio propagation as determined with half degree azimuthal resolution', Thesis, Stanford University.

TREHARNE, R.F. (1967): 'Vertical triangulation using skywaves', *Proc. Inst. Radio Electron. Eng.*, **28,** pp. 419–423

TREHARNE, R.F. (1981): 'Non-military applications of high frequency single station location systems', *J. Electr. & Electron. Eng. Australia*, **1,** pp. 87–92

WARRINGTON, E.M. and JONES, T.B. (1991): 'Measurements of the direction of arrival of HF sky wave signals and single site location using a seven element wide aperture (294m) interferometer array', to be published in Proc. IEE Pt. H, April 1991

Chapter 8

Wave interference effects for circular arrays

8.1 Introduction

We now come to the question of how much can be deduced about the structure of a wave-field from an analysis of wave interference effects. The observations might consist of the fluctuations of the indicated bearing of a WADF as a function of time (this Chapter) or a sequence of phase-difference measurements on various pairs of elements in an interferometer (Chapter 9).

Not surprisingly, the amount of information that can be derived with an orthodox DF, which provides only one indicated bearing at any one moment, proves to be very limited. In the numerical simulations for this type of problem, therefore, attention will be confined to two-ray wave-fields and the basic question is whether the ray azimuths can be found.

In order to distinguish between the zero-aperture bearing defined in Chapter 4 and the indicated bearing from a DF of finite aperture, we shall refer to the former as the tangent bearing; it is the direction normal to the tangent to a curve of constant phase at a specified point in the horizontal plane and is a property of the wave-field. A phase-measuring DF of finite aperture, centred on the same point, would produce an approximation to the same bearing but would average in some way (to be investigated below) over a finite segment of the curve. Thus, the 'segment bearing' for a finite aperture approaches the tangent bearing as the aperture approaches zero.

The total swing S_0 of the tangent bearing, as the phase difference φ between two rays varies from 0 to 2π, was found in Chapter 4 (eqn. 3) to be

$$S_0 = \frac{4A_1A_2 \cos \Delta_1 \cos \Delta_2}{(A_1^2 \cos^2 \Delta_1 - A_2^2 \cos^2 \Delta_2)} \tan \theta \qquad (1)$$

where A_1, A_2 are ray amplitudes, Δ_1 and Δ_2 are ray elevation angles and θ is *half* the difference of ray azimuths.

It might be expected that

(*a*) the swing of the segment bearing, S_a, would be small for very large apertures and would increase in magnitude towards that of the tangent bearing as the aperture is decreased, and

(*b*) the exact variation of swing with aperture would depend slightly on the processor, i.e. on the way in which the segment is sampled by the array elements and on the weighting given to amplitude and phase information.

These questions will be examined in this Chapter, by means of numerical simulations for the Wullenweber CDAA defined in Appendix 1 and for other types of DF.

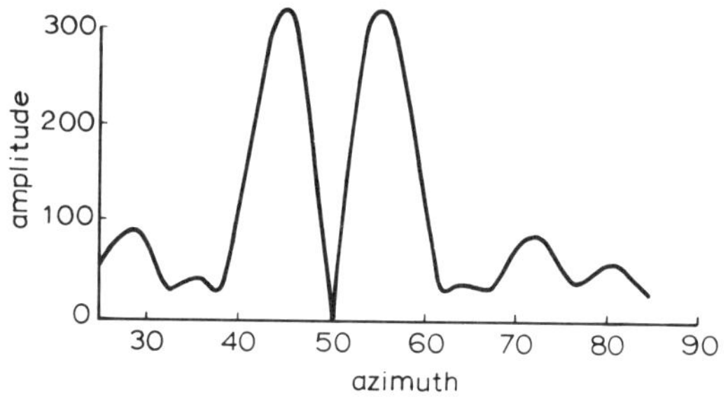

Fig. 8.1 Theoretical waveform for a single ray, 8·5 MHz, azimuth 50°, elevation 15°

8.2 Wullenweber array

8.2.1 Simulated waveforms

Fig. 8.1 shows part of the output of a spinning goniometer (corresponding to the 'difference' pattern shown in Fig. 1.3), for a single incident ray; in this and other computer simulations, the ray parameters and frequency are marked on the diagrams. The abscissa is pointing angle, i.e. the azimuth to which the axis of symmetry of the goniometer points at a particular moment of time. The ordinate is the amplitude of the goniometer output in arbitrary units. For a goniometer rotation rate of 600 r.p.m., 360° on the x-axis is equivalent to 0·1 s. The calculated output, which will be referred to as a waveform, is based on the assumption that the ray parameters do not change during one revolution. For convenience, only a portion of the waveform is shown, for a sector containing the ray.

Fig. 8.2 shows a calculated waveform for a two-ray situation, the parameters of which were adjusted to give a good match with an observed waveform (Fig. 8.3). Clearly, the match is not perfect and it is not claimed that the parameters chosen represent a unique or optimum solution to the problem of 'explaining' Fig. 8.3. But the comparison of Figs. 8.2 and 8.3 does suggest that some of the main features of the observed waveform, e.g.

the difference in height of the two maxima, and the partially-filled null, could be due to WI effects.

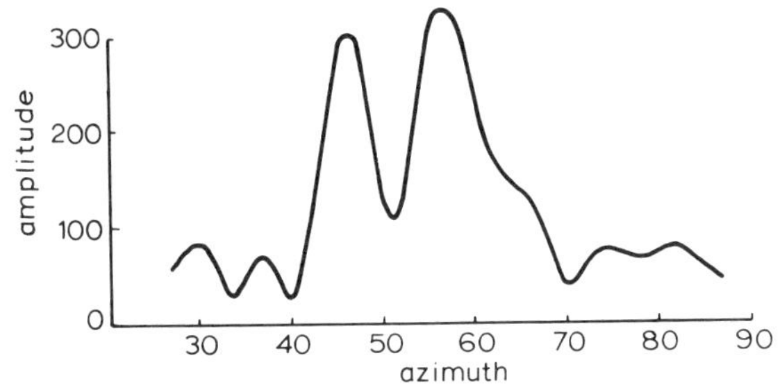

Fig. 8.2 Waveform for two-ray hypothesis, 8·5 MHz

	azimuth	elevation	rel. ampl.	rel. ph.
ray 1	52°	15°	2·5	0°
ray 2	58°	15°	1	110°

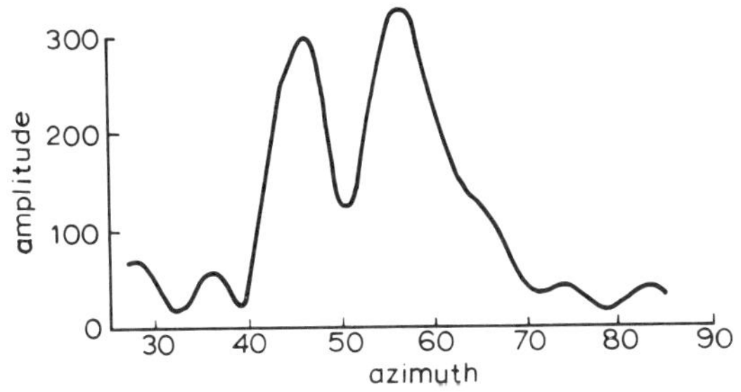

Fig. 8.3 Part of a low-difference waveform, live signal, 8·5 MHz

Warrington and Jones (1986) modelled the combined effect on the wavefield of several features of ionospheric propagation, and calculated the resulting waveforms for a smaller WADF of 24 elements. Because the calculated waveforms were very similar to those obtained in practice, the authors were confident that the modelling technique can be used to study the effects of changes in the array configuration or in the algorithms used to process the waveform.

8.2.2 Definition of segment bearing

The segment bearing or indicated bearing will now be defined rather arbitrarily as the pointing angle (to the nearest 0·1°) corresponding to the position of the principal minimum in the waveform, the principal minimum being the minimum ordinate between the 2 largest maxima of the pattern. When the calculated waveform is very badly distorted by WI from the perfect (one-ray) pattern, it is occasionally difficult to recognise the characteristic split-lobe or 'rabbit ears' of the difference pattern. Generally, however, the above definition provides a unique bearing that can be readily identified in the computer program, and is therefore a convenient choice in numerical studies.

The total processor in this case is therefore defined by the use of 32 doublets to sample the field, by the way in which the doublet outputs are combined to form a low-difference pattern and by the computer algorithm used to locate the principal minimum. Other algorithms, such as the selection of the maximum of the sum pattern, have been found to give very similar results to those described below.

8.2.3 Variation of segment bearing with phase

Suppose that the main beam of the sum pattern of the WADF can be approximately represented by the function cos $(D\theta)$, where θ is the azimuth angle from the beam axis and D is a function of frequency, aperture and elevation angle; D is large for a narrow beam. For two rays from the same elevation angle, and from azimuths of $(\eta + \alpha)$ and $(\eta - \alpha)$, the indicated bearing β is given by

$$\beta = \eta + \frac{1}{2D}\tan^{-1}\left\{\frac{(R^2 - 1)\tan 2D\alpha}{R^2 + 1 + 2R\cos\varphi\sec 2D\alpha}\right\} \tag{2}$$

where R is the amplitude ratio and φ is the phase difference. This expression, derived in Appendix 9-2 of Shirazi's thesis (1972), tends towards the standard expression for the tangent bearing (see eqn. 8 of Chapter 4) when $D \rightarrow 0$.

The values of β for $\cos\varphi = +1$ and -1 respectively are

$$\beta_1 = \eta + \frac{1}{2D}\tan^{-1} A$$

$$\beta_2 = \eta + \frac{1}{2D}\tan^{-1} B$$

where

$$A = \{(R^2 - 1)\tan(2D\alpha)\}/\{R^2 + 2R\sec(2D\alpha) + 1\}$$

$$B = \{(R^2 - 1)\tan(2D\alpha)\}/\{R^2 - 2R\sec(2D\alpha) + 1\}$$

The swing $(\beta_1 - \beta_2)$ is proportional to y, where

$$y = \tan^{-1} A - \tan^{-1} B$$

The value of R giving a maximum value of swing is found from the condition

$$\frac{dy}{dR} = 0$$

which yields the equation

$$R^4 - 2R^2\{1 + 2\sin^2(2\,D\alpha)\} + 1 = 0$$

The two values R_1 and R_2 giving maximum swing are therefore the positive roots of the equation

$$R^2 \pm 2R\sin(2D\alpha) - 1 = 0$$

Although theoretical expressions can give some insight into the importance of various factors, the need for major approximations can be avoided by using numerical simulations. As an example of this approach we now consider two rays with the parameters shown in Table 8.1, for a frequency of 13 MHz.

Table 8.1 Ray parameters

Parameter	Ray 1	Ray 2
Amplitude	2·0	1·0
Azimuth	90·0°	92·0°
Elevation angle	10·0°	30·0°

The indicated bearing, as a function of the phase difference between the rays at the centre of the array, is shown in Table 8.2.

Table 8.2 Indicated (segment) bearing as a function of phase

Phase difference (deg)	Indicated bearing (deg)
0	89·7
30	90·0
60	90·3
90	90·4
120	90·4
150	90·4
180	90·4
210	90·2
240	89·9
270	89·5
300	89·3
330	89·5

The bearing swing in this example is approximately 1·1° (from 89·3° to 90·4°). The mean of the tabulated values of indicated bearing is 90·0°, precisely that of the stronger ray.

Table 8.3 Tangent bearing as a function of phase

Phase difference (deg)	Indicated bearing (deg)
0	90·6
30	90·6
60	90·5
90	90·4
120	90·0
150	89·2
180	88·4
210	89·2
240	90·0
270	90·4
300	90·5
330	90·6

Table 8.3 shows the tangent bearing as a function of phase, calculated from eqn. 2 of Chapter 4. Note that the phase difference is now that at the point where the bearing is taken. The swing calculated from eqn. 1 is 2·2° and this value is seen to be consistent with Table 8.3.

The results from Tables 8.2 and 8.3 are compared in Fig. 8.4, with an appropriate adjustment of the phase origin of Table 8.2. The curves are

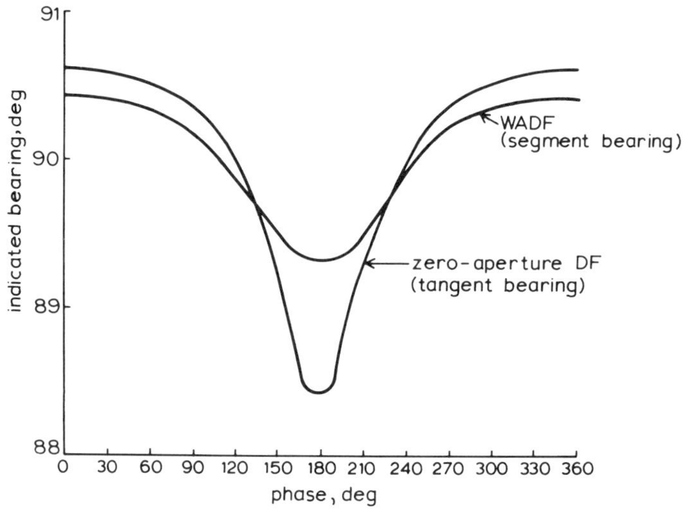

Fig. 8.4 Indicated bearing as a function of phase

seen to be of very similar shapes, but the segment bearing shows a smaller total swing than the tangent bearing, as expected. This reduction of swing is mainly due to a less extreme bearing for the antiphase condition (180°), where the tangent bearing is taken at the most kinked part of the wavefield; it is here that the value of aperture is most apparent. It should be noted that the curve for the tangent bearing is independent of frequency.

8.2.4 Variation of swing with amplitude ratio

Swing provides a convenient, if somewhat arbitrary, measure of the magnitude of WI effects in a given situation and of their nuisance value. It must, however, be emphasised that the swing is calculated for an artificial situation in which all ray parameters other than phase are held constant. From eqn. 1, we expect very large swings of the *tangent* bearing whenever the condition $A_1 \cos \Delta_1 = A_2 \cos \Delta_2$ is approached by the fading rays. To investigate this point further, we consider two rays from the same elevation angle and re-write eqn. 1 as

$$S_o = \alpha F$$

where α is the azimuthal separation of the rays and F is the function

$$F = \frac{2(A_1/A_2)}{\{(A_1/A_2)^2 - 1\}}$$

The variation of F with amplitude ratio near equality is shown in Table 8.4.

Table 8.4 Increase of swing near amplitude equality

A_1	A_2	A_1/A_2	F
1·0	0·5	2·0	1·33
1·0	0·6	1·67	1·88
1·0	0·7	1·43	2·74
1·0	0·8	1·25	4·44
1·0	0·9	1·11	9·47
1·0	1·0	1·0	∞
1·0	1·1	0·91	−10·48
1·0	1·2 etc	0·83 etc	− 5·45

If we now consider the swing S_a of the segment bearing on a WADF, we expect, on physical grounds, $S_a < S_0$ for identical ray conditions. Significant WI effects on a WADF, say $S_a > 1°$ for a typical value of 1° for α, therefore occur only if F is rather greater than 1. The peak of F is rather narrow; for example, the condition $F > 10$ requires that the amplitudes shall not differ by more than about 10%. The apparent amplitude of each of two fading modes is now affected by the vertical radiation pattern of the array; it is when the apparent amplitudes approach or pass through the condition of equality that very large swings can occur. However, at such moments, the goniometer waveform becomes so distorted that neither a human operator nor an automatic device is likely to be able to identify the bearing. For this reason, values of the amplitude ratio near unity have been excluded in computer simulations described below.

Two points of practical significance should be noted. Firstly, to obtain the azimuth of the stronger ray from Table 8.2 it is necessary to take the mean of all the indicated bearings; rejection of some bearings on criteria such as a distorted waveform, or low amplitude, might lead to a different result. Secondly, a rather steady indicated bearing which does not fluctuate by more than (say) ± 0·5° from its mean value must not be taken as evidence that only one ray is incident on the array: other rays of lower amplitude may well be present.

8.2.5 Variation of swing with other parameters

In Figs. 8.5 and 8.6 the swing is plotted as ordinate, against amplitude ratio as abscissa, for various sets of ray parameters and frequencies.

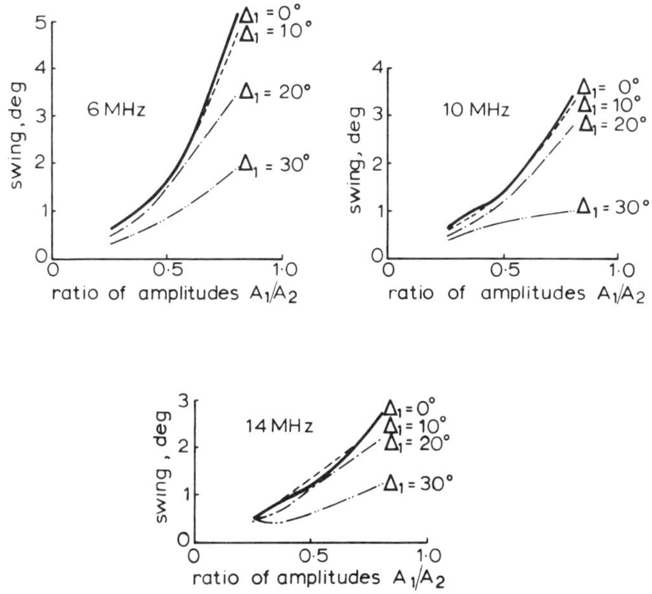

Fig. 8.5 Wave-interference swing for constant azimuth separation of 1° Elevation angle Δ_2 is constant at 20°

Fig. 8.5 shows curves for various values of Δ_1, with Δ_2 and the ray separation held constant. Fig. 8.6 shows curves corresponding to azimuth separations of 1°, 2° and 3°, for constant and equal elevation angles. The most striking features are the increase in swing with amplitude ratio and the decrease with increasing frequency. The spacing of the curves in Fig. 8.6 at low amplitude ratios is interesting: at 6 MHz the swing is proportional to α, whereas at 10 MHz and 14 MHz there is only a small increase in swing in going from $\alpha = 2°$ to $\alpha = 3°$. This effect is to be expected when the aperture of the array becomes significant compared with the scale of the kinked wave-field. An equivalent statement is that the beamwidth of the array becomes sufficiently narrow at the higher frequencies to provide a measure of rejection against one ray or the other, whatever the pointing angle.

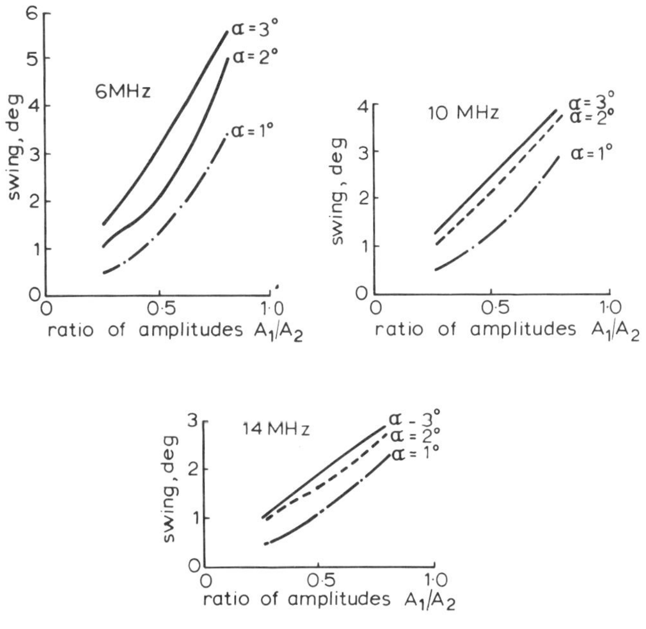

Fig. 8.6 Wave-interference swing for constant elevation angles
Constant elevation angles $\Delta_1 = 20°$, $\Delta_2 = 20°$
α is azimuth separation

8.3 Other types of DF

8.3.1 The Adcock

Bain (1953) considered the error γ of the indicated bearing of an Adcock with two incident rays. In his notation,

$$\tan 2\gamma = \frac{2s \sin \alpha \cos \varphi + s^2 \sin 2\alpha}{1 + 2s \cos \alpha \cos \varphi + s^2 \cos 2\alpha}$$

where α is the total separation in azimuth between the rays, φ is the phase difference, and

$$s = \frac{r \cos \Delta_2}{\cos \Delta_1}$$

$$r = \text{amplitude ratio}$$

$$= A_2/A_1$$

and γ is measured relative to the stronger ray. As a matter of interest, we shall now show that this expression is consistent with eqn. 2 above. We first replace α in Bain's expression by 2α, because this is the separation used by Shirazi, and also use R in place of s, thus implicitly assuming that $\Delta_1 \simeq \Delta_2$. With these changes,

$$\tan 2\gamma = \frac{2R \sin 2\alpha \cos \varphi + R^2 \sin 4\alpha}{1 + 2R \cos 2\alpha \cos \varphi + R^2 \cos 4\alpha}$$

It may then be shown that

$$\tan \{2(\gamma - \alpha)\} = \frac{(R^2 - 1) \tan 2\alpha}{1 + 2R \sec 2\alpha \cos \varphi + R^2}$$

which is identical with $\tan \{2(\beta - \eta)\}$ as derived from eqn. 2 with $D = 1$, the appropriate value for the Adcock pattern.

The mean-square-error relative to the stronger ray, averaged over all value of φ, is given approximately by

$$\mu_\sigma^2 = \frac{R^2 \, (\text{sep})^2}{2(1 - R^2)}$$

where sep has been written for the total separation in azimuth. Some values of this expression are tabulated in Section 8.3.3 for a particular example.

8.3.2. Doppler-style CDAA

In the same paper (1953), Bain gave theoretical expressions for WI effects on a cyclical DF designed by Earp and Godfrey (1947), in which a single antenna moves in a circle. The bearing is determined from the phase of the phase modulation imposed on a received signal by the circular movement.

A more practicable version of this system consists of a CDAA with fixed antennas spaced at equal intervals round the circle. The bearing is derived from phase difference measurements between successive pairs of elements. In a later paper (1956a), Bain computed the errors on a system of this type for various two-ray wave-fields and compared the error variance (i.e. mean-square-error relative to the stronger ray) for various types of DF system, namely, an Adcock, an Earp–Godfrey system with a rotating element and an Earp–Godfrey system with 36 fixed elements.

8.3.3 Comparison of Adcock and Doppler systems

One set of ray parameters used by Bain is shown in Table 8.5 and an extract from his error variance table is shown in Table 8.6.

Table 8.5 Ray parameters adopted by Bain

Parameter	Ray 1	Ray 2
Amplitude	1·0	0·8
Azimuth	0° (say)	α° (see Table 8.6)
Elevation angle	33·0°	33·0°

Table 8.6 Extract of error-variance results calculated by Bain

		Fixed-antenna cyclical system		
α deg	Adcock system deg^2	Diameter 1λ deg^2	Diameter 4λ deg^2	Diameter 10λ deg^2
1	0·89	0·88	0·81	0·60
2	3·56	3·43	2·65	1·41
3	8·00	7·41	4·79	1·94

Once again, the value of aperture becomes more apparent as the value of α increases. For a fuller explanation of the method of calculation and a complete tabulation of results, the original paper should be consulted.

8.4 Capture effect

Up to this point, we have used the term 'capture' as a convenient description of a tendency for the mean indicated bearing on a DF, averaged over a suitable period, to be that of the stronger ray. It has been shown that the Wullenweber, for example, tends to capture the strongest of two or more modes arriving from practically the same azimuth, but that the operator can usually discriminate against other transmissions, provided the azimuthal separation from the target signal is greater than the beamwidth. For example, given two transmissions within the receiver passband, the operator can in these circumstances obtain a good estimate of the azimuth of the strongest mode of either.

A Doppler system, however, exhibits an additional form of capture effect: it tends to capture the strongest signal within the passband and to suppress evidence of the existence of weaker signals. This phenomenon, which is closely related to the well known FM capture effect in modulation theory, is an advantage if the strongest signal is the wanted signal, but a major disadvantage otherwise.

Suppose that, with a Wullenweber system, the stronger of two close rays is an 'unwanted' ray and that the weaker is 'wanted'. Suppose also that the principal null of the goniometer pattern is placed, by means of a phase shift control loop, on the indicated bearing. Cooper (1973) has analysed a system of this kind and has shown that the signal/noise ratio, i.e. the ratio of the strengths of the wanted and unwanted rays, is enhanced by the system. If the relative phase of the inputs fluctuates rapidly, an integrator or lowpass filter in the control channel can be arranged to position the null at the mean-indicated bearing; provided the inputs are not of equal effective amplitude, the null will be steered to the precise azimuth of the larger ray, which will therefore be completely suppressed. Thus capture effect can be exploited to obtain improved reception of a signal.

8.5 One-ray interpretation of multiray fields

The indicated bearing on a CDAA is an example of a single-ray interpretation of what may be a field composed of n rays. Clearly, such a solution cannot be optimum, because the correct n-ray model must be capable of giving a more exact reconstruction of the signals in the antenna elements. Nevertheless, we can consider which of many single-ray interpretations gives the best fit in some least-squares sense and enquire whether a particular processor leads to this one-ray solution.

Suppose, for example, that the phases at N elements, relative to a reference element, are $\varphi_1, \varphi_2, \ldots, \varphi_N$ and that the phases calculated from a one-ray interpretation are $\varphi'_1, \varphi'_2, \ldots, \varphi'_N$. The least-squares-phase processor has been found if it can be shown that

$$\sum \{(\varphi_1 - \varphi_1')^2 + (\varphi_2 - \varphi_2')^2 + \ldots + (\varphi_N - \varphi_N')^2\}$$

is always minimised by the choice of ray DOA. In general, of course, the element phases are not known and the CDAA normally provides HAM only. Nevertheless, Bain (1956b) was able to show that the cyclical differential DF system of Earp and Godfrey does in fact produce this result. In other words, the system provides the optimum one-ray interpretation based on phase information alone, without amplitude weighting.

It can be argued that the most distorted parts of the phase field should be given a relatively low weight, because the phases of the element signals at these points of low amplitude are more affected by noise than elements elsewhere. Some simple form of amplitude weighting may therefore be desirable. Bain considered various mathematical functions that might be minimised for CDAAs. He showed that systems of the Wullenweber type are least-squares systems with some amplitude weighting; although the processor is not strictly optimum, he suggested that the amplitude weighting may give the Wullenweber an advantage over the Earp system.

8.6 Analysis of waveforms

8.6.1 Introduction

The output from a spinning goniometer, which will be referred to as the waveform, can be calculated for a set of specified ray parameters in a straightforward way. We now consider the converse problem, which is of more practical significance: can the ray parameters be correctly recovered from an analysis of a waveform? Any proposed method of analysis can be tested against a simulation such as Fig. 8.2, where the correct solution is known.

We shall ignore the question of the relationship between the magnitude of (say) the maximum of the waveform, measured perhaps in cm on a visual display, and the field strengths of the incident rays in μV/m; we are more interested in relative ray amplitudes than absolute strengths and more interested in the shape of the waveform than in its vertical scale. The shape of a two-ray waveform depends on six parameters: two azimuth angles, two elevation angles, one relative amplitude and one phase difference. For a three-ray field there are 10 independent parameters: three azimuth angles, three elevation angles, two relative amplitudes and two phase differences. Each additional ray adds four further unknowns.

A circular array with a highly directive radiation pattern that is rotated in azimuth is, of course, primarily intended for the determination of ray azimuth angles. If these can be found the average amplitudes of the (fading) rays can probably be estimated; and phase is usually of little interest. Because the observed waveforms are not very sensitive to small changes of (low) elevation angles, some quite different form of array may be preferred for accurate VAM. The main emphasis in this Section will therefore be on HAM.

8.6.2. Analysis of single waveforms

The waveform to be analysed might have been recorded in analogue form, e.g. on tape or by photographing a display, or in digital form, from an array of known properties, at a specified frequency. The shape of the waveform may be slightly distorted by noise and measurement errors, and affected by any change of parameters during the rotation; the number of incident rays is unknown. It will be assumed that whatever the nature of the initial record, the waveform can be converted to a digital form for computer analysis.

Case (i) *Rays well separated in azimuth*

Suppose, as an example, that only two rays are incident on the array, from two transmitters on quite different bearings. The value of the comparatively high directivity of a WADF now becomes apparent. As the main beam of (say) a sum pattern rotates, it 'looks' at each ray separately. Provided the sidelobe level is satisfactorily low, it should be possible to determine for each ray an indicated bearing that is little affected by the presence of the other. Even if each ray exerts a slight 'pull' on the bearing of the other, this effect can easily be corrected by some suitable form of successive approximations. In effect we can use an iterative method, starting with excellent first approximations to the ray azimuths and relative amplitude estimated by simple inspection of the waveform. Shirazi (1972) referred to this method as feature extraction and showed that it works well, i.e. converges very rapidly to the correct solution, when the ray separation exceeds the width of the main lobe from null to null.

Case (ii) *Rays close in azimuth*
When the rays are only a few degrees or less apart in azimuth, the problem becomes more difficult. This situation occurs when two or more modes arrive from the same transmitter.

Numerical experiments with simple iterative methods of search for the unknown ray parameters soon reveal three difficulties:

(*a*) The fit of a trial solution to an observed waveform can generally be improved (as expected) by adding another ray, because this adds four extra degrees of freedom to the solution.
(*b*) Quite different ray sets can be found that give acceptable fits to the waveform.
(*c*) Iteration in six or more unknowns in a computer program is very slow and unsatisfactory; because of (*a*) and (*b*), the program will not necessarily converge to the correct solution, i.e. the input parameters used in constructing a simulated waveform.

In physical terms, we are trying to resolve the distorted wave-field from one space sample over a limited aperture. During the brief sampling period ($\ll 0{\cdot}1$ s) in which the array is 'looking' at the rays, we can visualise the curves of constant phase as being frozen in shape. From a waveform that departs significantly from the one-ray pattern we can safely infer that these curves are not straight lines, but the information available is not sufficient for successful resolution of the field.

In the light of these rather fundamental difficulties, the analysis of a single waveform does not seem to be a profitable line of attack.

8.6.3 Analysis of sequences of waveforms

Clearly, we need to make use of time sampling as well as space sampling. A sequence of waveforms, taken on a given target transmitter over a period of seconds or minutes, must form the raw material for the analysis. The question is then whether some simple model of the wavefield is consistent with the sequence.

As a specific example, suppose that a set of 100 scans over the full 360° is taken over a period of 10 seconds with a goniometer spinning at 10 rev/s, and that the waveforms show some progressive change of shape over the period, arising from complex fading of the rays. The mean of the 100 indicated bearings provides a good first estimate of the azimuth of the strongest ray (if one is dominant throughout) and a good starting point for any iterative search for the azimuths of this and other rays. The next step might be to select (say) every 10th waveform for a more detailed analysis. It is not necessary to analyse the complete scan over 360°; we are more interested in a sector of (say) 36° wide, i.e. 1/10th of the scan, centred on the indicated bearing. The term waveform should now be understood to refer to the corresponding portion of the full curve. It therefore takes 0.01s to observe a waveform, the selected waveforms are spaced at intervals of 1s and the total observation period is 10 s.

The aim is to explain the observations, as defined above, in terms of a simple model of the wave-field. The model is based on the following explicit assumptions:

(*a*) the ray DOAs may be regarded as constant over the total observation period,
(*b*) other ray parameters may change between waveforms, but can be treated as constant for any one waveform,
(*c*) the simplest explanation, i.e. the minimum number of rays giving a satisfactory fit to all 10 waveforms within a tolerance set by signal/noise ratio, is the solution sought.

It must be accepted that any solution derived from these simplifying assumptions is based on an idealised model of the wave-field; assumptions (*a*) and (*b*), for example, are not likely to be precisely true in practice. Nevertheless, such a solution could provide a good starting point for any process of subsequent refinement.

The assumptions stated explicitly here are often made implicitly in classical DF experiments. Any form of time averaging is based on the assumption that the azimuth of arrival of (say) a single ray has a mean value that is worth observing, even though the instantaneous azimuth exhibits small fluctuations about this value and the ray amplitude fluctuates with time.

Analysis of p.d.f.
Archer (1985) extended his previous analysis (Shirazi, 1971) of two close rays subject to Rayleigh fading by making better approximations to the sum and difference patterns of a circular WADF. The resulting p.d.f. of the indicated bearing from a sample of waveforms is very similar in character to the curves shown in Fig. 4.5 of Chapter 4. The ray parameters can in principle be extracted when the shape of the p.d.f. is known, using published curves (Archer, 1989). Hudson (1986) showed how two rays could be resolved from the outputs of four fixed beams rather than the continuous waveforms, each beam shape being represented by a third-order Taylor series.

Exploitation of QUMP
If all the waveforms are of the same shape, with a deep principal minimum in a constant position and fairly symmetrical humps on each side of it, the natural deduction is that only one ray is incident on the array. The calculation of the mean azimuth is then trivial.

More commonly, some or all of the waveforms will be noticeably distorted compared with the simple one-ray shape. Suppose, however, that in a two-ray field, some of the waveforms happen to coincide with moments when one of the rays (ray *A*) has faded to a very low level compared with the other (ray *B*). The waveform should then show little distortion compared with a theoretical one-ray pattern for the appropriate elevation angle, Δ_B, and the azimuth α_B can be determined directly. Conversely, when *B* fades to a low level, α_A can be determined.

To express the criterion of little distortion in quantitative terms, it would be desirable to calculate a correlation coefficient between each observed waveform and the theoretical patterns for single rays at Δ_A and Δ_B, arriving from the indicated bearing. A coefficient greater than some specified value would then be used as an indication of QUMP conditions. However, Δ_A and Δ_B are generally unknown. Some measure of distortion that is not sensitive to elevation angle is therefore required; properties such as the symmetry of the waveform about the principal minimum, and the depth of the minimum, can be used.

While this simple approach may well be fruitful when a large sequence of waveforms is available, a few practical difficulties are worth noting. Firstly, we cannot be sure that one ray will undergo a deep fade at all elements of an array simultaneously: diversity effects across a WADF may be significant. Secondly, even in one-ray conditions the waveform may lack symmetry because of amplitude fading during the goniometer rotation and distributed noise tends to fill in the null. Thirdly, QUMP conditions, if they exist at all, will tend to be rare with three or more rays. If one ray is dominant throughout the observation period it is true that some fairly 'clean' waveforms may occur. But the azimuth of such a ray can in any case be estimated as the mean indicated bearing over the period.

Intersection method

Fig. 8.7 shows a superimposition of four difference waveforms caculated for a two-ray situation, with constant ray amplitudes and DOAs. The

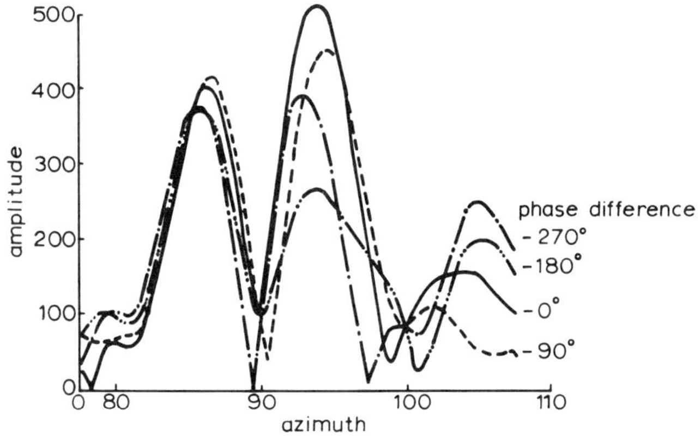

Fig. 8.7 Superimposed waveforms with two incident rays

	azimuth	elevation	rel. amp.
ray 1	90°	20°	2
ray 2	100°	30°	1

waveforms shown are for phase differences of 0°, 90°, 180° and 270°. A large separation of 10° in azimuth has been adopted for clarity of presentation.

As first pointed out by Shirazi (1972), these curves have common intersections at the azimuths of the two rays. The ordinate at the azimuth of the first ray is proportional to the amplitude of the second, because the first ray has been 'nulled', and vice versa. It is clear that such common intersections will arise only with two rays, that the ray amplitudes and DOAs must remain constant over a period in which the phase difference changes significantly and that departures from these conditions will blur the intersections. The restriction to constant amplitudes could be removed if ratio waveforms were available, as described below.

Ratio intersection method

The ratio of outputs, for a single ray, from two goniometers phased for different elevation angles is a function of (ray azimuth minus pointing angle) and of elevation angle, but not of ray amplitude. In effect, one goniometer can be used to normalise the output from the other goniometer to a constant ray amplitude.

Suppose that the rotor of a spinning goniometer carries two sets of capacitive plates, one set being used to form a low-angle difference pattern and the other set to form a high-angle difference pattern (see Appendix 1). For mechanical reasons these sets of plates are likely to be 180° apart, but fortunately will then possess the same axis of symmetry and will give zero outputs simultaneously when this axis points to a single ray. Suppose that the ratio of the amplitudes of the outputs can be calculated or displayed as a function of pointing angle.

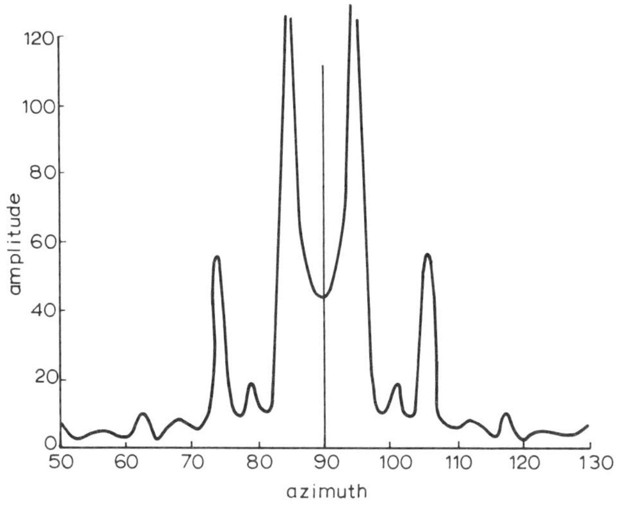

Fig. 8.8 Ratio of low/high waveforms. Single ray. Elevation 20°

Fig. 8.8. shows an example of such a ratio waveform in a one-ray field, for an elevation angle of 20° and an azimuth of 90°. It is worth emphasising that such curves are symmetrical and that the same shape is traced out on each scan, no matter how the ray amplitude varies; the shape is a function of Δ and could therefore be compared with a library of curves to give at least a crude measurement of Δ.

The presence of two or more rays is generally indicated by a departure from symmetry on individual scans and by a variation of shape from scan to scan. Superimposed waveforms for a two-ray situation intersect at the two-ray azimuths, as expected (Fig. 8.9). The presence of two rays only, on the same azimuth, is indicated by a sequence of waveforms of variable shape, but each one approximately symmetrical about that azimuth; the ratio at or near this pointing angle, where both goniometers give, ideally, a zero output, would be greatly affected by noise.

Further discussion of the intersection method and ratio intersection method can be found in Shirazi's thesis (1972). So far as the author is aware, neither has yet been tried in practice.

Elevation angles known

In Chapter 7 it was suggested that the exploitation of elevation-angle measurements would greatly assist mode resolution and identification. In the current discussion of waveforms, however, it has been assumed up to this point that the ray elevation angles are among the unknowns. It will now be shown that an important simplification of the two-ray problem

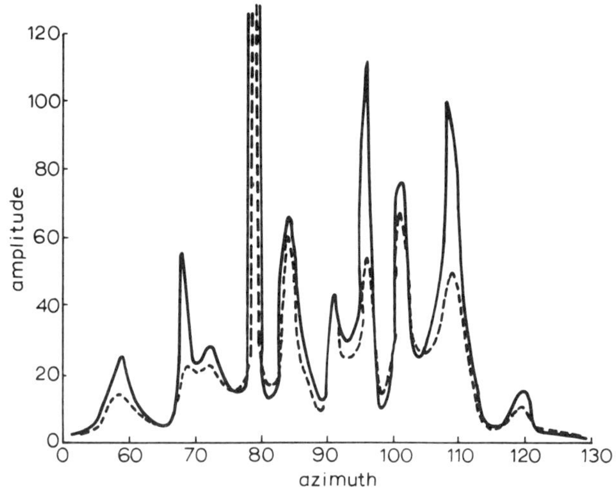

Fig. 8.9 Ratio curves for two rays on different azimuths

	azimuth	elevation	rel. ph.
ray 1:	90°	20°	0°
ray 2:	91°	40°	233·7°

______	rel. amp.	ray 1 = 10	ray 2 = 8
------	rel. amp.	ray 1 = 10	ray 2 = 10

Because the vertical array uses horizontally-polarised elements, whereas the CDAA uses vertical monopoles and because ray amplitudes vary with time, we must still regard the ray amplitudes as unknowns in the waveform analysis. However, if we choose to iterate in the two unknown azimuth angles α_1, α_2, it will be shown below that the amplitude estimates $\hat{A}_1$, $\hat{A}_2$ corresponding to each iteration $\hat{\alpha}_1$, $\hat{\alpha}_2$ can be immediately determined from the ordinates of the waveform at angles $\hat{\alpha}_1$, $\hat{\alpha}_2$. It is then only necessary to see if a value of the phase difference φ can be found that gives a satisfactory fit to the whole waveform.

The procedure is perhaps best explained with the aid of a numerical example, as shown in Table 8.7. An 'observed' difference waveform calculated for two incident modes is shown in column 2. The ray parameters used were actually $\alpha_1 = 90{\cdot}0°$, $\Delta_1 = 20{\cdot}0°$, $A_1 = 1{\cdot}0$, $\alpha_2 = 91{\cdot}0°$, $\Delta_2 = 30{\cdot}0°$, $A_2 = 1{\cdot}25$, but only Δ_1, Δ_2 are assumed known. The amplitude columns under the heading of theoretical waveforms show the calculated difference waveforms for one ray only, first at an elevation of 20° and then at 30°. In both cases the ray is assumed to be of unit amplitude, and to arrive from an arbitrary azimuth of 0°.

The azimuth of arrival of the stronger mode is likely to be somewhere near 90°, and the weaker within a few degrees of this value. We do not know whether the stronger mode is associated with $\Delta_1 = 20{\cdot}0°$ or $\Delta_2 = 30{\cdot}0°$. It would therefore be reasonable to iterate in both azimuth angles from (say) 85·0° to 95·0° and to regard, for example, $\hat{\alpha}_1 = 85{\cdot}0°$, $\hat{\alpha}_2 = 85{\cdot}1°$, as a different trial solution from $\hat{\alpha}_1 = 85{\cdot}1°$, $\hat{\alpha}_2 = 85{\cdot}0°$.

The first trial solution of $\hat{\alpha}_1 = 85{\cdot}0°$, $\hat{\alpha}_2 = 85{\cdot}0°$ can be rejected immediately, because it would give a perfect null at 85·0°, a feature not present in the observed waveform. In order to illustrate the calculation of amplitude estimates, we take as our next iteration $\hat{\alpha}_1 = 85{\cdot}0°$, $\hat{\alpha}_2 = 86{\cdot}0°$, although in practice an iteration step of 0·1° might be adopted.

The observed amplitude at 85·0° must be entirely due to the waveform of the second mode, 1° away from its null. Hence,

$$\hat{A}_2 = \frac{48{\cdot}99}{14.04}$$

$$= 3{\cdot}49$$

Similarly

$$\hat{A}_1 = \frac{57{\cdot}24}{21{\cdot}75}$$

$$= 2{\cdot}63$$

These are the amplitudes that *must* be adopted if a good fit to the observed waveform is to be obtained at 85·0° and 86·0°. The fit at other azimuths will depend on the phase difference between the modes. The next step is

Table 8.7 'Observed' and calculated waveforms

Observed waveform		Theoretical waveforms					
Az	Amp	Δ = 20·0°			Δ = 30·0°		
deg		Az	Amp	Phase	Az	Amp	Phase
80·0	11·01	−10·0°	2·94	314·8°	−10·0°	14·67	187·1°
81·0	15·02	− 9·0	7·39	7·9	− 9·0	21·02	167·6
82·0	19·07	− 8·0	10·23	359·0	− 8·0	24·48	149·1
83·0	25·68	− 7·0	9·92	322·6	− 7·0	24·98	127·1
84·0	36·67	− 6·0	14·88	268·0	− 6·0	24·85	99·5
85·0	48·99	− 5·0	27·00	241·7	− 5·0	26·73	71·2
86·0	57·24	− 4·0	38·44	230·3	− 4·0	29·61	49·7
87·0	56·84	− 3·0	43·03	224·3	− 3·0	29·79	36·1
88·0	46·08	− 2·0	37·38	220·8	− 2·0	24·66	28·1
89·0	27·52	− 1·0	21·75	218·9	− 1·0	14·04	23·8
90·0	14·04	0·0	0·00	–	0·0	0·00	–
91·0	27·19	1·0	21·75	38·9	1·0	14·04	203·8
92·0	41·01	2·0	37·38	40·8	2·0	24·66	208·1
93·0	46·03	3·0	43·03	44·3	3·0	29·79	216·1
94·0	42·32	4·0	38·44	50·3	4·0	29·61	229·7
95·0	33·20	5·0	27·00	61·7	5·0	26·73	251·2
96·0	23·07	6·0	14·88	88·0	6·0	24·85	279·5
97·0	15·71	7·0	9·92	142·6	7·0	24·98	307·1
98·0	14·08	8·0	10·23	179·0	8·0	24·48	329·1
99·0	17·84	9·0	7·39	187·9	9·0	21·02	347·6

therefore to multiply the theoretical waveforms by 2·63 and 3·49 respectively and then to calculate the ray-phase differences necessary at (say) three widely-separated points on the observed waveform if the observed amplitudes are to be explained. If inconsistent phase values are calculated, the solution can be rejected and the program proceeds to the next iteration. If the two phase values agree to within a few degrees, further points on the observed waveform can be tested.

Experiments with a computer program based on the above procedure gave the following results against toy problems. When no noise was present, the program gave the correct answer, plus typically one or two incorrect answers, from a single waveform. When two or more waveforms, constructed for the same DOAs but different relative amplitudes or phases, were analysed, the incorrect answers were readily eliminated. When noise, or a weak third mode, was added to the problem it was necessary to analyse many waveforms in order to eliminate the false solutions.

In this as in other iterative methods, much computer time can be saved, and some false solutions avoided, if good initial estimates of the ray parameters can be used as the starting point. In this connection, it is instructive to refer back to Fig. 8.4 and to Tables 8.1 and 8.2 in Section 8.2.3. The mean indicated bearing gives an excellent first approximation to α_1, and good estimates of Δ_1 and Δ_2 are now assumed to be available. From Table 8.2, the maximum bearing is 90·4° and the minimum bearing is 89·3°, and the corresponding phase differences (0° and 180°) are known; it is possible in principle to deduce the remaining ray parameters (A_1/A_2 and α_2) from these two bearings. In practice, various difficulties are likely to be encountered: other weak rays, noise and amplitude fading will all tend to produce departures of the curve in Fig. 8.4 from the simple shape shown, and the variation of indicated bearing with time may look rather different from the curve of indicated bearing against phase. Nevertheless, if a curve of approximately this shape is encountered, it should be possible to make an intelligent guess at the azimuth of the weaker ray and the amplitude ratio.

8.7 Summary

Various possible attacks on the problem of determining ray parameters from the output of a WADF have been considered, and it may be helpful to summarise the conclusions before proceeding to a discussion of the interferometer.

The behaviour of the segment bearing under conditions of WI is very similar to the behaviour of the tangent bearing, as described in Chapter 4. The main features such as capture of the indicated bearing by a dominant ray, swing etc. are again encountered in simulated problems. The main effect of the finite aperture of a WADF is to reduce the magnitude of the swing; the reduction becomes very significant if the ray separation is comparable with, or larger than, the beamwidth.

For a two-ray field with one ray dominant, the indicated bearing might typically fluctuate within about ±0·5° of a mean value. From a study of many simulations it seems clear that this mean value is extremely close to the azimuth of the dominant ray, even when the ray-elevation angles differ considerably. If the effective ray amplitudes pass through equality from time to time, the situation is more complicated and much larger swings can occur. The mean indicated bearing now depends on the proportion of the observation period for which each ray was the stronger.

Instead of confining our attention to indicated bearings, we can look at a single waveform, or a sequence of waveforms. The methods of analysis then available include the exploitation of capture, QUMP, swing analysis, intersections of waveforms, intersections of ratio waveforms, and VAM from other sources.

Not surprisingly, no one method emerges as sufficiently powerful to resolve any type of wave-field from a limited sample of noisy data. It seems unlikely that a unique solution can be obtained from a single waveform and, as the complexity of the problem increases (more rays and/or more noise) so a larger number of waveforms must be analysed if a reliable solution is to be obtained. Given the limited aperture of a typical WADF compared with the scale of the wave-field distortions, this need for time-sampling as well as space-sampling is in no way surprising. The theory given above indicates that a variety of methods can be used as aids to the intelligent interpretation of practical records.

References

ARCHER, M.J. (1985): 'Probability of direction indications for two unresolved signals', *Proc. IEE.*, **132F,** pp. 90–93

ARCHER, M.J. (1989): 'Determination of the directions of closely-spaced signal sources by a statistical method', Passive DF Colloqium, IEE Digest No. 1989/16, pp. 8/1–4

BAIN, W.C. (1953): 'Calculation of wave interference errors on direction finder employing cyclical differential measurements of phase'. *Proc. IEE.*, **100,** Pt III, pp. 253–260

BAIN, W.C. (1956a): 'Possible errors of a particular wide-aperture direction finder', *Proc. IEE.*, **103C,** pp. 313–324

BAIN, W.C. (1956b): 'The theoretical design of direction finding systems for high frequencies', *Proc. IEE.*, **103B,** pp. 113–119

COOPER, D.C. (1973): 'Zero-steering antenna system for receiving a signal close to the direction of strong interference'. *Electron. Lett.*, **9,** pp. 140–141

EARP, C.W., and GODFREY, R.M. (1947): 'Radio direction finding by the cyclical differential measurement of phase'. *J. IEE.*, **94,** Pt. IIIA, pp. 705–721

HUDSON, J.E. (1986): 'Simple, optimum resolution of close sources by maximum likelihood estimation using a sufficient statistic', Colloquium on direction finding systems in their operating environment, IEE. Digest No. 1986/10, pp. 6/1–4

SHIRAZI, S.M.S.A. (1972): 'Signal processing for high frequency direction finding'. PhD thesis, Electronic and Electrical Engineering Department, University of Birmingham

WARRINGTON, E.M., and JONES, T.B. (1986): 'Computer modelling of ionospheric errors induced in an HF DF system', Colloquium on direction finding systems in their operating environment, IEE. Digest No. 1986/10, pp. 1/1–4

Chapter 9

Wave interference effects for interferometers

9.1 Introduction

Interferometer systems have been widely used in radioastronomy and HF propagation research, and can take many forms. Sherrill (1971) reviewed the use of interferometers for the latter purpose and described 13 different configurations (his Fig. 7). Plan views of three of these are shown in Fig. 9.1, which is diagrammatic only. Fig. 9.1*a* shows an array of eight horizontal dipoles used by a team at the Weapons Research Establishment, Australia. Fig. 9.1*b* shows a configuration of five vertical monopoles used by the Southwest Research Institute in Texas. Fig. 9.1*c* shows a triangular arrangement of five vertical elements, built by a team at the University of Illinois.

The usual procedure for using arrays such as those shown in Fig. 9.1*a* and *b* is to perform a series of phase-difference (p.d.) measurements in a sequence such as

(*a*) Narrow spacing, *N*-*S* arm
(*b*) Wide spacing, *N*-*S* arm
(*c*) Narrow spacing, *E*-*W* arm
(*d*) Wide spacing, *E*-*W* arm.

Measurements *a* and *b* determine DOA relative to the *N*-*S* arm and *c* and *d* determine DOA relative to the *E*-*W* arm. The measurements at the narrow spacing serve to remove any ambiguities in the 'fine' measurements at the wider spacings.

In co-operative propagation experiments it is often possible to use a pulsed transmitter and to make measurements on time-resolved modes. For example, Ross, Bramley and Ashwell (1951) used a four-element interferometer with crossed 100 m arms to determine the arrival angles, in azimuth and elevation, of the modes of a pulsed signal. An individual mode could be selected for study by time-gating and thus the interferometer was effectively used in single-ray conditions.

An interferometer should also work well for a single ray (plane wavefront) accompanied by random Gaussian components symmetrically arranged around the ray direction. Denisov and Doroshenko (1983) showed that the maximum probability estimate of the sine of the bearing angle relative to the broadside direction can be found from a weighted sum of the total phase differences.

We shall discuss here some of the problems involved in using an interferometer to make measurements on a CW signal in multiray conditions. Performance now depends on factors such as the exact sequence of operations, the speed with which it is carried out, the number

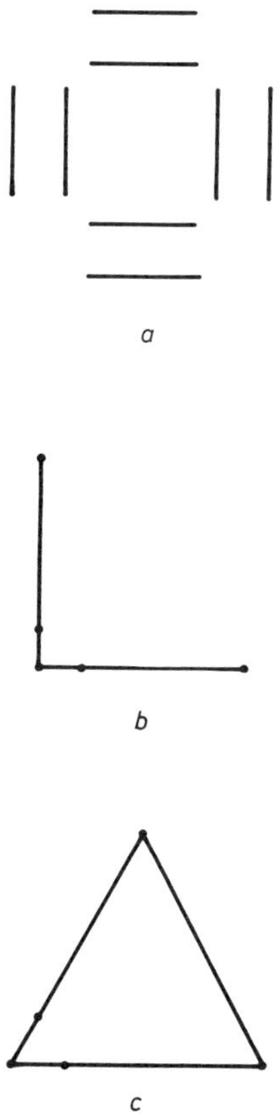

Fig. 9.1 Plan views of interferometers
(*a*) WRE system
(*b*) SWRI system
(*c*) University of Illinois system
Based on Fig. 7 in a paper by Sherrill, *Radio Sci.*, **6**, pp. 549–566, May 1971, copyrighted by American Geophysical Union

of times it is repeated to obtain a time average, and the methods used to resolve ambiguities and to combine observations from separate arms. The optimum combination of processor and array configuration is by no means clear, and a full discussion of all possible permutations is beyond the scope of this book. All that will be attempted here is an introduction to the types of calculations that are needed in the interpretation of observations and the analysis of accuracy.

9.2 Parameters used for numerical examples

We shall consider phase-difference (p.d.) measurements made with pairs of elements from the array shown in Fig. 9.1*b*; this array configuration was used in experiments described by Dodge and Martin (1969).

It will be assumed that the sum-and-difference twin-channel technique (see Section 7.7.1) is used, first to measure the p.d. for the narrow spacing (AB) and then for the wide spacing (AC) on a given arm.

The two-ray wave-field model used in numerical simulations is shown in Table 9.1. The frequency adopted throughout is 6 MHz.

Table 9.1 Ray parameters used in simulations

Ray	Amplitude	Arm angle
1	1·0	20·0°
2	1·5	30·0°

In this table, arm angle is the angle between the ray and a line drawn through the phase centres of the elements. The phase difference between the rays, measured at element *A*, will be specified for each example.

9.3 The RF ellipse

For a given ray p.d., the signal amplitudes at two elements are not generally equal and the c.r.t. display is an ellipse. We shall refer to this as the RF ellipse, although in practice it is usually IF signals that are displayed.

The ellipse is of constant shape and orientation for as long as *all* the ray parameters remain constant. Its equation can be derived theoretically, or points on the ellipse can be calculated in a suitable computer program, for a specified wave-field. This is, of course, the procedure for *constructing* a toy problem. The solution starts with one or more ellipses as the raw material.

Table 9.2 shows some points (x, y) on the ellipse for a ray p.d. of 80°, and the narrow spacing. Here x is the sum signal and y is the difference signal phase-shifted by 90°.

Table 9.2 RF ellipse for ray p.d. of 80°, narrow spacing

RF phase	x	y
0°	2·96	2·64
18	2·78	2·52
36	2·33	2·14
54	1·64	1·56
72	0·80	0·83
90	−0·12	0·01
108	−1·03	−0·81

These points, plotted in Fig. 9.2, form a narrow ellipse; the slope of the major axis is about 41°. The p.d. between elements deduced from this slope (i.e. based on a one-ray assumption) is 82° and the corresponding arm angle is 28·8°, fairly close to that of the stronger ray.

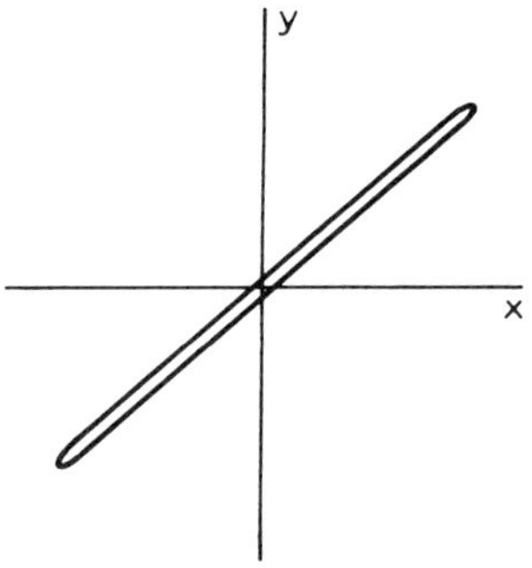

Fig. 9.2 RF ellipse, narrow spacing

The ratio of wide to narrow spacings is 150 m/13 m, i.e. 11·54, and the expected p.d. between the wide elements, again on the basis of a simple one-ray interpretation, is therefore about 946°. Some points on the ellipse calculated for the true two-ray situation are shown in Table 9.3. These points lie on a fatter ellipse, with the major axis running from about 122° to 302° (see Fig. 9.3).

The most plausible extrapolation from narrow to wide spacing is obtained by rewriting 122° as (360° + 122°) and doubling this to find the apparent p.d.; this procedure gives 964°, in reasonable agreement with the extrapolated value. The arm angle based on this p.d. is found to be 26·8°.

This calculation shows the problem of extrapolation is not entirely straightforward and that (not surprisingly) an incorrect answer is obtained if a one-ray interpretation is applied to what is really a two-ray situation. The elliptical nature of each display provides a visual warning that two or more rays are present. The calculation has been described in some detail in the hope that the distinctions between RF phase, ray p.d. and element p.d. are now clear.

Table 9.3 RF ellipse for ray p.d. of 80°, wide spacing

RF phase	x	y
0°	−0·79	0·20
18	−0·03	−0·97
36	0·73	−2·04
54	1·42	−2·91
72	1·97	−3·50
90	2·32	−3·74
108	2·45	−3·63

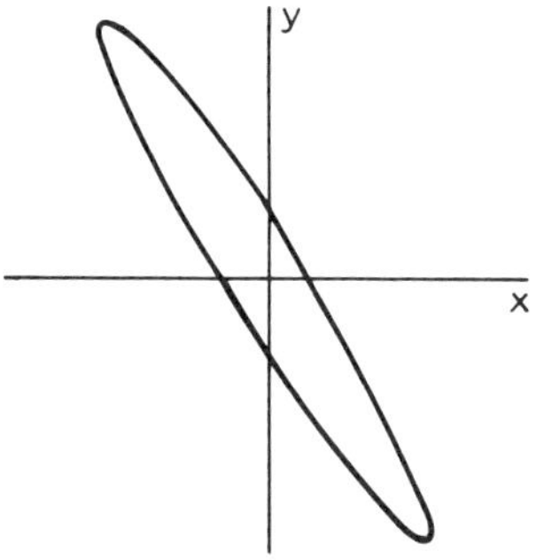

Fig. 9.3 RF ellipse, wide spacing

9.4 The two-ray parallelogram

9.4.1 Envelope of ellipses

If we now change the ray p.d. to some new value, but keep the arrival angles constant, new ellipses are obtained for the narrow and wide spacings. Interpretation of this pair of ellipses in isolation generally leads to exactly the same difficulties as those noted above. However, the envelope of the ellipses at one spacing for all possible ray p.d.s can be used to deduce the ray arm angles. For two rays this envelope is a parallelogram, the sides being parallel to the line displays obtained with either ray alone.

This important result, which has been exploited in some interferometer systems, will now be proved. Let A, B be the two-ray amplitudes, θ_1 and θ_2 the corresponding arm angles and α the p.d. between the rays at the first element. At a particular moment of time t the signal in element one can be written

$$S_1 = A \cos \omega t + B \cos (\omega t + \alpha)$$

where ω is the frequency in rads/s. The signal in the second element is

$$S_2 = A\cos(\omega t + \varphi_1) + B\cos(\omega t + \alpha + \varphi_2)$$

where

$$\varphi_1 = 2\pi d\cos\theta_1/\lambda$$

$$\varphi_2 = 2\pi d\cos\theta_2/\lambda$$

The sum signal, $(S_1 + S_2)$, to be displayed on the X plates of the c.r.t. is

$$X = 2A\cos\left(\omega t + \frac{\varphi_1}{2}\right)\cos\frac{\varphi_2}{2} + 2B\cos\left(\omega t + \alpha + \frac{\varphi_2}{2}\right)\cos\frac{\varphi_2}{2}$$

The difference signal, $(S_1 - S_2)$, is shifted in phase by $+\pi/2$ and displayed on the Y plates:

$$Y = A\{\sin(\omega t + \varphi_1) - \sin\omega t\} + B\{\sin(\omega t + \alpha + \varphi_2) - \sin(\omega t + \alpha)\}$$

$$= 2A\cos\left(\omega t + \frac{\varphi_1}{2}\right)\sin\frac{\varphi_1}{2} + 2B\cos\left(\omega t + \alpha + \frac{\varphi_2}{2}\right)\sin\frac{\varphi_2}{2}$$

These equations define one point (x, y) on the ellipse. We can obtain an equation independent of A by writing

$$y - x\tan\left(\frac{\varphi_1}{2}\right) = 2B\cos\left(\omega t + \alpha + \frac{\varphi_2}{2}\right)\left\{\sin\frac{\varphi_2}{2} - \cos\frac{\varphi_2}{2}\tan\frac{\varphi_1}{2}\right\}$$

Clearly, if B were zero, we would obtain the straight-line display

$$y = x\tan\frac{\varphi_1}{2}$$

The maximum and minimum values of $y - x\tan\left(\frac{\varphi_1}{2}\right)$, when B is not zero, occur when $\cos\left(\omega t + \alpha + \frac{\varphi_2}{2}\right)$ is $+1$ and -1. Note that this is the only term involving t and that it must reach each of these limiting values, once only, as ωt varies from 0° to 360°. Thus, the ellipse touches the two parallel boundary lines defined by

$$y = x\tan\frac{\varphi_1}{2} + c$$

$$y = x\tan\frac{\varphi_1}{2} - c$$

where

$$c = 2B\left\{\sin\frac{\varphi_2}{2} - \cos\frac{\varphi_2}{2}\tan\frac{\varphi_1}{2}\right\}$$

Similarly, by obtaining a relationship independent of B, we arrive at two other parallel boundary lines that the ellipses must touch:

$$y = x \tan \frac{\varphi_2}{2} + d$$

$$y = x \tan \frac{\varphi_2}{2} - d$$

where

$$d = 2A \left\{ \sin \frac{\varphi_1}{2} - \cos \frac{\varphi_1}{2} \tan \frac{\varphi_2}{2} \right\}$$

Now c and d depend on the ray amplitudes A and B, but not on the p.d., α. Provided A and B remain constant, but α varies, all the ellipses will touch the same parallelogram, once on each side. Thus the parallelogram defined by the four boundary lines is the envelope of the ellipses. The slopes of the sides are tan $\varphi_1/2$ and tan $\varphi_2/2$, i.e. the slopes of the straight-line displays for either ray alone.

9.4.2. Diagonals of the parallelogram

For two values of the ray p.d., say α_1 and α_2, the corresponding ellipses collapse to straight lines which are the diagonals of the parallelogram. The values of α_1 and α_2 can be found from the condition that the ratio y/x must be constant for any point on the diagonal, independent of t. A constant ratio of Y to X requires that

$$\cos \left(\omega t + \frac{\varphi_1}{2} \right) = \pm \cos \left(\omega t + \alpha + \frac{\varphi_2}{2} \right)$$

Hence

$$\alpha_1 = \tfrac{1}{2} (\varphi_1 - \varphi_2)$$

$$= \frac{\pi d}{\lambda} (\cos \theta_1 - \cos \theta_2)$$

$$\alpha_2 = \tfrac{1}{2} (\varphi_1 - \varphi_2) + \pi$$

$$= \frac{\pi d}{\lambda} (\cos \theta_1 - \cos \theta_2) + \pi$$

The important points to notice are

(*a*) that the collapse of the ellipse to a straight line does *not* signify QUMP conditions

(*b*) that the slopes of the diagonals do *not* provide direct information about the arm angles.

9.5 More than two rays

The theory given above can be generalised to n rays; the envelope of the ellipses is now a polygon with $2n$ sides, made up of n pairs of parallel lines with slopes determined by the ray arrival angles.

Clearly, a 'clean' parallelogram or polygon is obtained only if the ray amplitudes remain constant while the p.d. varies. The side lengths of the parallelogram, for example, are proportional to A and B and will therefore vary as A and B fade. A parallelogram display is probably best built up on a storage tube over a period of a few seconds, which may perhaps be long enough for phase difference to vary over several cycles while amplitudes remain sensibly constant. A polygon may take rather longer to form, depending on the number of rays.

9.6 Practical implementation

The use of the sum and difference polygon display for mode resolution was pioneered at the University of Illinois by Bailey and McClurg (1963). It is an attractively simple procedure that is capable, in principle, of providing information about the number of modes, their average amplitudes and their DOAs.

However, it is the author's impression that interpretation of a *sequence* of such observations is sometimes difficult in practical multiray conditions. Suppose that measurements are made first at a narrow spacing, then at a wide spacing, on an *NS* arm and that similar measurements are then made on an *EW* arm. In effect, we are space-sampling and time-sampling the wave-field, but a space sample is based on only two elements and diversity effects may be significant, particularly at the wide spacing. Sufficient time must be allowed to build up an envelope on the display. In order to resolve ambiguities, sort out arm angles and pair up the *E-W* and *N-S* arm angles for each ray, it is necessary to adopt a simplified model of the wave-field (constant amplitudes, constant DOAs, varying phases) which may not always be satisfactory. Whereas a single goniometer waveform on a Wullenweber can be obtained in 0·01 sec (over 36°), a period of several seconds is presumably required to build up a polygon.

These difficulties can be ameliorated in various ways. Dodge and Martin (1969) displayed all four polygons on the same storage tube, using an antenna-selector switch to cycle round the elements. They used a simple analogue simulator, with knobs calibrated in terms of frequency, azimuth and elevation angles, to assist in interpretation of the display. The absence of clear polygons on the display provides a visual warning of an unfavourable situation. Nevertheless, the impression remains that this method is sensitive to the wave-field model, and will not work in all circumstances. When one ray is dominant in a multiray situation, it appears from the results of simulation studies quoted by Sherrill (1971) that the DOA deduced (on a one-ray interpretation) from a time-averaged display tends to be that of the dominant ray. Thus the interferometer can exhibit a capture effect, similar to that described for the CDAA in earlier chapters.

9.7 Wavefront testing

Treharne (1967, 1973) and his group at WRE, Australia adopted the philosophy that the interferometer is basically a tool for the measurement of the DOA of a single ray; they relied on relative fading between rays to produce moments of QUMP when the DOA of the remaining ray was measured. Thus, with two rays *A* and *B*, the DOA of *A* is measured when ray *B* fades to a low level and vice versa. The DOAs measured at such moments, and interpreted in terms of single-ray model, should then fall into two distinct populations. The probability of obtaining QUMP conditions must decrease greatly when the number of rays exceeds two. In (say) a three-ray field it is possible for two rays to cancel each other at a particular point (i.e. to have the same amplitudes and to be in antiphase) but it is questionable whether this condition would apply over a large array.

The moments of QUMP can be selected by some method of wavefront testing. The idea is to detect that a given set of element signals is consistent with the hypothesis of a single plane wavefront, i.e. a single ray. In practice, some tolerance must be allowed in any numerical test because of noise, measurement errors and the coning associated with a single 'ray'. Various types of test can be devised; those involving as many elements as possible and utilising both phase and amplitude information are the most reliable, i.e. give the fewest false indications. Some of the situations that can give rise to false indications will now be briefly discussed.

Firstly, we must note the possibility of two or more rays having different DOAs but the same arm angle. A special example is that of a number of modes arriving from the same azimuth but different elevation angles, in a vertical plane perpendicular to the arm. The usual sum-and-difference display *for that arm* will then be a stable straight line, but the presence of more than one ray would be readily inferred from the elliptical display for a perpendicular arm.

A standard technique is to apply a 'bright-up' pulse to a visual display only when the wavefront test is satisfied. Suppose, for example, that a test of phase coincidence (PC) is used in conjunction with the sum-and-difference phase meter, so that the display is only visible when the signals in the *X* and *Y* channels differ in phase by less than some small tolerance. With two rays, *A* and *B*, we must in general expect four lines to appear at various moments on the c.r.t.: two are the diagonals of the parallelogram and the others correspond to ray *A* alone and ray *B* alone respectively. Because the diagonals do *not* correspond to moments of QUMP, PC in this case gives two false angles and two correct angles (on the tube). The two false angles occur at moments when the signals in the two antennas happen to be equal in amplitude, even though two rays are incident on each; this condition tends to be transitory and is said to be easily distinguished by an operator from the true QUMP condition.

It is helpful to recall at this point the shapes of the curves of constant phase (CP) and of constant amplitude (CA) for a two-ray wavefield, as discussed in Chapter 3. The CPs can approximate to straight lines over comparatively large apertures ($>1\lambda$). Clearly, measurements must be made at a minimum of three well-spaced elements to detect whether the CPs are kinked. Measurements of amplitude at two elements are generally sufficient to detect that amplitude is not constant over the whole horizontal plane, but the two elements might happen to lie on a CA. Because the CAs in a two-ray field are straight lines, the use of several collinear elements in an amplitude test does not overcome this objection. However, the use of three or more *non* collinear elements should lead to very few false alarms.

The general conclusion to emerge from this qualitative discussion is that there are some pitfalls in the use of simple wavefront tests, and that special circumstances can be found in which such tests fail, but as the severity of the test increases, so the possibility of false alarms can be made very small.

Modern digital techniques allow more sophisticated tests to be used. Rice (1982) reported the probability of observing various degrees of phase-front non-linearity with two or three rays incident from different directions on the CRC linear array; the measure of non-linearity employed was the r.m.s. deviation from the best straight-line fit to the measured phases along the array. The bearing accuracy achieved when the deviation was below a suitable wavefront linearity threshold was comparable to that expected on single modes. Warrington and Jones (1991) tried two separate criteria, namely the agreement of angles determined from different element spacings and the near equality of amplitudes in the elements. Significant reductions in the spreading of the individual angle measurements and in the occurrence of erroneous peaks in the distributions were obtained with either criterion.

9.8 Combination of observations from two arms

So far we have been mainly concerned in this Chapter with the problem of determining the arm angles for two or more rays, from observations made on one arm of an interferometer. We now consider the accuracy of a DOA obtained from measurements on two orthogonal arms. For this purpose it is sufficient to consider a single ray arriving from azimuth α, elevation Δ. Suppose that the arms are *N-S* and *E-W* and that the electrical spacing between the pairs of elements used is a, where $a = 2\pi d/\lambda$. The correct p.d.s φ_1, φ_2 for the two arms are

$$\varphi_1 = a \cos \alpha \cos \Delta$$

$$\varphi_2 = a \sin \alpha \cos \Delta$$

Hence

$$\tan \alpha = \varphi_2/\varphi_1$$

$$\cos \Delta = \frac{1}{a}\sqrt{(\varphi_1^2 + \varphi_2^2)}$$

We now consider the effects of phase errors $\delta\varphi_1$, $\delta\varphi_2$ on the determination of α and Δ. By differentiation,

$$\delta\varphi_1 = -a \sin \alpha \cos \Delta\delta\alpha - a \cos \alpha \sin \Delta\delta\Delta$$
$$\delta\varphi_2 = a \cos \alpha \cos \Delta\delta\alpha - a \sin \alpha \sin \Delta\delta\Delta$$

Elimination of $\delta\Delta$ yields

$$\delta\alpha = \frac{\delta\varphi_2 \cos\alpha - \delta\varphi_1 \sin\alpha}{a \cos\Delta}$$

If $\delta\varphi_1$ and $\delta\varphi_2$ are independent errors with the same standard deviation σ_φ, the corresponding s.d. σ_α of α is

$$\sigma_\alpha = \frac{\sigma_\varphi}{a \cos\Delta}$$

Thus, the HAM is comparatively well determined at low-elevation angles and badly determined for nearly vertical rays.

Similarly, it may be shown that

$$\delta\Delta = \frac{-\delta\varphi_1 \cos\alpha - \delta\varphi_2 \sin\alpha}{a \sin\Delta}$$

$$\sigma_\Delta = \frac{\sigma_\varphi}{a \sin\Delta}$$

and the elevation angle is comparatively badly determined at low angles. These results are fairly obvious from physical arguments, but it is sometimes useful to have the formal expressions.

It is sometimes stated that the errors in α and Δ are likely to be least when $\alpha = 0°$, $90°$ etc., greatest when $\alpha = 45°$, $135°$ etc and that the average error at $\alpha = 45°$ etc is $\sqrt{2}$ times that at $\alpha = 0°$ etc. However, the expressions for σ_α and σ_Δ derived above do not show such an 'arm effect'. It is true that the two arms are not generally of equal weight in the determination of α and Δ, but the combined weight appears to be independent of α. For a given arm, the relationship between arm angle θ and c.r.t. angle t is

$$t = \frac{a}{2} \cos\theta$$

Hence

$$\frac{dt}{d\theta} = \frac{-a}{2} \sin\theta$$

and

$$\frac{dt}{d\theta} \rightarrow 0 \text{ as } \theta \rightarrow 0$$

This very simple geometrical effect is sometimes referred to as 'flatness' of the display.

When two or more rays are present, it is necessary to identify the pairs of arm angles that correspond to each ray before the ray DOAs can be calculated. Some pairings can immediately be eliminated; for example, one ray cannot give both arm angles less than 45° for orthogonal arms. Relative mode amplitudes and fading behaviour are usually used to resolve any remaining uncertainties.

References

BAILEY, A.D. and McCLURG, W.C. (1963): 'A sum-and-difference interferometer system for HF radio direction finding', *IEEE Trans.*, **ANE-10,** pp. 65–72

DENISOV, V.P. AND DOROSHENKO, V.G. (1983): 'A priori uncertainty in the problem of statistical optimization of phase direction finders', Izv. vuz. Radioelektron, **26,** no. 11, pp. 86–88. English translation in Radioelectron. & Commun. Syst., **26,** no. 11, pp. 95–97

DODGE, C. and MARTIN, P.E. (1969): 'HF direction finding during multimode propagation conditions', 21st annual Southwestern IEEE conference and exhibition, San Antonio, 1969, (IEEE), Paper 15D

RICE, D.W. (1982): 'HF direction finding by wave front testing in a fading signal environment', *Radio Sci.*, **17,** pp. 827–836

ROSS, W., BRAMLEY, E.N. and ASHWELL, G.E. (1951): 'A phase-comparison method of measuring the direction of arrival of ionospheric radio waves', *Proc. IEE.*, **98,** pp. 294–302

SHERRILL, W.M. (1971): 'A survey of HF interferometry for ionospheric propagation research', *Radio Sci.*, **6,** pp. 549–566

TREHARNE, R.F. (1967): 'Vertical triangulation using skywaves', *Proc. Inst. Radio Electron. Eng.*, **28,** pp. 419–423

TREHARNE, R.F. (1973): 'Single station location: a brief survey of the development of a new system of high frequency direction finding', 14th National radio and electronics engineering convention, 20–24 August 1973, Melbourne, (IREE. Aust.)

WARRINGTON, E.M. and JONES, T.B. (1991): 'Measurements of the direction of arrival of HF sky wave signals and single site location using a seven element wide aperture (294m) interferometer array', to be published in Proc. IEE Pt. H, April 1991, pp. 121–130

Chapter 10

Wavefront analysis: the concept

10.1 Introduction

Signal processing in DF has, until recently, been almost entirely of an analogue nature. The spinning goniometer usually used in a Wullenweber system is an example: the inputs to the goniometer are the signals from a selected arc of elements and the output is fed through a single-channel receiver which determines the narrow frequency band to be examined.

The development of multichannel HF receivers in recent years has opened up the possibility of using instead some form of digital processing; the amplitudes and phases of the element signals at a selected frequency can be measured, recorded digitally and processed in a computer program. The experimental and mathematical techniques are referred to collectively as wavefront analysis (WFA). The aim is to resolve the wave-field into its component rays and to print out the ray parameters.

Direct analytic or iterative attacks on the equations relating the ray parameters to the measured signals become very unattractive when more than a small number of rays are present. However, a general method of solution, applicable to almost any form of antenna array, can be used with fading signals (Gething, 1971). In this chapter some simple examples of the technique will be given for idealised problems involving neither noise nor measurement errors, in order to explain the principles involved. The analysis of more realistic data is then discussed in Chapter 11.

10.2 Definitions

10.2.1 Input to the computer program

It will be assumed that a well-adjusted n-channel receiving and measuring system is available, where n is the number of elements in the DF array. 'Well adjusted' means that the channels are almost perfectly matched in gain and in phase-change characteristics and that any small remaining imbalance can be determined by a calibration procedure and allowed for in the computer program.

A simultaneous set of measurements of n amplitudes and n phases (relative to one element as the reference) will be referred to as a frame of data; it forms the basic unit of information for input to the program and can be regarded as one space sample or 'snapshot' of the field. The measurements can be repeated at regular intervals, at a rate such as one frame/sec, in order to obtain the benefits of time sampling. Given a group of (say) 100 successive frames, we have two main options: to calculate 100 separate solutions based on single-frame analysis and then take the mean, or to incorporate all frames in one multiframe analysis.

10.2.2 Wave-field model

A single-frame analysis is based on the assumption that the antenna signals can be explained in terms of a small number of discrete rays and that any residual component in each element can be ascribed to random noise and/or small measurement errors.

Multiframe analyses are based on the same assumption, but require two additional features in the model: that the rays exhibit fading in amplitude and/or phase, but that the ray DOAs remain sensibly constant over the observation period.

The ultimate test of whether either idealised model of the wave-field is a satisfactory approximation in any particular practical application must be based on the analysis of practical data. It will be shown that various checks on the validity of the assumptions can be made. The need to adopt a model is rather more clear and explicit in WFA than in other forms of DF, but the assumption of one ray arriving from a fixed azimuth is implicit to many conventional DF equipments.

10.2.3. Number of unknowns

At a given moment, each of the incident rays is defined by four parameters: its amplitude, phase, azimuth angle and elevation angle. The amplitude is that of the polarisation components to which the elements respond and the phase must be measured relative to some standard. If one of m incident rays is taken as the amplitude and phase reference, there are $(4m - 2)$ parameters that we wish to determine.

With n elements, one of which is used as a reference, the information from a single frame consists of $(2n - 2)$ numbers. In general, therefore, we require $n \geqslant 2m$ if a solution is to be obtained from a single frame. Bain (1956) pointed out that this rule does not hold for the special case of $m = 1$: the complete DOA of one ray cannot be determined with less than three elements. It should also be borne in mind that the signals in, for example, small horizontal loops on a vertical tower do not depend on the azimuth angles; the number of unknowns that could be determined with such an array is therefore $(3m - 2)$.

Over f frames, the number of variables to be determined is as follows:

	m (constant) elevation angles
	m (constant) azimuth angles
	$f(m-1)$ relative amplitudes
	$f(m-1)$ relative phases
Total	$2m + (2m-2)f$

The total number of measured relative amplitudes and phases is $(2n-2)f$. Suppose, as a specific example, that we wish to resolve a three-ray wave-field with a four-element array. This is impossible with a single-frame analysis, because there are 10 unknowns to be determined from six measurements. However, with three frames the number of unknowns (18) is just equal to the number of measurements, and for four or more frames the number of measurements exceeds the number of unknowns.

In a practical problem, the number of incident rays is unknown initially and there is a danger of obtaining an incorrect answer by forcing (say) a three-ray solution to what may really be (say) a four-ray problem. However, it will be shown that such a solution can generally be recognised as unsatisfactory.

10.3 Outline of solution method

The main problem is to determine the ray DOAs; when these have been found the ray amplitudes and phases can be calculated relatively easily.

Suppose that the signals from n elements are combined in a beam-forming network (BFN) with n inputs and one output. The elements can now be said to form an array, the radiation pattern of which is a function of the configuration of the elements *and* of the properties of the BFN. Suppose further that the only energy incident on the array consists of m rays ($m < n$) from fixed directions and that these directions happen to coincide with perfect nulls in the radiation pattern of the array. Then the BFN output will be zero at all times, no matter how the rays fluctuate in amplitude and phase.

In mathematical terms, the BFN forms a linear combination of the element signals $Z_1, \ldots, Z_n$ of the form

$$d_1 Z_1 + d_2 Z_2 + \ldots + d_n Z_n$$

where the ds are complex in general and represent the phase and amplitude changes introduced by the BFN. We therefore seek to synthesise in the computer a hypothetical BFN such that

$$\sum_{r=1}^{n} d_r Z_r = 0$$

The corresponding radiation pattern will then exhibit nulls at the ray arrival angles, and possibly elsewhere. If more than one BFN with the above property can be found, it should be possible to identify the true DOAs from the common nulls in two or more patterns.

10.4 Numerical examples

10.4.1 Introduction

The power and simplicity of the procedure outlined above, and its applicability to almost any form of array, become clear when numerical examples for various types of array are examined. Three numerical simulations for different situations will now be briefly described. For clarity of presentation, 'perfect' data will be used in these toy problems; that is to say, no noise will be added deliberately to the calculated antenna signals and the ray DOAs will be assumed not to vary at all with time.

10.4.2 Vertical tower, one-frame analysis

The simplest type of problem occurs with a one-dimensional array, such as the vertical tower of horizontal loops defined in Appendix 3. Suppose that we construct a three-ray problem. The antenna signals produced by each ray are of the form

$$S_r = A \cos \Delta \sin (\psi_r)$$

where

$$\psi_r = \frac{2\pi h_r}{\lambda} sin\ \Delta$$

and S_r is the signal in the loop at height h_r due to a ray of amplitude A arriving from elevation angle Δ. The cos Δ term represents the element pattern.

The ray parameters used to construct the problem are shown in Table 10.1. The phase of the first ray at the base of the tower is taken as the phase reference. The phasor sums of the signals in each loop are formed and resolved into X and Y components; thus

$$X_r = a_r \cos \varphi_r$$

$$Y_r = a_r \sin \varphi_r$$

where a_r is the amplitude of the total signal in the rth loop and φ_r is its phase. The calculated values of X and Y are shown in Table 10.2.

Table 10.1 Ray parameters used to construct 3-ray problem

Ray	Amplitude	Phase	Elevation Angle
1	1·00	0·0°	15·0°
2	2·00	123·4°	23·2°
3	0·70	234·5°	31·6°

Table 10.2 Components of loop signals

Loop	1	2	3	4	5	6	7	8
X	−0·581	−0·301	0·721	1·494	1·197	−0·015	−1·247	−1·726
Y	0·762	1·168	0·827	−0·286	−1·428	−1·503	−0·184	1·479

We now endeavour to form, from elements 1 to 7, an array with nulls at angles Δ_1, Δ_2, Δ_3. We choose to apply a symmetrical amplitude taper and form the combination

$$a(Z_1 + Z_7) + b(Z_2 + Z_6) + c(Z_3 + Z_5) + Z_4 = 0$$

where $Z_r = X_r + jY_r$ and a, b, c are amplitude taper coefficients. Note that in free space and with a single ray, the signal combinations $(Z_1 + Z_7)$, $(Z_2 + Z_6)$ etc would be in phase (or antiphase). The tapered array in free space must have symmetrical nulls, i.e. each null at Δ_s is matched by a null at $-\Delta_s$.

The subarray

$$a(Z_2 + Z_8) + b(Z_3 + Z_7) + c(Z_4 + Z_6) + Z_5 = 0$$

would have the *same* radiation pattern in free space. Above a horizontal plane reflector (not necessarily a perfect reflector), the subarrays will both exhibit nulls at angles Δ_s.

We therefore solve the following equations for a, b, c:

$$a(X_1 + X_7) + b(X_2 + X_6) + c(X_3 + X_5) + X_4 = 0$$

$$a(X_2 + X_8) + b(X_3 + X_7) + c(X_4 + X_6) + X_5 = 0$$

$$a(Y_1 + Y_7) + b(Y_2 + Y_6) + c(Y_3 + Y_5) + Y_4 = 0$$

$$a(Y_2 + Y_8) + b(Y_3 + Y_7) + c(Y_4 + Y_6) + Y_5 = 0$$

The solution obtained from the first three equations is

$$a = -0{\cdot}118$$

$$b = +0{\cdot}418$$

$$c = -0{\cdot}822$$

and this set of values satisfies the fourth equation.

The radiation pattern of the tapered subarrays composed of elements 1 to 7 and 2 to 8 can now be examined; they are found to have nulls at the angles shown in Table 10.3.

Table 10.3 Nulls in sub-array patterns

Array 1–7	15·0		19·8	23·2	31·6		42·6	
Array 2–8	15·0	15·7		23·2	31·6	32·8		54·2

The only common nulls are at 15·0, 23·2 and 31·6. Thus the correct angles have been found.

In moving from subarray (1–7) to subarray (2–8) we take a space-step within the wave-field at a particular moment of time. It is possible to solve this problem by the space-stepping technique only because the elements are uniformly spaced and the two subarrays (in free space) are therefore identical. Note that, to find three elevation angles, we must have three degrees of freedom in the taper coefficients. Because only *relative* taper affects an array radiation pattern, the centre element is given an arbitrary taper of unity. The problem has been solved on the basis of a three-ray hypothesis; this hypothesis is supported by the fact that the taper solution satisfies all four equations and by the existence of three common nulls, no more and no less, in Table 10.3. The final confirmation would be to show that rays at these angles all have non-zero amplitudes and that the antenna signals reconstructed from the solution are in excellent agreement with the initial data.

This method of solution was discovered independently by Dr. J. Kelso (1972) and S. Wylie, and is often referred to as a Prony solution, because of the underlying analysis technique. It can be readily generalised to more elements and more rays; if the physical ideas of space stepping and symmetrical amplitude taper are borne in mind, the form of the linear relationships for any new situation can be written down immediately. The method does not impose any restriction on the ground reflection coefficient, other than that the ground must be a horizontal plane of uniform properties; it can therefore be applied to vertically-polarised as well as horizontally-polarised elements.

Table 10.4 Mode parameters

	Frame 1	Frame 2	Frame 3
Mode 1			
Azimuth	17·0°	17·0°	17·0°
El. angle	20·0°	20·0°	20·0°
Amplitude	1·00	1·15	1·21
Phase (radians)	0·0	0·15	5·20
Mode 2			
Azimuth	17·0°	17·0°	17·0°
El. angle	35·0°	35·0°	35·0°
Amplitude	0·50	0·51	0·77
Phase (radians)	1·24	6·56	5·96

Table 10.5 Signal components in antenna elements

	Frame 1	Frame 2	Frame 3
X_1	0·055	0·634	−0·284
Y_1	−0·501	−0·865	−0·639
X_2	0·509	1·133	0·519
Y_2	−0·130	0·153	−0·612
X_3	−0·793	−1·077	0·309
Y_3	0·631	1·153	1·563
X_4	1·068	1·317	1·752
Y_4	1·023	0·835	−0·921
X_5	0·299	0·371	1·726
Y_5	1·415	1·145	0·123
X_6	1·393	1·245	1·025
Y_6	0·468	0·248	−1·469
X_7	−0·242	−0·603	−1·944
Y_7	−1·401	−1·512	−0·342
X_8	0·841	1·452	0·583
Y_8	−0·104	−0·293	−1·256

10.4.3 8-element CDAA, multi-frame analysis

We now consider a circular array of eight uniformly-spaced elements in a circle of diameter 150 m, illuminated by two modes from the same azimuth angles. The mode parameters obtained from a computer simulation of fading from frame to frame are shown in Table 10.4. The components of the antenna signals, calculated for a frequency of 10 MHz, are shown in Table 10.5. Here X_1, Y_1 are the components of the signal in element one on azimuth 0°, X_2, Y_2 are the components of the signal in element two, on azimuth 45°, and so on. No allowance has been made for the elevation pattern of each element, but this simplifying assumption does not affect the basic technique, which depends on the control of *array* nulls.

If only one ray were incident, the following wavefront tests would be satisfied:

$$X_1^2 + Y_1^2 = X_2^2 + Y_2^2 = X_3^2 + Y_3^2 \text{ etc.}$$

$$\tan^{-1}\frac{Y_1}{X_1} + \tan^{-1}\frac{Y_5}{X_5} = \tan^{-1}\frac{Y_2}{X_2} + \tan^{-1}\frac{Y_6}{X_6}$$

$$= \tan^{-1}\frac{Y_3}{X_3} + \tan^{-1}\frac{Y_7}{X_7} \quad \text{etc}$$

i.e.

$$\frac{Y_1X_5 + Y_5X_1}{X_1X_5 - Y_1Y_5} = \frac{Y_2X_6 + Y_6X_2}{X_2X_6 - Y_2Y_6} \quad \text{etc}$$

The first of these is an equal-amplitude test, the second compares different estimates of the phase of the ray at the centre. Because these tests are *not* satisfied by the data in Table 10.5, more than one ray must be involved. We now seek complex-taper coefficients d_1 to d_8 such that

$$\sum_{1}^{8} d_r Z_r = 0$$

where $$Z_r = X_r + jY_r$$

Writing $$d_r = a_r + jb_r$$

we obtain two equations

$$\sum (a_r X_r - b_r Y_r) = 0$$

$$\sum (a_r Y_r + b_r X_r) = 0$$

Each frame therefore gives two equations in 16 unknowns, a_1 to a_8 and b_1 to b_8. From a group of eight frames or more we can therefore obtain a solution. Because only two rays are involved, some of the coefficients prove to be indeterminate and can be given arbitrary values. In Table 10.6, coefficients obtained from a group of eight frames are shown.

The data given in Table 10.5 for the first three frames are sufficient to enable the reader to verify that the linear relationships are satisfied by these values.

Table 10.6 Taper coefficients from 8-frame analysis

Antenna	a	b
1	4·0	0·0
2	0·0	0·0
3	0·0	0·0
4	0·0	−4·40587
5	0·0	−6·12356
6	0·0	4·09225
7	0·0	−8·26603
8	0·0	1·0

To locate all the nulls or null contours in the corresponding radiation pattern it would be necessary to calculate the pattern at small intervals of azimuth and elevation. For simplicity, we shall now suppose that the correct azimuth of arrival has been determined, perhaps with a spinning goniometer. It is now very simple to conduct a search for nulls at intervals of (say) 0·1° in elevation on azimuth 17°. Only 900 points have to be tested in the computer and the only nulls found are at 17·0° and 35·0°. From this result, and the fact that only four taper coefficients have non-arbitrary values, it is safe to deduce that this is a two-ray problem. Once the DOAs have been determined, the calculation of ray amplitudes and phases from the frame data presents no problem.

10.4.4 Discussion of numerical simulations

The two problems described briefly above are sufficient to illustrate several features of the WFA method. In the three-ray VAM problem (Section 10.4.2) the basic procedure was space-stepping in the wave-field, i.e. selecting subarrays 1–7 and 2–8 from a set of simultaneous measurements. The space step does not alter the subarray geometry (apart from its height above the ground) because the elements are uniformly spaced. The taper values found were real, not complex; this effect arises because the signals for a single ray differ in amplitude, but not in phase.

The other example was based on a multiframe analysis; we can exploit the power of time sampling as well as space sampling provided we are satisfied that DOAs are likely to be constant over the period of analysis. In this example, the *Z*'s were signals in the individual elements of an array. However, linear relationships can be found among signals from groups of elements and the basic WFA technique can be applied, for example, to the outputs from various beam-cophasal arrays. The output from a rotating goniometer is the output of a continually changing group of elements, but with the same group whenever the goniometer points to a particular azimuth. Suppose, for example, that the symbol Z_1 is used for the output when the goniometer points to azimuth 10°, Z_2 for azimuth 20° and so on. Then linear relationships can be formed from the *Z*'s which are satisfied for each revolution of the goniometer, provided

(*a*) the number of incident rays is smaller than the number of *Z*'s per revolution,
(*b*) all the ray parameters remain constant during a revolution,
(*c*) the DOA's remain constant from revolution to revolution.

With two-dimensional or three-dimensional arrays, the search for nulls involves both the azimuth and elevation co-ordinates. Although this necessity for two-dimensional search in the (α, Δ) plane is an unfortunate feature of the WFA method, it represents a very considerable simplification compared with a multidimensional iterative search over many ray parameters. The 2-D search can sometimes be shortened: mathematically, by locating null contours and moving along them; physically, by using auxiliary equipment such as a goniometer or a CDAA to give an indication of the most fruitful area of the (α, Δ) diagram for further examination.

With an array consisting of two linear subarrays of vertical monopoles in a cross or *L*-formation, each arm can be used separately and a search for nulls is then made in one co-ordinate only, namely arm angle. When the results from the two arms are combined, arm angles can be paired off and converted into azimuth and elevation angles. Exploitation of the WFA technique is therefore particularly simple with this type of array.

Although taper coefficients are complex in general, a small number of nulls (compared with the number of elements) can be controlled with amplitude taper alone. Phase shifters can then be used to produce other desirable features in the radiation pattern, such as a beam maximum in a specified direction (Vu, 1986; Vu and Rathinam, 1988).

10.5 Further one-dimensional theory

10.5.1 Introduction

As a further illustration of the application of WFA theory to a one-dimensional, uniformly spaced, array we now consider some of the theory developed by Cawsey (1972) for the vertical tower with eight horizontal loops. A variety of single-frame analysis methods will be described; the aim in each case is to form a polynomial in c, where

$$c = \cos 2\psi$$

and

$$\psi = \frac{2\pi h}{\lambda} \sin \Delta$$

For example, a three-ray problem should lead to a cubic in c, with roots c_1, c_2, c_3, corresponding to the three arrival angles Δ_1, Δ_2, Δ_3. The key problem is to form the correct polynomial; once this has been done, there is no difficulty in finding the roots by analytic or iterative methods. A good first approximation to the ray parameters can be refined by curve-fitting techniques.

The following trigonometric identities are required:

$$\begin{aligned}
\sin \psi &= \sin \psi \\
\sin 2\psi &= 2 \cos \psi \sin \psi \\
\sin 3\psi &= (2c + 1) \sin \psi \\
\sin 4\psi &= 4c \cos \psi \sin \psi \\
\sin 5\psi &= (4c^2 + 2c - 1) \sin \psi \\
\sin 6\psi &= (8c^2 - 2) \cos \psi \sin \psi \\
\sin 7\psi &= (8c^3 + 4c^2 - 4c - 1) \sin \psi \\
\sin 8\psi &= (16c^3 - 8c) \cos \psi \sin \psi
\end{aligned}$$

10.5.2 1-ray solution

The element signals are all in phase or antiphase when only one ray is incident on the tower. The elevation angle can be derived by fitting a sine curve to a plot of element amplitude against element height and this procedure could be carried out quite simply, even if the elements were not uniformly spaced.

With uniformly-spaced elements, two good initial estimates of c can be found analytically, from the odd and even elements respectively. Let the (real) element signals be represented by X_1 to X_8. Then it is easily seen from the trigonometric identities that

$$\frac{X_7 + 3X_5 + 3X_3 + X_1}{2(X_5 + 3X_3 + 2X_1)} = \frac{(8c^3 + 16c^2 + 8c)\sin\psi}{(8c^2 + 16c + 8)\sin\psi}$$

$$= c$$

$$\frac{X_8 + 4X_6 + 6X_4 + 4X_2}{2(X_6 + 4X_4 + 5X_2)} = \frac{(16c^3 + 32c^2 + 16c)\sin\psi\cos\psi}{(16c^2 + 32c + 16)\sin\psi\cos\psi}$$

$$= c$$

10.5.3 Two rays in phase

A small phase spread in a single frame of data may indicate the presence of one ray only, or of two or more rays that happened to be in phase at the moment the frame was taken. With two or more rays the amplitudes will not, in general, fall on a sine curve when plotted against height. We now consider the problem of determining the elevation angles of two in-phase rays. This should be regarded as a difficult case because the solution must be obtained from eight X components only, rather than the eight Xs and eight Ys.

For two rays with amplitudes A, B, angles ψ_1, ψ_2 and $c = \cos 2\psi_1$, $d = \cos 2\psi_2$, the signals in the even elements can be written in the following form

$$X_2 = \alpha + \beta$$

$$\alpha = 2A\cos\Delta_1 \sin\psi_1 \cos\psi_1$$

$$\beta = 2B\cos\Delta_2 \sin\psi_2 \cos\psi_2$$

$$X_4 = 2\alpha c + 2\beta d$$

$$X_6 = \alpha(4c^2 - 1) + \beta(4d^2 - 1)$$

$$X_8 = \alpha(8c^3 - 4c) + \beta(8d^3 - 4d)$$

We now form the following combinations of signals:

$$D = 4X_4^2 - 4X_2(X_2 + X_6)$$

$$= 16\alpha\beta\{2cd - (c^2 + d^2)\}$$

$$E = 2X_2(2X_4 + X_8) - X_4(X_2 + X_6)$$

$$= 16\alpha\beta\{(c^3 + d^3) - cd(c + d)\}$$

$$F = (X_2 + X_6)^2 - X_4(2X_4 + X_8)$$

$$= 16\alpha\beta\{2c^2d^2 - cd(c^2 + d^2)\}$$

Hence

$$\frac{E}{D} = -(c + d)$$

$$\frac{F}{D} = cd$$

and the equation

$$Dx^2 + Ex + F = 0$$

has the required roots, c, d. Note that D, E, F all depend on the amplitude product term $\alpha\beta$ and are all zero in a one-ray case.

A similar quadratic can be formed from the odd elements, with coefficients D', E', F' given by

$$D' = 4\{(X_3^2 - X_1^2) - X_1(X_3 + X_5)\}$$

$$E' = 2\{(X_3^2 - X_1^2) + X_7(X_1 + X_5) - X_5(X_3 + X_7)\}$$

$$F' = (X_1 + X_5)^2 - (X_3 - X_1)(2X_3 + X_5 + X_7)$$

10.5.4 Five rays from one frame

One frame of data from eight loops gives 14 independent numbers. In theory, these data should be just sufficient to give a solution for 13 independent ray parameters associated with five rays, plus one confirmatory check that the five-ray hypothesis is valid. Cawsey (1972) has shown how such a solution can be obtained.

The first step is to form new (complex) signal combinations G_0 to G_7 from the measured (complex) signals Z_1 to Z_8. The Gs are defined by the equations

$$G_0 = Z_1$$

$$G_1 = Z_2/2$$

$$G_2 = (Z_1 + Z_3)/4$$

$$G_3 = (2Z_2 + Z_4)/8$$

$$G_4 = (2Z_1 + 3Z_3 + Z_5)/16$$

$$G_5 = (5Z_2 + 4Z_4 + Z_6)/32$$

$$G_6 = (5Z_1 + 9Z_3 + 5Z_5 + Z_7)/64$$

$$G_7 = (16Z_2 + 14Z_4 + 6Z_6 + Z_8)/128$$

These expressions are chosen so that each G_r is proportional to $\cos^r \psi$.

The next step is to form six equations in six (real) amplitude taper coefficients a_0 to a_5. These equations (in complex form) are

$$a_0G_0 + a_1G_1 + \ldots + a_5G_5 = 0$$

$$a_0G_1 + a_1G_2 + \ldots + a_5G_6 = 0$$

$$a_0G_2 + a_1G_3 + \ldots + a_5G_7 = 0$$

One of these taper values is arbitrary, e.g. $a_5 = 1.0$; a solution for a_0 to a_4 can be obtained from five of the equations and checked against the sixth. It can then be shown that the roots x_1 to x_5 of the polynomial

$$a_0 + a_1 x + \ldots + a_5 x^5 = 0$$

are equal to $\cos\psi_1, \cos\psi_2 \ldots \cos\psi_5$ and the corresponding elevation angles Δ_1 to Δ_5 can be calculated.

In Appendix 10.1 of his paper (1972), eqn. 16, Cawsey gave a general expression for the signal combinations required for any number l of equally-spaced elements. It is, in the notation used here,

$$G_r = \frac{1}{2^{r-1}} \sum_{i=1}^{l} \frac{i(r-1)!\, Z_i}{\left(\frac{r+i}{2}\right)!\left(\frac{r-i}{2}\right)!}$$

where the summation includes all terms for which $(r - i)$ is positive and even.

10.6 Basic theorems

This chapter will be concluded with a summary of the theorems on which WFA is based. They may be stated formally as follows:

(*a*) When the number m of rays arriving at an array is less than the number n of elements, the signals $Z_1,\ldots,Z_n$ in the elements, measured simultaneously, will satisfy certain linear relationships of the form

$$\sum_{r=1}^{j} d_r Z_r = 0$$

where $(m + 1) \leqslant j \leqslant n$ and the ds are complex coefficients,

(*b*) The coefficients d_r depend on the ray arrival angles, but not on the ray amplitudes and phases,

(*c*) The same coefficients d_r, when interpreted as complex taper coefficients in an adding network, produce a radiation pattern with nulls at the ray-arrival angles,

(*d*) If more than one linear relationship can be found that is satisfied by the Zs, i.e. if $m \leqslant (n - 2)$, the corresponding radiation pattern will exhibit common nulls at the ray-arrival angles but will usually differ elsewhere,

(*e*) Adding networks of the form

$$\sum_{j=1}^{n} e_{rj} Z_j$$

can be synthesised for $r = 1,\ldots, m$, with a nonzero response to ray r but with nulls at all other ray arrival angles. The amplitude and phase of the rth ray can be determined using the rth network of this set.

These five statements may be regarded as formal results in antenna-array theory; apart from the implicit assumption of a finite number of discrete rays, they do not require the adoption of any particular wavefield model. The validity of the theorems is not restricted to the HF band. No assumptions have been made about array geometry and it is not even necessary for the elements to be of identical size, although they must respond to the same polarisation component. Only when the theorems are applied to multiframe WFA is it necessary to make further assumptions about the wave-field.

References

BAIN, W.C. (1956): 'The theoretical design of direction finding systems for high frequencies', *Proc. IEE.*, **103B,** pp. 113–119

CAWSEY, D. (1972): 'Numerical methods for wavefront analysis', *Proc. IEE.*, **119,** pp 1237–1242

GETHING, P.J.D. (1971): 'Analysis of multicomponent wavefields', *Proc. IEE.*, **118,** pp. 1333–1338

KELSO, J.M. (1972): 'Measuring the vertical angles of arrival of HF skywave signals with multiple modes', *Radio Sci.*, **7,** pp. 245–250

VU, T.B. (1986): 'Simultaneous nulling in sum and difference patterns by amplitude control', *IEEE. Trans.*, **AP–34,** pp. 214–218

VU, T.B., and RATHINAM, M. (1988): 'Direction finding with null steering at baseband using digital signal processors: a computer simulation', Proc. AP–S 88, Internat. symposium digest: antennas and propagation, Syracuse 6–10 June, **1,** pp. 254–257 (New York, IEEE.)

Chapter 11

Wavefront analysis using imperfect data

11.1 Introduction

The numerical simulations used in Chapter 10 to illustrate some aspects of WFA theory all involved 'noiseless' data, generated in a computer program for idealised conditions such as perfect ground reflection and constant DOAs. The aim was to demonstrate a mathematical technique, by means of which correct answers for the ray parameters could be obtained.

Some of the problems that arise in the analysis of imperfect WFA data will now be briefly discussed. When the input data to a computer program are contaminated by noise and measurement errors it is no longer reasonable to expect a unique and correct answer from a minimum number of frames. Other potential sources of error include departures of the behaviour of the real wave-field from the assumed model, and differences between the real and computed patterns of the array; clearly, a measuring instrument of known properties is required if accurate measurements are to be made.

Techniques for the statistical analysis of samples of imperfect data can, of course, be tested against simulated problems to which known amounts of noise, measurement errors, angle fluctuations etc. are added in a series of controlled numerical experiments.

The ability of an array to resolve a given wave-field must eventually break down when the signal/noise ratio becomes sufficiently unfavourable. In a problem involving (say) two strong rays and one weak ray, it is likely to be the weak ray that is the first to be lost from the solution as the noise level increases. In these circumstances, a partially-correct solution might be obtained.

11.2 Linear relationships

11.2.1 Geometrical interpretation

The antenna signals $(Z_1, \ldots, Z_n)$ on a given frame of data can be regarded as defining a point in n-dimensional space. From f frames of data we can obtain f such points, which tend to be coincident if there is no relative fading between rays, but well separated, in some sense that could be defined quantitatively, when significant fading is present. The search for a linear relationship satisfied by all frames is equivalent to the search for a 'plane' of $(n - 1)$ dimensions or fewer on which all the points lie. The effect of noise and/or small DOA fluctuations is to introduce some scatter into the positions of the points.

The determination of the 'best' linear relationship is a standard statistical problem that is treated, for example, in Kendall and Stuart (1979). As a specific example, we now consider the problem of finding an unknown set of taper coefficients, a_1 to a_8, for the eight-loop array. Each frame of data gives two equations:

$$a_1X_1 + a_2X_2 + \ldots + a_8X_8 \simeq 0$$

$$a_1Y_1 + a_2Y_2 + \ldots + a_8Y_8 \simeq 0$$

It is convenient to drop the distinction between the Xs and the Ys and to assume that all Xs and Ys are normally distributed with the same variance. In effect, f frames of X and Y data can be regarded as $2f$ frames of X data. The distance d_r of a point $(X_1 \ldots X_8)$ from the plane defined by the a_js is given by

$$d_r^2 = \frac{(\Sigma\, a_j\, X_j)^2}{\Sigma\, a_j^2}$$

A reasonable definition of a least-squares solution would be one that minimises Φ, the sum of the $2f$ squared distances obtained from f frames. The eight conditions

$$\frac{d\Phi}{da_j} = 0 \qquad j = 1, \ldots, 8$$

unfortunately lead to nonlinear equations. An alternative approach is to minimise

$$\theta = \sum^{2f} \left\{ \sum^{8} (a_j\, X_j) \right\}^2 - \mu \sum^{8} a_j^2 \tag{1}$$

where μ is an 'undetermined multiplier', regarded as constant. The eight conditions

$$\frac{d\theta}{da_j} = 0 \qquad j = 1, \ldots, 8$$

now lead to eight linear equations in nine unknowns, a_1 to a_8 and μ. The rth equation is

$$C_{r1}a_1 + C_{r2}a_2 + \ldots + C_{r8}a_8 = \mu a_r$$

where

$$C_{r1} = C_{1r} = \sum^{2f} (X_1\, X_r)$$

If we now use small *c*s to denote covariances, following Kendall and Stuart, we have

$$c_{r1} = \frac{1}{2f} C_{r1} \qquad \text{etc.}$$

The nine unknowns can be reduced to eight by division by (say) a_8 and solving for the relative tapers b_1 to b_7, where

$$b_1 = a_1/a_8 \quad \text{etc.}$$

We then have eight equations in eight unknowns, the *r*th equation being

$$c_{r1}b_1 + c_{r2}b_2 + \ldots + c_{r7}b_7 + c_{r8} = \mu' b_r \tag{2}$$

where $\mu' = \mu/2f$. In the final equation, the right-hand side is μ'.

It would seem natural to eliminate μ', in order to obtain equations involving the *b*s and (known) covariances only. However, this leads back to the nonlinear equations mentioned above. An alternative is to eliminate the *b*s and solve for μ'. This process appears to lead to an 8th-order equation in μ'. For each of the eight roots it is then possible to find a 'set' of *a*s and *b*s. The 'set' we require is that corresponding to the value of μ with the smallest magnitude; this can be seen from eqn. 1 where, with perfect data, we require $\mu = 0$ to minimise θ.

The polynomial in μ', the eigenvalues of the matrix, can be obtained from the condition that the determinant

$$D = \begin{vmatrix} c_{11} - \mu' & c_{12} & \ldots & c_{18} \\ c_{12} & c_{22} - \mu' & \ldots & c_{28} \\ \ldots & \ldots & \ldots & \\ c_{18} & c_{28} & \ldots & c_{88} - \mu' \end{vmatrix}$$

must vanish if eqns. 2 are to give consistent, nonzero solutions for the *b*s.

The recommended procedure is therefore as follows:

(*a*) Find the smallest solution for μ' from the condition $D = 0$.
(*b*) Substitute this value of μ' in eqns. 2 and solve these linear equations for the *b*s.
(*c*) Locate the nulls of the tapered arrays defined by the set of *b*s.

Note that the observations have been combined into what is often called the (signal) correlation matrix, in which the *ij*th term is proportional to $(X_i X_j)$ summed over all frames. In most problems, this is a complex matrix.

11.2.2 Physical interpretation

It has already been pointed out that the process of forming linear relationships in the computer is equivalent to combining the antenna signals in a hypothetical beam-forming network (BFN). With perfect data, and *m* array nulls coinciding perfectly with the ray arrival directions, the BFN can be described as a zero output network. With imperfect data, we search instead for a minimum output network. Such a network may still define a radiation pattern with fixed, infinitely deep nulls, but fluctuations in ray DOAs and the incidence of noise from other directions lead to the nonzero output. The aim must be to position the nulls, by means of amplitude and phase tapers, to minimise this output.

In order to make a fair comparison of the outputs obtained with different taper sets, it is necessary to normalise the average magnitudes of the tapers in some way; otherwise, a set of amplifiers will tend to produce a bigger output than a set of attenuators. It is for this reason that a condition such as

$$\sum^{n} a_i^2 = 1$$

has to be applied to an amplitude taper set a_i.

11.2.3 Mathematical generalisation

In Subsection 11.2.1 it was implicitly assumed that we were seeking one 'best' plane through a set of points obtained from f frames of data. Thus it is possible to solve for a maximum of seven rays with eight loops, given at least four frames of data. But if the number of rays is $(n - 2)$ or fewer, where n is the number of elements, it would be possible with perfect data to find many sets of linear relationships. For example, with r rays, any set of $(r + 1)$ or more elements should lead to such a relationship. Space stepping within a given frame can also produce other relationships.

It is already clear that equations such as those in Subsection 11.2.1. can be more compactly expressed in matrix notation and that the covariance matrix of the measured data and its latent roots (eigenvalues), particularly the smallest, play an important role in the theory. Working in the matrix notation defined in Appendix 5, we now consider how to analyse noisy data from any array of sensors. The method of eigenanalysis introduced briefly in this subsection is one of the main themes in Chapters 12 and 13, where modern developments in superresolution algorithms are reviewed. The problem is to decide whether the observed points lie on or near several planes of $(n - 1)$ dimensions or less. A plane is a satsifactory fit if the 'error' represented by the distance of a point from the plane is consistent with the estimated noise level.

The antenna signals $\mathbf{Z}j$ on frame j are related to the array steering matrix $\mathbf{M}$, complex mode amplitudes $\mathbf{W}j$ and noise components $\mathbf{E}j$ by equation (A5.1) of Appendix 5:

$$\mathbf{Z}j = \mathbf{M}\,\mathbf{W}j + \mathbf{E}j$$

Each $\mathbf{Z}j$ represents one point n-dimensional space. Since we cannot define a plane by one point, we must analyse a sample of frames to exploit relative fading, which causes a spreading of the points.

As an initial approach, we consider the distance of each point $\mathbf{Z}j$ from a plane of dimensions $(n - 1)$, that is, we solve for the maximum possible number of rays. Let $\mathbf{X}$ denote an arbitrary n-vector of unit length. Then the product $\mathbf{X}^T\mathbf{Z}\mathbf{X}$ is the projection of $\mathbf{Z}$ in the direction defined by $\mathbf{X}$. For f

points defined by f frames of data, the sum of squared deviations normal to a best-fit plane is proportional to

$$P_N = \mathbf{X}^T\hat{\mathbf{R}}\mathbf{X}$$

where the symbol P_N is chosen to suggest noise power.

The n eigenvalues $\mu_1, \mu_2 \ldots \mu_n$ of $\hat{\mathbf{R}}$ are real numbers satisfying the equation

$$\det|\hat{\mathbf{R}} - \mu\mathbf{I}| = 0$$

The covariance matrix can be represented by its eigenvalues and eigenvectors γ_i according to the following equation:

$$\hat{\mathbf{R}} = \sum_{i=1}^{n} \mu_i \gamma_i \gamma_i^H$$

With the eigenvalues listed in descending order, μ_n is the smallest and is the minimum value of $\mathbf{X}^T\hat{\mathbf{R}}\mathbf{X}$. Although all eigenvectors are of unit length, it is convenient to use the term 'large eigenvector' as an abbreviation for 'an eigenvector corresponding to a large eigenvalue' and similarly for the small eigenvalues and eigenvectors. The smallest eigenvector of $\hat{\mathbf{R}}$ leads to the angle solutions through the usual mechanism of searching for nulls in the corresponding array radiation pattern (step (c) of Subsection 11.2.1).

Instead of solving for the maximum number of rays, $(n - 1)$, we can solve for planes of lower dimensionality. It is natural to seek the 'signal subspace' projection vector $\mathbf{Y}$ of unit length which minimizes the noise power P_N and maximizes the total signal power P_s summed over all rays:

$$P_s = \mathbf{Y}^T \hat{\mathbf{R}} \mathbf{Y}$$

Although the true number of arriving rays is initially unknown, the eigenvalues tend to fall into two groups, the r larger ones corresponding to rays and the $(n - r)$ smaller ones (which are approximately equal) representing noise; with suitable conventions on relative amplitudes they are equal to σ^2, as defined in Appendix 5. The number of rays, r, can therefore be deduced from the number of large eigenvalues. Unfortunately, the dividing line between the ray eigenvalues (each of which contains a noise component) and the noise eigenvalues is rarely clear-cut with real data, and the noise could of course swamp a small ray eigenvector.

When the number of rays has been estimated from a study of the eigenvalues, the ratio of total signal power to total noise power can be found. The total noise power is proportional to n times the average noise eigenvalue. The total signal power is proportional to the sum of the large eigenvalues with the noise contributions removed. Hence,

$$\mathrm{SNR} = 10 \log_{10} \left\{ \frac{\Sigma \text{ large} - \dfrac{r}{(n - r)} \Sigma \text{ small}}{\dfrac{n}{(n - r)} \Sigma \text{ small}} \right\}$$

$$= 10 \log_{10} \left\{ \frac{\Sigma \text{ large} - \dfrac{r}{n} \Sigma \text{ all}}{\Sigma \text{ small}} \right\}$$

where the summations extend over the large, small or all eigenvalues.

The matrix procedure outlined above was derived by D. V. Mardle and J. H. Ellis in the course of research on wavefront analysis and was described by the author in June 1978 (Gething, 1978, 1987). It appears to be very similar to a matrix inversion procedure suggested by Owsley (1971) and to the theory underlying the MUSIC algorithm (Schmidt, 1979) considered in the next chapter. Ellis (1980) and Schmidt (1981) later gave fuller treatments of the geometrical interpretation.

An alternative approach that is sometimes suggested is to use a 'singular value decomposition' of $\hat{\mathbf{R}}$. The ordinary theory of least-squares can be expressed in terms of singular value decompositions, as emphasised by Good (1969).

Table 11.1 Ray parameters used to construct problem

Ray	Elevation angle	Mean amplitude
1	8·5	2
2	16·1	1
3	28·1	1
4	37·4	4

11.3 Example of eigenvalues: vertical tower

As a simple example of a clear-cut division between large and small eigenvalues, the results obtained for a four-mode simulated problem may be quoted. The data analysed consisted of 50 frames with a signal/noise ratio of 20 dB, the frequency was 12 MHz, and the ray elevation angles and mean amplitudes are shown in Table 11.1. The eigenvalues found, listed in decreasing order, were

131·8
49·4
14·8
6·3
0·3
0·2
0·2
0·2

and it seems reasonable to identify the first four as produced by real modes. The corresponding angles found were 8·8°, 15·3°, 29·6°, 37·8°. The angular errors, ranging from 0·3° to 1·5°, are probably fairly typical for a problem of this nature. The signal/noise ratio estimated from the eigenvalues is

$$10 \log_{10} \left\{ \frac{202{\cdot}3 - 4/8 \times 203{\cdot}2}{0.9} \right\}$$

i.e. about 20·5 dB.

11.4 Later steps in the solution

11.4.1 Introduction

It is not always possible to arrive at a correct solution, from one or more frames, for the number of incident rays and their DOAs. This is the critical part of the analysis; provided it is successfully accomplished, the subsequent calculations of other ray parameters are relatively straightforward. The remaining steps in the analysis are

(*a*) the calculation of ray amplitudes and phases for each frame,
(*b*) the reconstruction of theoretical antenna signals from the solution and a comparison of observed and theoretical signals,
(*c*) possible refinement of the solution, e.g. by curve-fitting,
(*d*) the comparison of solutions obtained from different groups of frames and/or by different analysis methods, and the exercise of human judgement.

Naturally, if a completely incorrect solution for the ray DOAs is obtained initially, because of the limited resolving power of the array or the poor quality of the measured data, the subsequent calculation of ray amplitudes etc. will be unreliable.

11.4.2 Calculation of complex amplitudes

As an example of the application of a least-squares method to the calculation of complex ray amplitudes on a single frame of data, we now consider a three-ray problem on the eight-loop vertical array. It will be supposed that the angles of arrival Δ_A, Δ_B, Δ_C of the rays have been correctly determined from (say) 50 frames of data, and that a computer routine for the formation and solution of least-squares normal equations is available; this will be referred to as l.s.

Let A represent the amplitude of ray A on a particular frame and φ_A its phase. Then if we can find A_x and A_y, where

$$A_x = A \cos \varphi_A$$

$$A_y = A \sin \varphi_A$$

and similarly for B_x, B_y, C_x, C_y the problem is solved. The basic equations relating these components to the antenna signal components ($X_1 \ldots X_8$, $Y_1 \ldots Y_8$) are

$$\begin{array}{c}
S_{1A}\,A_x + S_{1B}\,B_x + S_{1C}\,C_x = X_1 \\
S_{2A}\,A_x + S_{2B}\,B_x + S_{2C}\,C_x = X_2 \\
\cdot \quad \cdot \quad \cdot \quad \cdot \quad \cdot \\
S_{8A}\,A_x + S_{8B}\,B_x + S_{8C}\,C_x = X_8 \\
S_{1A}\,A_y + S_{1B}\,B_y + S_{1C}\,C_y = Y_1 \\
\cdot \quad \cdot \quad \cdot \quad \cdot \quad \cdot \\
S_{8A}\,A_y + S_{8B}\,B_y + S_{8C}\,C_y = Y_8
\end{array}$$

where

$$S_{1A} = \cos \Delta_A \sin\left(\frac{2\pi h_1}{\lambda} \sin \Delta_A\right)$$

$$S_{2A} = \cos \Delta_A \sin\left(\frac{2\pi h_2}{\lambda} \sin \Delta_A\right)$$

etc.

These S coefficients are calculated as the first step in the procedure. A very simple method of solution is now available, namely to call l.s. 100 times, i.e. for two sets of eight equations on each of 50 frames. However, this method is wasteful of computer time, because the same matrix of S coefficients is being inverted each time. A more elegant and quicker method is to invert this matrix once only, then multiply the inverse by the antenna signals for each frame. Assuming for a moment that the inversion has been accomplished, we can write

$$A_x = e_{1A}X_1 + e_{2A}X_2 + \ldots + e_{8A}X_8$$
$$A_y = e_{1A}Y_1 + e_{2A}Y_2 + \ldots + e_{8A}Y_8$$
$$B_x = e_{1B}X_1 + e_{2B}X_2 + \ldots + e_{8B}X_8$$

etc.

for each frame.

Now e_{1A} is the value of A_x when $X_1 = 1, X_2 = X_3 = \ldots = X_8 = 0$. Similarly, e_{2A} is the value of A_x when $X_2 = 1, X_1 = X_3 = \ldots = X_8 = 0$ and e_{1B} is the value of B_x when $X_1 = 1, X_2 = X_3 = \ldots = X_8 = 0$. If we call l.s. for a hypothetical frame on which $X_1 = 1, X_2 = X_3 = \ldots = X_8 = 0$, the subprogram returns solutions for A_x, B_x, C_x which are equal to e_{1A}, e_{1B}, e_{1C} respectively. All the e coefficients can be determined from a total of eight hypothetical frames. Once these have been found, the calculation of A_x etc for the real frames involves only simple multiplications and additions.

Each set of e coefficients, such as e_{1A}, e_{2A} ... e_{8A}, can be interpreted physically as follows. It defines a BFN which produces nulls on the directions of all rays except one (A in this case) and a normalised response of unity to this ray. Any small errors in the original determinations of the DOAs will clearly produce corresponding small errors in the calculated amplitudes.

It is perhaps worth noting that a BFN that leads to selection of one ray, and rejection of all others, would be likely to give high-quality reception of a wanted signal, free from the effects of wave interference and multipath distortion.

11.4.3 Reconstruction of antenna signals

One objective of the amplitude calculation is to arrive at an estimate of mean-ray amplitudes and hence to gauge the relative importance of various contributions to the wave-field; phases are usually of less interest. However, both must be calculated on a frame-by-frame basis in order to test the 'fit' of the solution to the original measured data.

Suppose that the measured signal components in a given antenna s on a given frame f are X_{sf}, Y_{sf} and that the reconstructed signals from the solution are X'_{sf}, Y'_{sf}. Then $\{(X_{sf} - X'_{sf})^2 + (Y_{sf} - Y'_{sf})^2\}$ is a measure of the unexplained energy in that element. The unexplained energy will be referred to, for brevity, as noise, but it might arise from a combination of many physical causes in practice, e.g. measurement errors, undetected rays and angle fluctuations of the detected rays, as well as external and internal noise.

A comparatively large noise component in one element, persisting over many frames, provides a warning of a possible physical fault in that element or receiver channel. A comparison of signal and noise components over many frames gives an estimate of average signal/noise and hence of the reliability of the answer. Clearly, a solution giving a very poor fit to the original data should be regarded with considerable suspicion, particularly if the signal/noise ratio estimated from the fit is substantially worse than would be expected from other evidence. For these reasons, the examination of noise residuals should be regarded as an essential part of the WFA procedure.

11.4.4 Comparisons of several possible solutions

As has been stressed at several points in the preceding discussion, the number of incident rays is generally unknown initially. Nor can we rely on the eigenvalues to give a clear and correct indication of the number of rays when noisy data are analysed. But some analysis methods, particularly those involving space stepping, are based on an initial hypothesis about the number of rays; that is to say, the best tactics for a three-ray analysis may be quite different from the best tactics for a two-ray analysis.

With near-perfect data, this problem is not serious; if (say) a three-ray analysis is applied to a two-ray problem, the usual outcome is a three-ray solution but with a near-zero amplitude for the 'false' ray, which can therefore be discarded. With noisy data, however, the situation is less clear-cut. As the number of rays in the solution is increased, so the fit of the solution to the original data can be improved because of the extra degrees of freedom.

Suppose that a noisy three-ray toy problem is constructed and that various possible solutions are obtained involving two rays, three rays, four rays etc. Then we might hope to find a very marked improvement in fit in going from two rays to three, but only minor improvements in going from three to four, four to five etc. However, if the third ray is comparatively weak, the improvement in going from two to three rays may also be comparatively small. Some suitable test of significance must be adopted to test the reduction of the apparent total noise with increasing number of rays.

It is usually necessary to adopt some empirical rule initially, then to refine it as experience is gained in the interpretation of practical data. An example of an empirical rule is as follows:

(*a*) calculate the total noise energy e_1, e_2, e_3 for one-, two- and three-ray solutions, summed over all elements and all frames of a group,

(*b*) now multiply e_1, e_2, e_3 by arbitrary weights of 1·0, 1·4, 1·7 respectively,
(*c*) choose the solution with the smallest weighted noise energy.

Sometimes a solution for a given number of rays can be obtained by more than one method: for example

(*a*) single frame, Wylie–Kelso (Prony) method,
(*b*) single frame, Cawsey method,
(*c*) multiframe.

Method (a) requires no assumption about the ground-reflection coefficient, whereas (b) and (c) assume perfect reflection. Method (c) requires that any angle fluctuations should be small. Because of these different underlying assumptions, but the similarity of the searches for linear relationships, it is reasonable to describe the methods as semi-independent. Clearly, confidence in the solution is increased if semi-independent methods give nearly identical answers. Similarly, given a group of (say) 200 frames, one would hope to obtain consistent answers from (say) four separate groups of 50 frames each, or from the odd-numbered and even-numbered frames. Ultimately, all the internal and external evidence must be reviewed and human judgement exercised in order to reject implausible answers. A good example of this procedure can be found in a paper by Clarke and Tibble (1978), who made a comparison of the results from several methods of analysis of elevation angle data obtained on a controlled oblique path; oblique ionograms and azimuth-angle measurements were also available over the same path.

11.5 Systematic errors

Systematic errors in angle determinations can arise if oversimplifying assumptions are made in calculating the radiation pattern of an array for a specified BFN. If, for example, we assume a perfect ground reflection, the elevation angles of the nulls will be different from the true positions for the real ground reflection coefficient; the effect may be comparatively small for an excellent reflector (as defined in Section 6.5.4) but is systematic.

The effect of a ground reflection is essentially linear, in the sense that the complex amplitude of the reflected ray is directly proportional to the complex amplitude of the incident ray. The same remark applies to mutual-coupling effects; the voltage at the terminals of any one element is usually represented as a linear function of the currents in all elements. Linear relationships between the antenna signals are therefore expected even when such effects are significant. The error arises only when we apply the coefficients determined from the real signals to an oversimplified model of the antenna array.

For the eight-loop vertical array above perfect ground it is theoretically possible, with noise-free data, to find linear relationships involving real coefficients only, because for any one ray all the loop signals are in phase or antiphase. For imperfect reflection, the rigorous calculation would involve

(*a*) the determination of complex taper coefficients,
(*b*) the calculation, with these coefficients, of the positions of the array nulls above real ground.

It was shown in Section 6.5.4 that the effect of imperfect (but 'good') ground reflection for this type of array is closely approximated by a perfect reflection at a depth δh below the surface of the ground, where δh is a function of the ground constants and the frequency, but is approximately independent of Δ. This result suggests that the imaginary parts of the taper coefficients found from the linear relationships should be very small or zero, and that there is no objection to the simpler procedure of searching for the best set of real coefficients. In the calculation of the position of the nulls, the height of each loop should be increased by δh. If the values of the ground constants are not known precisely, the adopted value of δh will be slightly in error; however, numerical calculations suggest that the errors in elevation angles produced by the assumption that an 'excellent' ground is a perfect reflector (i.e. $\delta h = 0$) are not likely to exceed a few tenths of a degree (Gething, 1973).

Space-stepping analysis methods of the Wylie–Kelso (Prony) type (see Section 10.4.2) are generally independent of the ground reflection coefficient and a nonzero value of δh produces no error, either in the calculation of the taper coefficients or in the location of the nulls.

In effect, the nulls of the tapered array in free space are used for the angle measurements and the only requirement is for a horizontal reflecting plane of uniform properties. Two points are worth noting. Firstly, if the reflection is good enough to justify the assumption $\delta h \simeq 0$, it is also reasonable to assume that a loop at zero height would give a zero signal in any wave field. We can then use an extra subarray and an extra space step in the analysis, incorporating this hypothetical element. Secondly, as Kelso (1972) showed, the ground reflection coefficient at the ray arrival angles $\Delta_1, \Delta_2, \ldots$ can be determined from the original data, provided the number of incident rays is not too large; in effect, the direct and reflected components of each ray are treated as separate rays reaching the array from $+\Delta_1$ and $-\Delta_1$, $+\Delta_2$ and $-\Delta_2, \ldots$ respectively and their complex amplitudes are determined. The ratio of reflected amplitude to incident amplitude gives the complex reflection coefficient at $\Delta_1, \Delta_2, \ldots$ etc. From these values, the ground conductivity and permittivity can, in principle, be determined. However, the calculation appears to depend on small differences between signals and therefore requires very high-quality data if reliable results are to be obtained.

11.6 Angle-free linear relationships

A given set of amplitude tapers, applied to the loops on the vertical tower, usually produces a vertical radiation pattern exhibiting comparatively sharp nulls at one or more elevation angles. However, some taper sets can be found that produce very broad nulls in elevation. Suppose that a relationship can be found such that the BFN output is 30 dB or more down on the loop signals for a ray arriving from any elevation angle, i.e. from any angle between 0° and 90°. A linear relationship of this kind will be called angle-free. The BFN will give a near-zero output for any combination of rays.

A broad null extending from 0° to 90° in elevation at a lowish frequency might become a null sector extending from 0° to 60° at the 30 dB level at some higher frequency. The corresponding linear relationship is then angle-free up to 60°. If it is satisfied, within a suitable tolerance, by measured data, the presumption is that there are one or more rays arriving from below 60°. If it is not satisfied, either there is a fault in the measuring equipment or there is some incident energy from above 60°.

Various types of taper solutions, such as

(*a*) Dolph–Chebȳshev (Cawsey, 1974)
(*b*) Binomial, with alternate coefficients reversed in sign (Gething and Haseler, 1974)

for linear arrays in free space can lead to very broad nulls. They can be used as a starting point from which angle-free linear relationships for the eight-loop array are derived. The connection between an array in free space and the vertical tower above a perfect reflector is as follows.

Suppose the loops are at heights nh above the reflector, with $n = 1$ to 8, and that the signals are Z_1 to Z_8. We first calculate the signals G_1 to G_8 in a hypothetical array with elements at 0·5 h, 1·5 $h, \ldots,$ 7·5 h. They are given by

$$G_s = \frac{1}{2\cos\left(\frac{\psi}{2}\right)}(Z_{s-1} + Z_s)$$

for a single ray defined by

$$\psi = \frac{2\pi h}{\lambda}\sin\Delta$$

Now consider the 16-element uniformly-spaced array formed by this array and its image. The pattern (above ground) of the hypothetical eight-element array, with taper coefficients $a'_1 \ldots a'_8$ will be identical with that of the 16-element array with taper coefficients $-a'_8, -a'_7 \ldots -a'_1, a'_1 \ldots a'_8$. The array output V is given by

$$V = \sum_1^8 a'_s G_s$$

$$= \frac{1}{2\cos\left(\frac{\psi}{2}\right)} \sum_1^8 (a'_s + a'_{s+1}) Z_s$$

with

$$a'_9 = 0$$

Hence

$$2V\cos\left(\frac{\psi}{2}\right) = \sum_1^8 a_s Z_s$$

where $$a_s = a'_s + a'_{s+1}$$

and the taper coefficients required, a_s, can be simply calculated from any set a'_s that gives a broad null in free space.

The a'_s can be the taper coefficients of a Dolph–Chebŷshev array, reversed in sign from element to element. The radiation pattern then defines a broad null-sector, the depth of which is defined by a constant sidelobe level within the sector. The taper set for eight elements and a 34 dB level can be found from equations given by Jasik (1961). It is

$$(-254{\cdot}7,\ 230{\cdot}2,\ -187{\cdot}2,\ 135{\cdot}8,\ -86{\cdot}4,\ 46{\cdot}8,\ -20{\cdot}3,\ 6{\cdot}3)$$

The a_s are therefore

$$(-24{\cdot}5,\ 43{\cdot}0,\ -51{\cdot}4,\ 49{\cdot}4,\ -39{\cdot}6,\ 26{\cdot}5,\ -14{\cdot}0,\ 6{\cdot}3)$$

These values define a first checking function,

$$F_1 = \sum_{j=1}^{8} a_{1j}\, Z_j \simeq 0$$

The coefficients for F_1, plus those for two other checking functions F_2 and F_3, are shown in Table 11.2. They have been normalised so that

$$\sum_{j=1}^{8} a_{rj}^2 = 1 \qquad r = 1, 2, 3$$

and are approximately orthogonal

$$\sum_{j=1}^{8} a_{rj}\, a_{sj} \simeq 0 \qquad r \neq s$$

Table 11.2 Coefficients for three checking functions

Loop	F_1	F_2	F_3
1	0·244	0·393	0·468
2	−0·429	−0·491	−0·265
3	0·513	0·234	−0·313
4	−0·493	0·172	0·471
5	0·395	−0·449	−0·037
6	−0·264	0·466	−0·412
7	0·141	−0·298	0·436
8	−0·062	0·115	−0·175

These functions provide valid checks up to a maximum value of $f \sin \Delta$, where f is the signal frequency; this and the maximum value of Δ at two values of f are shown in Table 11.3.

Table 11.3 Angle ranges over which checking functions can be used

Loop	F_1	F_2	F_3
Max $f \sin \Delta$	11·4	9·4	7·6
Max Δ at f = 10 MHz	90°	70°	49°
Max Δ at f = 17 MHz	42°	34°	27°

The fact that $F_1 \simeq 0$ for many problems implies that the eight loops provide no more than seven independent measurements of the wavefield in these circumstances. $F_2 \simeq 0$ shows further redundancy and implies that no more than six independent samples are available. When $F_3 \simeq 0$ there cannot be more than five. By manipulation of F_1, F_2, F_3 it is possible to obtain linear relationships not involving a specified element, or a specified pair of elements. Hence it is possible, in principle, to

(*a*) identify one or two faulty channels
(*b*) correct up to three faulty channels if their identity is known, using data from five (or more) good channels.

In practice the effects of noise would restrict the number of errors that would be detected and corrected reliably.

Suppose that, at even lower frequencies and elevation angles, the number of independent measurements falls to four. In effect, the signals in the odd-numbered loops (say) can be predicted fairly accurately if the signals in the even-numbered loops are known. We then expect linear relationships among sets of three neighbouring loops such as 2, 3, 4 etc.

The existence of angle-free linear relationships arises from the limited resolving power of the array. The possibility of applying simple preliminary checks to raw data is to be welcomed, but the effect of redundancy is to reduce the number of rays that can be found from practical data. For example, any attempt to solve for the maximum of five rays from a single frame appears to require a totally unrealistic signal/noise ratio, in excess of 70 dB. The topic of resolving power is discussed more generally in Section 11.8.

11.7 Masquerading

11.7.1 Description

Suppose that the signals induced in an n-element array by a particular wavefield A are denoted by the vector $\mathbf{Z}_A$ and that another wavefield B would give rise to vector $\mathbf{Z}_B$. Suppose further that $\mathbf{Z}_A$ and $\mathbf{Z}_B$ are identical within a tolerance set by the noise level. Then we shall say that field A can masquerade as field B and vice versa. The term masquerading was chosen deliberately to suggest a close connection with the well known problem of aliasing (Betts, 1970) in signal analysis. In effect, we can choose plausible noise vectors $\mathbf{n}_A$, $\mathbf{n}_B$ such that

$$\mathbf{Z}_A + \mathbf{n}_A = \mathbf{Z}_B + \mathbf{n}_B$$

This equation will be written in the abbreviated form

$$\mathbf{Z}_A \, M \, \mathbf{Z}_B$$

If A and B are composed of several rays, these rays can be split into subsets; hence we can write equations of the form

$$(\mathbf{Z}_C + \mathbf{Z}_D) \, M \, (\mathbf{Z}_E + \mathbf{Z}_F)$$

from which other relationships follow, e.g.

$$\mathbf{Z}_C \, M \, (\mathbf{Z}_E + \mathbf{Z}_F - \mathbf{Z}_D)$$

$$(\mathbf{Z}_C - \mathbf{Z}_E) \, M \, (\mathbf{Z}_F - \mathbf{Z}_D)$$

etc.

Any negative amplitudes implied by such equations do not create any difficulty of interpretation; it is only necessary to change the relevant ray phase by 180° to recover a positive amplitude.

A specific example of masquerading is given in Appendix 4, for the vertical array of loops, with two incident modes at elevation angles of 6·0° and 10·0°. It is shown that, at 6 MHz, any combination of these rays can be masqueraded by rays at 8·6° and 15·2°, with appropriately calculated amplitudes and phases. This masquerading can continue indefinitely, and many other equivalences can be found.

11.7.2 Discussion

Masquerading occurs chiefly at low frequencies and low elevation angles and is a manifestation of inadequate resolving power. On this physical interpretation it cannot be avoided by alterations in the number and dispositions of the elements within a given aperture; it can arise with irregular, as well as regular, spacings.

Superdirectivity can, in principle, always be attained and the rays correctly resolved if the effective signal/noise is sufficiently high, but practical obstacles to the achievement of more than a few dB of superdirectivity are known to be formidable and cost factors tend to increase exponentially (Hansen, 1981). Superdirectivity, even if it is attained, is not synonymous with superresolution, which forms the subject of Chapters 12 and 13.

Because masquerading can continue from frame to frame, the collection of many frames of data does little to ameliorate the problem at low frequencies and low elevation angles. Because some of the ray parameters change from frame to frame, the data from f frames cannot be directly averaged and the improvement in signal/noise ratio is less than the factor of $\sqrt{f}$ that simple statistical theory might suggest.

The phenomenon of masquerading is, of course, very closely related to the angle-free linear relationships discussed in Section 11.6.

11.8 Resolving power

11.8.1 Physical factors

It has been demonstrated, by means of numerical simulations, that the

WFA method of solution leads to a correct determination of the ray parameters in favourable circumstances, i.e. a small number of well-separated rays and a comparatively high signal/noise ratio. As the problem is gradually made more difficult in various ways, e.g. by decreasing the ray separation and increasing the noise level, there comes a point at which the analysis fails to resolve the rays correctly. Such a breakdown must occur at some point with any analysis method and arises largely from the limited resolving power of the array.

The ability to resolve multiray wave-fields depends on several physical factors: the effective aperture of the array in wavelengths for the ray DOAs, the number of elements, the signal/noise ratio, the number of rays and their separations. It also depends on the number of frames used in the analysis. It must also depend in some way on ray amplitudes, because it is clearly difficult to recognise weak rays in the presence of strong rays. It should also be noted that, in a multiframe analysis, fading is helpful but that large fluctuations in ray DOAs are unhelpful. Lastly, resolving power depends to some extent on analysis tactics, e.g. on the way in which subarrays are chosen for space-stepping and on the combinations of space-stepping and time-stepping adopted. We are, of course, mainly interested in the resolving power that can be attained with optimum tactics.

A general theoretical expression for resolving power, applicable to any array, any wave-field structure and any analysis method, would be a very complicated function of the factors mentioned and would not be easy to interpret. A more profitable approach is to derive semi-empirical expressions for particular arrays, based on simple theory and tested against a variety of toy problems.

11.8.2 Accuracy target

We discuss first the related problem of determining the DOA of a *single* ray, given (say) a DF array with one main lobe. Clearly, the signal/noise ratio required if the DOA is to be determined to (say) 0·1° is much greater than that required for a 1° accuracy.

Consider, for example, a lobe with a 10° beamwidth to 3 dB points; this is not unreasonable for a horizontal or vertical array with an aperture of a few wavelengths. The pattern can be approximately represented as cos (9θ), where θ is the angle from the lobe axis. The signal/noise ratio required to determine a ray DOA to 1° is about 40 dB; for 0·1° we require about 80 dB. As expected from information theory, a high signal/noise can be traded for an increase in (angular) information.

11.8.3 Vertical array of loops

We have seen that, in principle, five rays can be found from a single frame of data. However, this result applies only if we have 14 independent measurements from the eight loops with which to determine the 13 unknown ray parameters. The existence of angle-free linear relationships (Section 11.6), particularly at low frequencies and low-elevation angles, means that the amount of available information in practical situations is less than the maximum.

Numerical tests show that, at about 10 MHz, a signal/noise ratio of about 70 dB or better is required if a correct solution to a five-ray problem is to be obtained. For more realistic signal/noise ratios of the order of 20 or 30 dB, it is not possible to solve for more than three rays with any reasonable hope of success.

For a three-ray solution, there is some choice of tactics. For example, several space-steps can be made with comparatively small subarrays such as (1, 2, 3, 4), (2, 3, 4, 5), etc; each subarray yields a cubic in cos 2ψ. However, bearing in mind that the signal in any element can be fairly accurately predicted if those in the elements on each side of it are known, each subarray does not provide four independent samples of the wave-field. It seems better to derive one cubic from elements 1, 3, 5, 7 and another from elements 2, 4, 6, 8; numerical tests confirm that these are the best tactics (Cawsey, 1972).

For a frequency of 6 MHz, the *lowest* angle that can be reliably determined is as shown in Table 11.4.

Table 11.4 Minimum angle, single-frame analysis at 6 MHz, vertical tower

Signal/noise ratio, dB	Minimum angle
20	11·0°
30	7·5°
40	5·0°
50	4·0°

In a two-ray problem at 6 MHz, 30 dB, the higher angle should be at least 21·5° if it is to be reliably determined. A minimum separation of about 3° at 30° can also be resolved. For further details, the paper by Cawsey (1972) should be consulted.

11.8.4 Circular array

For a 12-element circular array with a diameter of 150 m, it is again doubtful whether more than about 4 independent samples of the wave-field are obtained at low HF. The elevation angle that can be reliably determined for various signal/noise ratios at 6 MHz is shown in Table 11.5.

Table 11.5 Minimum elevation angle, single-frame analysis at 6 MHz, circular array

Signal/noise array, dB	Minimum angle
20	25·0°
30	20·5°
40	17·0°
50	14·5°

In a two-ray problem at 6 MHz, 30 dB, the higher angle should be at least 34° if it is to be reliably determined. A minimum separation of about 3° at 40° elevation can also be resolved.

11.8.5 Multiframe analyses

As noted in the discussion of masquerading (Section 11.7.2), the precision of a solution from f frames does not increase as rapdily as $\sqrt{f}$. Because we are dealing with fading rays, the number of unknowns to be found unfortuntely increases almost proportionately to the number of frames. As an example, consider the problem of determining three rays with measurements made at four (independent) sampling points. The ratio of the number of unknowns to the number of measurements is shown in the final column of Table 11.6.

Table 11.6 Three rays and four elements

No. of frames	No. of measurements	No. of unknowns	Ratio
1	6	10	1·67
2	12	14	1·17
3	18	18	1·00
4	24	22	0·92
etc	etc	etc	etc
20	120	86	0·72

The problem can be solved from a minimum of three frames and beyond this number the accuracy of the solution should gradually increase. But note that the ratio of unknowns to measurements $(6 + 4f)/6f$, rapidly approaches 0·67 as f becomes large. Moreover, if the sampling period is made too long, the structure of the wave-field may change significantly; for example, new sources of interference may produce additional rays.

In summary, multiframe methods of analysis are probably only slightly superior to single-frame methods applied to all available frames, for those problems where the aperture and the number of elements are barely adequate, and do not greatly extend the range of problems that can be successfully resolved.

11.8.6 Geometrical interpretation

Suppose that a linear relationship involving all n elements of an array can be found. Then the observations lie on a hyper-plane of $(n - 1)$ dimensions in n-dimensional space. The problems posed by angle-free linear relationships, masquerading and limited resolving power can all be discussed in terms of this geometrical model. For example, the checking relationships $F_1 = 0$, $F_2 = 0$, $F_3 = 0$ discussed in Section 11.6 all represent 7-d hyper-planes in 8-d space. The fact that practical data lie on or near all these planes means that the eight dimensions can effectively be reduced to five (or less).

It may be helpful to visualise points in normal 3-d space. Ideally, one hopes to find that these lie on a unique 2-d plane. However, if the points lie almost on a straight line, the equations defining the plane (e.g. by means of the vector components of its normal) are very ill-conditioned. Fading between rays from frame to frame can help to increase the spread of the points in space and hence to lead to a more reliable determination of the plane. However, if the only effect of fading is to increase the spread along a straight line, the improvement may be negligible.

11.9 Conclusions on the analysis of imperfect data

This Chapter has been devoted to a discussion of some of the many problems that arise in the statistical analysis of imperfect data. As expected, the infinite resolving power associated with perfect, noiseless measurements no longer applies; the practical resolving power is limited by signal/noise ratio and aperture. The existence of angle-free linear relationships, the merging of rays in some numerical examples and the phenomenon of masquerading all arise from this limitation.

In practice, we may wish to use a DF array under a variety of propagation conditions, some of which might produce closely-spaced rays. Thus, some problems might require a degree of superdirectivity for their solution that cannot be attained. One objective of any analysis method must be to recognise such problems, which produce ill-conditioned equations and hence large standard deviations for the ray parameters.

There are many common-sense precautions and checks that can be applied in the course of the analysis. If the number of incident rays is greater than the maximum that can be solved with a given number of elements, satisfactory linear relationships will not be found. A completely incorrect solution for the ray parameters will (except in the case of masquerading) give a poor fit to the original data. Solutions from various groups of frames can be examined for consistency. The putative solution from a given set of data can be examined to see whether it would be classified as a resolvable problem. The signal/noise ratio estimated from the fit of the solution to the original data can be compared with estimates based on other evidence. When a plausible solution has been obtained by a routine analysis, it can be refined by iterative methods.

The number of rays that can be dealt with by a given array is theoretically determined from the number of elements; in practice, elements can be regarded as giving independent information only if they are (say) $\frac{1}{2}\lambda$ or more apart. With closer spacings, it is unwise to attempt a solution for the maximum number of rays. From arrays with apertures of a few wavelengths it seems possible to solve for two or three rays, provided the signal/noise ratio is of the order of 20–30 dB. This is a true signal-to-distributed-noise ratio, in the sense that a strong ray from an interfering source should be correctly resolved in the solution. It should also be borne in mind that the basic amplitude and phase measurements can be made in a comparatively narrow bandwidth, e.g. a few hundred Hz. Thus the required signal/noise does not seem unreasonable.

Wavefront analysis provides a convenient and simple method of calculating ray parameters, when the wave-field is resolvable. The necessary equations for almost any form of array can be written down as soon as the underlying physical picture of adjustable nulls is understood. The search for linear relationships is readily adapted to a statistical approach in which the 'best' planes through sets of points are sought. We have supposed that both amplitude and phase data are available, but in some circumstances it is possible to deduce ray parameters from amplitude data alone, according to Cottony (1971).

WFA methods provide insight into the relative merits of space-stepping and time-stepping and suggest ways in which they can be combined. The exploitation of relative fading, in multiframe methods, leads to a modest increase in the range of problems that can be solved. This statement refers to a comparison of WFA methods, which seem to the author to provide the only valid attack on multiray problems.

References

BETTS, J.A. (1970): 'Signal processing, modulation and noise' (English Universities Press Ltd.), Section 6.3

CAWSEY, D. (1972): 'Numerical methods for wavefront analysis'. *Proc. IEE,* **119,** pp. 1237–1242

CAWSEY, D. (1974): 'Linear antenna arrays with broadened nulls'. *Electron. Lett.,* **10,** pp. 291–292

CLARKE, R.H., and TIBBLE, D.V. (1978): 'Measurement of the elevation angles of arrival of multicomponents h.f. skywaves'. *Proc. IEE,* **125,** pp. 17–24

COTTONY, H.V. (1971): 'Processing of information available at the terminals of a multi-port antenna'. 1971 International symposium on antennas and propagation, Sendai, Japan. (Tokyo, Japan Inst. Elect. and Comms. Engrs.) pp. 9–10

ELLIS, J.H. (1980): 'Fundamental limitations of wave-front analysis by least-squares', *Proc. IEE.,* **127H**, pp. 219–227

GETHING, P.J.D. (1973): 'Concept of phase centre of an array applied to elevation-angle measurements'. *Proc. IEE.,* **120,** pp. 541–543

GETHING, P.J.D. (1978): 'Radio direction finding and the resolution of multicomponent wave-fields' (Peter Peregrinus), Subsection 11.2.2

GETHING, P.J.D. (1987): 'Optimum solution for multiple emitter location', *IEEE Trans.,* **AP-35,** p. 876

GETHING, P.J.D., and HASELER, J.B. (1974): 'Linear antenna arrays with broadened nulls', *Proc. IEE.,* **121,** pp. 165–168

GOOD, I.J. (1969): 'Some applications of the singular decomposition of a matrix'. *Technometrics,* **11,** pp. 823–831

HANSEN, R.C. (1981): 'Fundamental limitations in antennas', *Proc. IEEE.,* **69,** pp. 170–182

JASIK, H. (1961): 'Antenna engineering handbook' (McGraw–Hill), pp. 2–20 to 2–24

KELSO, J.M. (1972): 'Measuring the vertical angles of arrival of HF skywave signals with multiple modes', *Radio Sci.,* **7,** pp. 245–250

KENDALL, M.G., and STUART, A. (1979): 'The advanced theory of statistics', 4th edition, vol. 2 (Macmillan)

OWSLEY, N.L. (1971): 'Source location with an adaptive antenna array', Naval Underwater Systems Center Report NL-3015

SCHMIDT, R.O. (1979): 'Multiple emitter location and signal parameter estimation', Proc. RADC Spectrum Estimation Workshop, Griffiths AFB, Rome, N.Y., 3–5 October, pp. 243–258; republished in *IEEE Trans.*, **AP-34,** 1986, pp. 276–280

SCHMIDT, R.O. (1981): 'A signal subspace approach to multiple emitter location and spectral estimation', PhD thesis, Stanford University, CA

Chapter 12

Superresolution algorithms

12.1 Introduction

12.1.1 Meaning of superresolution

The word superresolution is now widely used in place of the term wavefront analysis and the phrase 'the resolution of multicomponent wavefields'. It is used to describe techniques of signal processing that can be applied in many different fields such as radar, sonar, radio astronomy, seismology and spectral analysis, as well as radio direction finding.

Some hundreds of papers on superresolution algorithms designed to determine the angular spectrum of the arriving rays have appeared in the literature over the last ten years, many based on concepts described in earlier chapters. In this chapter we describe the main categories of algorithms and their performance against simulated problems. Further developments and refinements of the processing techniques are discussed in Chapter 13. The aim in both chapters is to describe and interpret the important results rather than to reproduce mathematical detail from the research papers.

12.1.2 Angular spectrum

The output of the estimator is usually a graph of the angular spectrum, analogous to an array radiation pattern. Fig. 12.1 (diagrammatic only) shows the kind of plots that might be produced over an azimuth sector for a two-ray problem. In Fig. 12.1*a* the two arrival angles are represented by the two main maxima in the pattern; we postpone until later the discussion of how to determine the number of humps that represent genuine ray arrival angles.

Alternatively, minima in the pattern corresponding conceptually to array nulls can be used, as in Fig. 12.1*b*. Because noise and measurement errors lead to minima rather than perfect nulls, the reciprocal of the generating function for Fig. 12.1*b* can be plotted to provide sharp cusps at the arrival angles (Fig. 12.1*c*). This form of presentation is now generally favoured.

Kay and Demeure (1984) rightly pointed out that visual inspection of

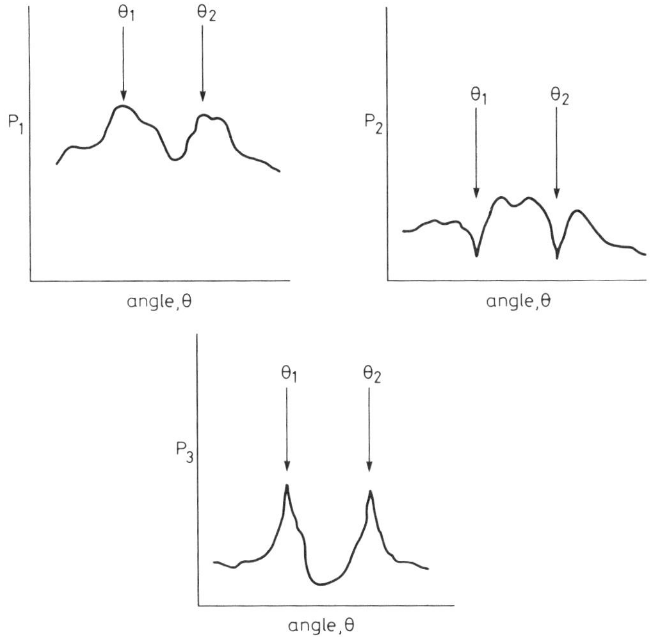

Fig. 12.1 Three types of spectral estimator P for two-ray problem
a use of maxima
b use of mimima
c use of inverted minima (cusps)

diagrams showing features such as sharp cusps is not a scientifically-valid means of assessing the resolution of a spectral estimator. In our example, Figs 12.1 *a*, *b* and *c* may be simple transforms of each other, with exactly the same resolution.

Once the arrival angles have been determined, the calculation of ray amplitudes is straightforward. The amplitude of the *i*th ray is found, in effect, by forming a beam with nulls on all the rays except the *i*th.

12.1.3 Definition of coherence

Suppose that two rays arriving at an array have real amplitudes $A_1(t)$ and $A_2(t)$ respectively at time t and that the corresponding phases at the same moment are $\varphi_1(t)$ and $\varphi_2(t)$. If the fluctuations with time of $A_2(t)$ are perfectly correlated with those of $A_1(t)$, i.e. if the correlation coefficient is $+1$, and if the same condition applies to the phase fluctuations, the arriving rays are said to be coherent. Note that coherence as so defined is not directly related to the coherence ratio defined in Section 2.4 for the power ratio of the specular and diffracted components of a single ionospheric mode.

Perfect coherence represents a difficult case for many resolution algorithms, in that the advantages of complex fading cannot be exploited; the two rays may not be resolved, particularly if they are closely spaced in arrival angle. The spectral estimator is likely to produce instead a peak at an intermediate angle, approximately a weighted mean determined by the amplitudes of the two rays. Partial coherence (correlation) between rays tends to increase the sampling time required to achieve the same resolution performance as with incoherent rays; seem for example, some of the results given in a paper by R. L. Johnson (1988). Local reflections at the receiving site and jammers can both produce a condition of coherence; they also increase the total number of unknowns to be determined.

A fuller discussion of the conditions under which coherence might be encountered, and of the spatial smoothing technique used to reduce its adverse effects, will be found in Chapter 13.

12.1.4 Context

We shall concentrate in this and the next chapter on the problem of narrowband direction finding. We shall, however, refer to sensors rather than antenna elements, and rays rather than modes, in order to preserve some generality.

Narrowband here means that the receiver bandwidth Δf_0 is very small compared with the carrier frequency f_0 and hence the radiation pattern is effectively constant across the band; the condition is discussed in more detail by R.L. Johnson and Sherrill (1982). Hudson (1981) pointed out that a single modulated carrier shows a lack of coherence across an aperture greater than c/B, where c is the velocity of propagation and B is the bandwidth; the narrowband approximation is valid if the aperture is very much smaller than c/B. For a receiver bandwidth of 6 kHz, and an array length of 100m, the aperture is $0{\cdot}002c/B$ where c is the velocity of radio waves in free space; hence the condition is clearly satisfied.

We shall assume that the direction finder is a fixed wide-aperture system of known location and measurement characteristics, but mention will be made of the literature on sensors of unknown or variable position. The usual wave-field model of a finite number of rays arriving from constant directions will continue to be adopted, with the explicit recognition that real data will be affected by distributed external noise with no clear directional properties, and by receiver noise.

12.2 Overview of solution methods

12.2.1 Covariance matrix of the measured signals

In Chapter 11, a number of problems that arise in the analysis of imperfect data were discussed and illustrated with numerical examples. The standard statistical criterion of least squares cannot easily be directly applied, because the problem is highly non-linear in the relationships between the measurements and the angle parameters to be determined. The number of unknowns is also very high: for example, a four-ray problem sampled on 1000 frames involves not only four azimuth angles and four elevation angles, both assumed constant, but 3000 relative amplitudes and 3000 relative phases. Iterative search over 6008 unknowns is not attractive and other methods are normally preferred. However, maximum-likelihood non-linear optimisation is possible, according to Wang and Kaveh (1985); Nickel (1987) also described parametric target model fitting (PTMF) as a possible option in a radar context.

A mathematical generalisation was briefly introduced in Section 11.2.3, based on the eigenanalysis of $\hat{\mathbf{R}}$, the covariance matrix of the measured signals **Z** defined in Appendix 5. Most high resolution methods of analysis use this matrix as the initial input.

With an appropriate noise model (to be discussed in Chapter 13) the basic equation to be solved is eqn. A5.2 of Appendix 5.

$$\hat{\mathbf{R}} = \mathbf{M}\ \mathbf{S}\ \mathbf{M}^{H} + \mathbf{U} \qquad (1)$$

where **M** is the matrix of steering vectors, **S** is sometimes called the signal correlation matrix and **U** is the noise matrix.

There is no simple, direct method of solution. $\hat{\mathbf{R}}$ represents the measurements. **M** contains the known geometry of the sensor array, and any imposed taper, entangled with the unknown arrival angles to be determined. **S** contains the complex ray amplitudes which, even if not of intrinsic interest, are needed eventually to identify spurious angle solutions by means of their very low amplitudes. Noise components and measurement errors are by definition unknown in detail, even when estimates of statistical parameters are available.

It is not surprising that a variety of attacks on this difficult equation have been suggested, involving different weighting factors, projections, approximations and extrapolations of data. In order to arrive at an overview of the various algorithms, we begin by following the procedure already introduced in Section 11.2.3 and project eqn. 12.1 on to some undefined unit vector **X**. This time, however, we choose to interpret the resulting scalar quantities as sums of squared voltages (strictly, dot products of pairs of voltages) rather than squared distances from a plane.

The resulting equation can be written as a power budget:

$$P_{\mathrm{M}} = P_S + P_N$$

where P_M represents the power of the measured voltages in the sensors, P_S is the power of the signals accepted by the array and P_N is the noise power. It is clear from the physical interpretation that this equation should be valid for one frame or any sample of frames. It is normal to start from a multi-frame version of the matrix $\hat{\mathbf{R}}$, in order to exploit relative fading, to achieve some statistical smoothing and to reduce the computing load as compared with a multiplicity of single-frame solutions.

The number of frames is sometimes referred to as the time–bandwidth product, on the assumption that each frame represents an independent sample of data. Successive frames are usually assumed to be unrelated 'snapshots'. However, in principle, successive frames can be obtained from the outputs of a tapped delay line attached to each element, thus preserving a precise phase relationship between frames; discussions of this possibility may be found in papers by Wax, Shan and Kailath (1984) and Gabriel (1989).

12.2.2 Statistical criteria

In each of the main categories of solution methods to be described in Section 12.3, an initial statistical criterion is adopted. The usual procedure is, not surprisingly, to maximise P_S or to minimise P_N; there is no difference in principle between these criteria with P_M fixed. In the Maximum Likelihood Method, the same condition is expressed as a requirement to minimise the normalised power:

$$\frac{P_M}{P_S} = 1 + \frac{P_N}{P_S}$$

A different criterion is adopted in the Maximum Entropy Method; in effect, the array nulls are adjusted so that P_S is near zero and $\hat{\mathbf{R}}$ becomes as noise-like and structureless as possible.

12.3 Spectral estimators

12.3.1 Maximum Likelihood Method

Properties of maximum likelihood have been claimed for a wide variety of processing schemes by proponents of the schemes. However, the Maximum Likelihood Method (MLM) is usually taken to refer to a criterion applied by Capon (1969, 1970) to the analysis of seismic data, giving an improvement in resolution by a factor of about four over conventional analysis. It is based on an implicit model of a wave-field consisting of a single wanted arriving signal plus distributed noise.

Conceptually, the steering weights for a specified direction-of-look are adjusted to maintain constant processing gain in that direction but to minimise the total power output from the array, thus minimising the contributions to the output power arising from noise and from any signals arriving from other directions. The direction-of-look is defined by a column vector **A**, which is the form that the array matrix **M** would take for a single ray from that direction.* We therefore consider the projection of P_M on to the 'trial' direction, subject to the constraint of holding P_S constant. For convenience the caret over $\hat{\mathbf{R}}$ will be omitted in this section, but with the understanding that **R** is determined from a finite sample of frames. The projected covariance matrix in the steering direction is $\mathbf{A}^{\mathrm{H}}\ \mathbf{R}\ \mathbf{A}$. Constrained minimisation (D. H. Johnson 1982) leads to the condition, quoted by R. L. Johnson and Miner (1986), that the following spectral estimator P_{MLM} should be maximised:

$$P_{MLM} = \frac{1}{\boldsymbol{A}^{H}\, \boldsymbol{R}^{-1}\, \boldsymbol{A}}$$

The steering angle giving the highest value of this estimator function can be identified with the supposed single arriving ray. More generally, but less

*The symbol **A** is used interchangeably in some papers for both the array matrix for r rays and the steering vector for a single ray. It seems less confusing to use two separate symbols.

rigorously, other maxima in the plot of the estimator are assumed to correspond to other, weaker, rays.

The signal covariance matrix **R** and its inverse $\mathbf{R}^{-1}$ are calculated only once from the collected data, but **A**, $\mathbf{A}^H$ and P_{MLM} have to be computed at small angular intervals (as for an array radiation pattern) in order to locate the maxima accurately. Considering for a moment the azimuth-only case, we are effectively measuring bearings with a rotating sum beam-pattern that varies in shape as it rotates. For this reason the technique is sometimes called the Scanned Array Beam (SAB) method.

Versions of MLM have been developed in which the initial rather unsatisfactory assumption of a single arrival is abandoned. Hudson (1985*a*) used an iterative method to solve for two arrivals, which can be coherent. Brandwood (1989) devised an Iterative Scanning Nulled Beam method, the basic principle of which is to scan the sensor array a number of times over the whole region containing arriving rays, putting a null on the strongest ray each time. The position of the nulls can be corrected subsequently.

Ziskind and Wax (1988) also used an iterative technique, which they called Alternating Projection; they solve first for a single source, which leads to a projection matrix update equation to find the next strongest source, and so on. Several iterations are required for the angle estimates to converge to stable values. This method also can be applied to coherent signals.

Scott (1987) showed how the calculation of **R** could be avoided by using a trapezoidal factorisation method instead. Oh and Un (1989*a*) suggested an efficient iterative algorithm, based on transforming Hermitian forms into inner products of vectors, to reduce the computational complexity of Alternating Projection. A paper by Stoica and Sharman (1990) can be recommended for a clear exposition of the mathematical basis of maximum likelihood methods. The authors suggested a new method of direction estimation (MODE) in which the taper coefficients obtained from eigenanalysis appear to be adjusted to satisfy a maximum likelihood criterion; MODE performed better than MUSIC, especially on highly-correlated sources.

Johnson and Miner described two variants of MLM: the Adaptive Angular Response (AAR) developed by Borgiotti and Kaplan (1979), and independently by Lagunas-Hernandez and Gasull-Llampallas (1984) to measure power density rather than power in the spectrum; and the Thermal Noise Algorithm (TNA) suggested by Gabriel (1980). The corresponding estimators are

$$P_{AAR} = \frac{\boldsymbol{A}^H \boldsymbol{R}^{-1} \boldsymbol{A}}{\boldsymbol{A}^H \boldsymbol{R}^{-2} \boldsymbol{A}}$$

and

$$P_{TNA} = \frac{1}{\boldsymbol{A}^H \boldsymbol{R}^{-2} \boldsymbol{A}}$$

The performance of MLM and other methods against simulated problems is discussed in Section 12.4.

12.3.2 Maximum Entropy Method

The Maximum Entropy Method (MEM) developed by Burg (1972) and others is designed, as the name implies, to maximise the entropy of the output from the processing algorithm, subject to certain constraints. Some of the early key papers are reprinted in a book edited by Childers (1978).

Whereas a conventional filter designed to 'clean up' a signal attempts to remove as much noise as possible (Fig. 12.2*a*), the MEM algorithm may be regarded as a filter that attempts to whiten its output spectrum, i.e. to make it as noise-like and structureless as possible (Fig. 12.2*b*). In so doing it places nulls on all the signals in the input spectrum. The transfer function of the filter is used to determine the characteristics of the signals that have been removed.

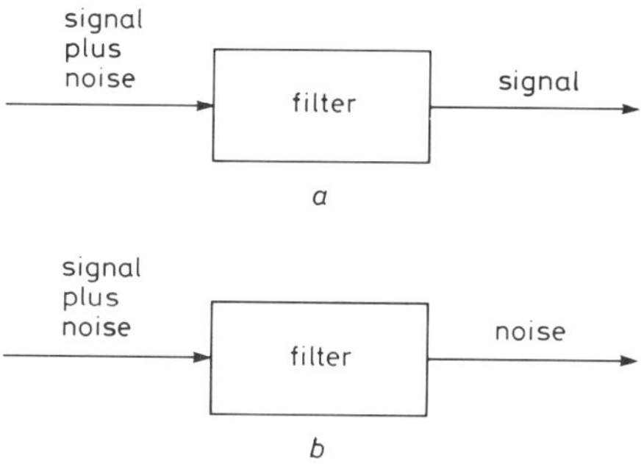

Fig. 12.2 Comparison of filters
a conventional filter to 'clean up' signal
b maximum entropy filter to make output as noise-like as possible

The MEM philosophy should be advantageous when more is known initially about the noise characteristics than about the arriving rays; loosely, a criterion of noise structurelessness is preferred when little or nothing is known about signal structure. R. L. Johnson and Sherrill (1982) stated that the object of the maximum entropy spectral estimator is 'to be maximally non-commital about which member of the family of signals we have observed.'

The literature on MEM draws heavily on research in related fields of autoregressive (AR) and autoregressive moving average (ARMA) processes and time series analysis, as well as filter theory. An AR process of order n is defined (Friedlander and Porat 1989*a*) as a stationary process in which subsets of length $(n + 1)$ from a set of N consecutive measurements $(y_1, y_2, \ldots y_N)$ satisfy a difference equation of the form

$$y_t = - \sum_{k=1}^{n} a_k \, y_{t-k} + u_t$$

where $\{u_t\}$ is stationary zero-mean Gaussian white noise with variance σ^2; the standard estimation problem is then to find σ^2 and the unknown coefficients $a_1, a_2, \ldots$ from the measurements. ARMA models are pole-zero models where the poles describe the signals and the zeros are due to the presence of noise (D. H. Johnson, 1982). Conventional maximum entropy spectral analysis (MESA) generates an all-pole spectrum; Benjamin (1980) showed how the process could be generalised to allow for spectra with zeros as well as poles.

Burg's method of spectral analysis is equivalent to a least mean-square error linear prediction with a discrete-time all-pole model (Van den Bos, 1971). Linear predictive models can be used to extrapolate a time series forwards or backwards in time, outside the time window in which it was collected. Similarly, the spatial window represented by a finite array aperture can be extended, with the aim of enhancing resolution, by forward–backward linear prediction (FBLP) techniques that predict the signals expected in additional elements at each end of (say) a linear array.

MESA can be extended to multi-dimensional spectra and non-uniformly spaced sampling points, but does not lend itself to a closed-form solution; iterative search techniques for the spectral components are normally required. Whereas AR and MEM may be regarded as identical in one dimension, they are distinct methods in the multidimensional case and MEM appears to require the solution of a non-linear optimisation problem (McClellan, 1982). In view of the fact that other estimators give better resolution in the DF context than MEM, a full treatment of these topics is outside the scope of the present chapter. The quoted paper by McClellan and another by Lang and McClellan (1982) provide a good introduction to multidimensional analysis, with references to other relevant papers.

The largest value of the 'interval' $(s - r)$ between sensors r and s in a uniformly-spaced array is $(n - 1)$, where n is the number of sensors. The voltage products for this interval occur only in the first and last columns (and first and last rows) of the covariance matrix; these columns, which can be denoted as $\mathbf{R}_{.1}$ and $\mathbf{R}_{.n}$, give the full benefits of the available aperture. One form of the MEM spectral estimator is calculated from the first column of the inverse of $\mathbf{R}$; see, for example, Nickel (1987):

$$P_{MEM} = \frac{(\boldsymbol{R}^{-1})_{11}}{\mid \boldsymbol{A}^H (\boldsymbol{R}^{-1})_{.1} \mid^2}$$

In performance tests on MEM and other estimators, R. L. Johnson and Miner (1986) used an alternative solution method proposed in Lang and McClellan's paper; the details may be found in either paper.

12.3.3 Eigenanalysis methods

The concept of searching for the best linear relationships among the sensor signals by decomposing the signal covariance matrix into its eigenvectors and eigenvalues was introduced in Chapter 11. Papers from the early

1970s onwards describe various aspects of eigenanalysis methods. The best known now is MUSIC, which is described in the next subsection. A brief survey of related work is given here.

Owsley (1971) used a matrix inversion method of modal decomposition and proposed an algorithm suitable for digital computation. Pisarenko (1972) studied the characteristics of spectral estimators formed from various weighted means of the eigenvalues of the covariance matrix and also described (Pisarenko 1973) a solution method based on the smallest eigenvalue and the roots of the corresponding polynomial equation. When Pisarenko harmonic decomposition (PHD) is applied to angle measurements, r rays produce, in the noise-free case, a matrix of $(r + 1) \times (r + 1)$ elements with one eigenvector orthogonal to the r ray vectors. The corresponding eigenvalue is zero in the noise-free case and σ^2 with noise, where σ^2 is the variance of the noise distribution. For a uniformly-spaced linear array the angles of arrival can be determined from the roots of a polynomial, a procedure based in effect on a well-known result in standard antenna array theory (Schelkunoff, 1943).

Eigenanalysis methods were being used in system control theory from the mid-1960s according to Stoica and Nehorai (1988) and for the study of adaptive array theory by the mid-1970s; see, for example, Reed, Mallett and Brennan (1974) and Gabriel (1976). Gabriel referred to the eigenvalues falling near 1.0 as non-unique, but those greater than a threshold value of 1.10 as unique and useful.

Gething (1978) recognised the importance of eigenanalysis for radio direction finding, referred briefly to the geometrical interpretation and the need to distinguish between the signal and noise eigenvalues, and illustrated the process with a numerical example; Ellis (1980) later gave a fuller treatment of the geometrical interpretation.

Van Blaricum and Mittra (1978) also discussed applications of eigenanalysis, arriving at a very similar result to Pisarenko. Van Blaricum and Mittra's approach provided the basis of the method used by Lai and Dyson (1981) for the determination of directions of arrival and is, in effect, used in some current eigenanalysis methods that solve for the maximum number of rays.

Reddi (1979) used the signal subspace rather than the signal nullspace (noise subspace) used in MUSIC in his minimum norm method of calculating the estimator. Bienvenu and Kopp (1980, 1981) explored various properties of eigenanalysis solutions, including the distinction between signal eigenvectors and noise eigenvectors.

R. L. Johnson and Sherrill (1982) distinguished between two main types of eigenanalysis:

(*a*) Principal component analysis (PCA), designed to eliminate coherent components in the covariance matrix
(*b*) Pisarenko's method of harmonic decomposition (PHD).

The PCA method requires the calculation of array beam patterns and an iterative search for nulls. The PHD method is as outlined above and, for uniform spacing, the angles are calculated from the roots of a polynomial.

This is a much faster procedure than searching in arrival angle, even though iterative routines must be used to solve fifth order polyomials and above. Stoica and Nehorai (1988) analysed the statistical performance of the PHD method.

12.3.4. MUSIC

MUSIC is an acronym for Multiple Signal Characterisation. The MUSIC procedure, proposed by Schmidt (1979, republished 1986), is now easily the best-known and most frequently-cited method of superresolution. A clear and thorough treatment of MUSIC may be found in Schmidt's thesis (1981). The philosophy is very similar to that described in Section 11.2, but is worth repeating.

In essence, the signals measured in each element of an n-element array are seen as defining an n-dimensional space. Each single frame of data then represents a single point in this space. The collection of many frames of data on fading signals leads to a spread of points. The existence of linear relationships for perfect data, when the number of arriving rays is less than n, would cause the point to lie on hyperplanes of dimension $(n - 1)$ or less. With imperfect data, the requirement is to search for hyperplanes that minimise the squared error vectors perpendicular to them. The fit can be tested for various hypotheses about the number of arriving rays, a parameter that is initially unknown. The fit of the solution to the original data will always improve as the assumed number of rays is increased; however, the accuracy of the angle determinations may decrease. For example, given 20 sensor elements we can choose to solve for 19 rays, but some of the determined angles may be spurious and others may be less accurate than if a smaller number of rays had been assumed.

Schmidt showed that a better procedure than solving for the maximum number of rays is to partition the n-dimensional space into a signal (i.e. ray) subspace and an orthogonal noise subspace, sometimes referred to as the signal nullspace. The computational approach, starting from the covariance matrix $\mathbf{R}$, requires the determination of the eigenvalues and eigenvectors of this matrix. A threshold on the eigenvalues is then set, such that the K eigenvalues larger than the threshold produce eigenvectors which span the signal subspace and all smaller eigenvalues produce eigenvectors which span the noise subspace; the value of K is selected by applying a χ^2 test to successive hypotheses $K = 0, 1, 2 \ldots (n - 1)$. The first acceptable hypothesis establishes the best estimate of r, the true number of rays.

The matrix $\mathbf{R}_N$ of eigenvectors spanning the noise subspace is then formed and used to calculate the spectral estimator

$$P_{MUSIC} = \frac{1}{\boldsymbol{A}^H \boldsymbol{R}_N \boldsymbol{R}_N^H \boldsymbol{A}}$$

The use of $\mathbf{R}_N$ to combine all the noise eigenvectors is equivalent to looking for common nulls in two or more radiation patterns and rejecting non-common nulls as spurious solutions. Alternatively, the estimator could be calculated from the signal subspace matrix, which in principle contains the same information; however, Schmidt gave reasons for preferring the noise subspace matrix.

The arrival angles are determined from the K highest cusps in the estimator function. These peaks may be thought of as the inverse of minima in a composite radiation pattern which, with perfect data, would contain nulls at the arrival angles of the rays. There should be no spurious peaks and the angle estimates are statistically unbiased. Once the angles have been found, the calculation of amplitudes and other ray parameters is straightforward, as in other methods.

Schmidt's thesis (1981) and papers (1979, 1983, 1986) provide the best source of information on the underlying theory and should be studied for a full appreciation of the many attractive features and wide applicability of MUSIC. It can, for example, be applied to arrays of irregular configuration and to any polarisation component of the arriving rays. Two key features distinguish MUSIC from other eigenanalysis methods:

(*a*) The combined use of two or more noise eigenvectors in the estimator to determine angles

(*b*) The use of cusps rather than minima in the spectral plot to indicate angles.

A practical problem is that the number of arriving rays has to be determined by setting a threshold between small noise eigenvalues and larger signal eigenvalues. The angle determinations are likely to be in error if the number of arrivals is wrongly estimated; underestimation generally produces much greater errors than mild over-estimation. An additional problem is the computing load, particularly if the angle search has to be conducted in both azimuth and elevation. Additional procedures are required to deal with two or more coherent arrivals.

These problems are not unique to MUSIC but occur in various combinations and degrees of severity in other methods as well.

12.3.5 ESPRIT

The acronym ESPRIT stands for Estimation of Signal Parameters via Rotational Invariance Techniques. The procedure was suggested by Paulraj, Roy and Kailath (1986) and is claimed to possess a number of advantages compared with MUSIC.

To understand the physical basis of the mathematical procedure, consider an irregularly spaced array consisting of elements A, B, C ... and suppose that an identical array is placed at A', B', C' ... as in Fig. 12.3. The same linear displacement vector $\mathbf{d}$ applies to each element pair AA', BB', CC' etc. Interferometer-type nulls for each pair of elements AA', BB', ... can now be placed on one of a number of arriving signals by forming a doublet with a null in that direction. We will suppose that the signal in element A' is shifted in phase by φ, where

$$\varphi = \frac{2\,\pi\,|\,\mathbf{d}\,|\,\cos\theta}{\lambda}$$

and subtracted from the signal in A.

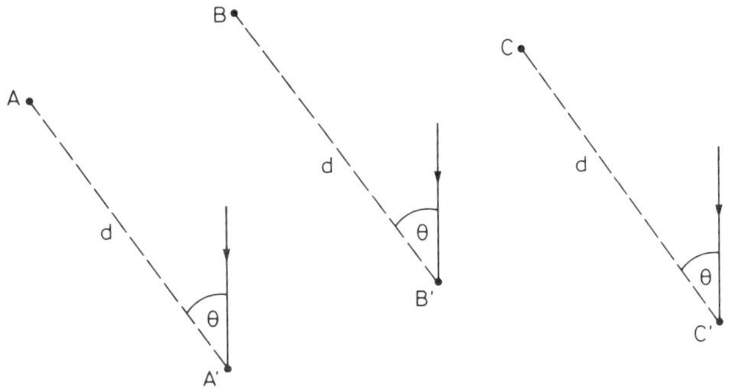

Fig. 12.3 Two congruent arrays of sensors
'Pairwise identical' doublets AA', BB', CC' formed from the two arrays, A, B, C and A', B', C'. One ray only shown, arriving from cone angle θ.

Applying the same phase shift and subtraction process to the pairs BB', CC' etc. we obtain a set of doublet output signals from which (ideally) any contribution from the ray arriving from θ has been eliminated.

It follows that the rank of the signal covariance matrix for the doublet signals will be one less than the rank of the matrix for the array, A, B, C, or for the array A', B', C'. For r arrivals we expect to find r phase shifts $\varphi_1, \ldots \varphi_r$ for which this reduction in rank occurs. This is the key to the ESPRIT procedure. The elements A, A', B, B', C, C' ... may be regarded as forming the total array, which contains 'pairwise identical' elements drawn from the two sub-arrays.

A full treatment, including the effects of external noise, was developed by Paulraj, Roy and Kailath (1986) and Roy and Kailath (1989) in terms of matrix pencil theory. They showed that the number of doublets n must exceed the number of arriving rays r. A diagonal $r \times r$ matrix Φ is introduced, with diagonal elements $e^{i\varphi_1}$, $e^{i\varphi_2}$, ... $e^{i\varphi_r}$ that are functions of the phase shifts to be determined. Each matrix element represents one of the phasor rotations (i.e. phase shifts) without change of phasor amplitude applied to the signals from array elements A', B', ... This interpretation of Φ helps to explain why the term 'invariant rotation' is used in the acronym.

Certain features of the solution can be understood without reference to the mathematical details. First, the arrival angles of rays in the plane of the array, say azimuth in Fig. 12.3, can be determined directly as soon as the phase shifts are known, without the need to search a radiation pattern. Secondly, there may be ambiguities if the interferometer spacing d is greater than λ. Thirdly, we can determine only cone angle relative to the 'interferometer arm' AA' for each arriving ray, rather than azimuth and elevation angles separately; it is only at zero elevation angle that cone angle becomes equal to azimuth angle relative to the arm. Fourthly, it is not necessary to know the array geometry or radiation pattern provided the displacement vector between the two sub-arrays is accurately known.

Paulraj, Roy and Kailath claim significant advantages in computer speed and storage requirements for ESPRIT as compared with MUSIC. They also regard the ability to work without array calibration as very attractive in some applications. However, the following disadvantages should be noted:

(*a*) The array design must be such that the 'pairwise identical' condition is satisfied
(*b*) The effective halving of the number of array elements by combining them into doublets reduces the maximum number of rays that can be resolved
(*c*) The angles of arrival are determined from phase shifts for a comparatively small displacement (i.e. narrow aperture) of less than λ
(*d*) A planar array determines cone angle relative to the displacement vector, rather than azimuth and elevation angles separately

ESPRIT, like MUSIC, encounters difficulties in resolving coherent rays. Ouibrahim, Weiner and Sarkar (1988) described a further development of ESPRIT to overcome this problem, by using a 'moving window' operator in place of the phase shift (delay) operator to form the matrix pencil. Liu and Zou (1988) proposed an iterative approach to source covariance matrix rotation, in order to overcome the same problem. Other refinements discussed by Zoltowski and Stavrinides (1989) include a 'total-least-squares' version, TLS–ESPRIT, a version called PRO–ESPRIT that exploits the redundancy of the two sub-arrays and a combination of the two.

12.3.6 Other developments from MUSIC

A number of variants and offshoots of MUSIC and other eigenanalysis techniques have been developed in recent years. Some methods are restricted to particular forms of array, such as uniform linear. Brief summaries of some of the important papers are given here.

D. H. Johnson and de Graaf (1982) described an adaptive beamforming scheme based on an eigenvector solution, which they claimed to be less sensitive to a wrong hypothesis about the number of arriving rays than MUSIC. Tufts, Kumaresan and Kirsteins (1982) suggested that the best approximation of rank r to the signal covariance matrix should be sought, using a singular value decomposition. Kumaresan and Tufts (1983) described a 'minimum-norm' method involving a different weighting of the eigenvectors as compared with MUSIC to find a vector spanning the whole noise subspace; the method is limited to uniformly-spaced linear arrays. Their algorithm contains a typographical error (absence of a minus sign), as noted by Jeffries and Farrier (1985).

Barabell (1983) considered several methods of reducing the signal-to-noise ratio required for successful resolution. One method involved examining the roots of the polynomial obtained for a uniformly-spaced array. Another used the properties of the signal–space eigenvectors to define a pole–zero spectrum function.

Hudson (1985*b*) used what he called a pragmatic algebraic approach, in which a least-squares eigenvector method was used to 'handle excess antenna elements.' The same author (1985*c*) suggested that an early step in finding the right steering weights for an adaptive array should be to determine the ray arrival angles. The same suggestion was made by Friedlander and Porat (1989*b*), as the first part of a two-step procedure for placing nulls on interfering signals. Recursive updating of solutions to take account of new data was considered by Schreiber (1986), again in the context of adaptive arrays; he suggested three algorithms that improve, either in speed or accuracy, on previous procedures. These procedures make it much more practical to use methods based on the inverse of the covariance matrix in real-time applications. Kim and Un (1989) considered the application to interference cancellation of a minimum eigenvector method with reduced degrees of freedom.

Bienvenu and Kopp (1980) suggested that the noise parameters should be determined first, in order to improve the angle determinations. Le Cadre (1989) used ARMA noise modelling to describe noise correlations by a small number of parameters, with the same objective.

MUSIC was used to solve a two-dimensional radar problem, involving range as well as angle, by Spielman, Paulraj and Kailath (1986). Du Fort (1983) applied a form of eigenanalysis to beam power outputs sampled with a single-channel receiver and obtained comparatively slow convergence to satisfactory results. Mayhan and Niro (1987) also made use of the fact that MUSIC can be applied to beam outputs instead of element outputs; for clustered targets, a significant reduction in processing requirements can be obtained by selecting a few neighbouring beams that cover the sector of interest, rather than many elements. Their paper discusses the effects of uncertainty in beam shape.

The property of orthogonality between noise–space eigenvectors and the signal direction vectors may not apply exactly to a sample covariance matrix, particularly if the sample is small and the noise level is high. The variance of a projected eigenvector was considered by Sharman and Durrani (1986), who then formed a joint log likelihood function to be minimised over all the direction vectors in the array manifold. This procedure, referred to as Likelihood MUSIC by the authors, has a higher computational cost than the basic method but the performance is significantly better with small samples.

Brandwood (1987*a*, *b*) suggested a method of noise–space projection that avoids the need for eigenanalysis. The method leads to a large reduction in complexity and computing requirements with very little loss of performance. Reilly (1987) split the signal covariance matrix into an $(n \times n)$ unitary matrix and an $(n \times n)$ upper triangular matrix in order to obtain a fast decomposition, and also found only a very small performance penalty.

Lee, Er and Evans (1989) solved for the maximum number of rays, i.e. $(n - 1)$ from n elements, and then distinguished between genuine and spurious nulls in a second stage of processing, using either artificial intelligence or optimum beamformer techniques to set the threshold level for a genuine null; the separation is, however, difficult with wideband sources. Starting from the other end, Oh and Un (1989*b*) solved first for one ray, then for two, three ... but removing the influence of the rays already found in an iterative refinement procedure; their simulations suggest that the threshold SNR for successful solution can be substantially reduced compared with MUSIC.

Haberman and Griffiths (1989) applied model-based AR methods to the signal–space eigenvectors for a uniformly-spaced array and found, from limited simulations, that the method may give slightly better results than MUSIC at low SNR.

The joint posterior P.d.f. of the signal parameters and the noise covariance matrix was studied by Reilly, Wong and Reilly (1989), who then integrated to remove the effects of the latter. A significantly improved performance compared with MUSIC was observed, particularly with regard to the threshold SNR at which a valid solution could be obtained.

Farrier (1989) considered the eigenvalues of the matrix

$$\mathbf{R}_r = [(\mathbf{I} - \mathbf{H}(\mathbf{H}^H \mathbf{H})^{-1} \mathbf{H}^H) \mathbf{R}_s]$$

where $\mathbf{R}_s$ contains the signal components and $\mathbf{H}$ contains the r signal vectors. If only r_1 of these vectors are correct, the matrix will have $(r - r_1)$ 'large' eigenvalues and r_1 'small' noise eigenvalues. By a process of weighting and adjustment, the direction vectors can be improved. The method was found to give a 10 dB improvement in the noise threshold compared with MUSIC.

12.3.7 Relationships between methods

Not surprisingly, some of the methods of superresolution reviewed in this chapter are closely related to each other and become identical in particular limiting cases.

Burg showed that the Capon MLM spectrum is a smoothed version of the MEM spectra for different filter lengths (see Childers (1978), which contains reprints of some of the earlier key papers). We would therefore expect the MLM estimate to exhibit a higher degree of statistical stability but lower spectral resolution than the MEM estimate (R.L. Johnson and Sherrill, 1982). Gabriel (1980) stated that TNA appears to possess an interesting combination of both MESA and MLM pattern characteristics.

Nickel (1988) suggested that the MUSIC method can be interpreted as a Capon-type method, but based on a covariance matrix computed from the eigenvectors. This reconstructed matrix corresponds to an infinite SNR; hence (on this argument) the superior resolution of the MUSIC pattern compared with the Capon pattern.

Nickel also showed that the MUSIC method can be interpreted as the weighted average of Kumaresan–Tufts (KT) patterns (see Section 12.3.6) of different filter lengths, i.e. the average of KT patterns generated by subarrays consisting of 1 to n elements. The limit of the MEM pattern for infinite SNR is the KT pattern. The various relationships discussed by Nickel are shown in Fig. 12.4, based on a diagram in his paper.

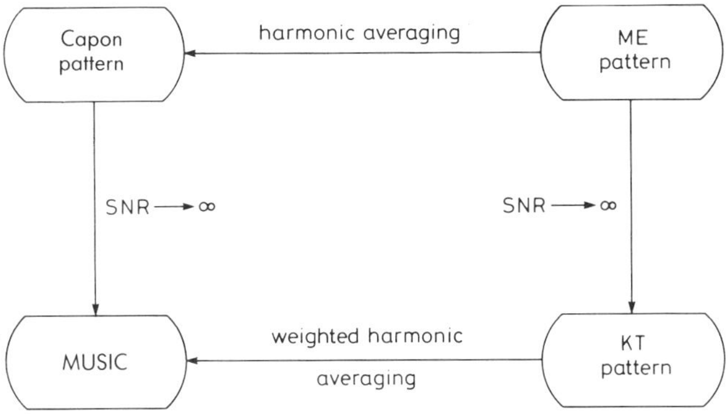

Fig. 12.4 Relationships between methods
Based on Fig. 1 in a paper by Nickel, Proc. IEE Pt.F., ***135,*** *1988, pp. 7–10*

A unified approach to the Prony, Pisarenko and matrix pencil methods is being developed by Ouibrahim (1989). He states that each depends on the examination of the dependence/independence of the same set of data vectors. However, the methods are known to perform differently, particularly at low SNR.

12.4 Performance of algorithms

12.4.1 Sources of information

We now come to a consideration of the performance of each of the categories of algorithms described above, and of other methods such as the Prony solution illustrated in Section 10.4.2. The performance parameters of most signifiance are resolving power and the accuracy of the calculated directions of arrival. We are interested in both comparative performance, when two or more methods are available for a given range of problems, and the absolute accuracy of the calculated angles of arrival. The relevant literature contains results based on theoretical arguments, solutions to simulated problems and practical measurements on live signals.

Simulations are used for a number of distinct purposes: to demonstrate that a proposed algorithm works correctly in favourable circumstances; to check theoretical results, for example on the reduction of angle variance with increasing sample size; and to compare different processing schemes against problems with known 'true answers'. Comparisons of accuracy and resolving power for two or more methods may be found in a number of papers; typical results are reviewed below. Resolving power is usually assessed on problems involving two rays arriving from nearly the same direction; the literature contains results for various methods, sensor arrays, and combinations of noise components and additional rays. The effects of aperture can be removed to some extent, in order to facilitate comparisons, by following the usual convention of expressing the minimum resolvable separation as a fraction of a beamwidth. However, such a comparison appears to give insufficient weight to the value of a narrow beam in rejecting unwanted energy from other directions.

As the noise level in a simulated problem is increased, we can reasonably expect the variances of the determined angles to increase. This process is not necessarily smooth or gradual; there may be a critical level of SNR at which the solution method breaks down completely and produces a 'wild' answer. The explanation may lie in an incorrect determination (underestimation) of the number of arriving rays, or in the problem of masquerading described in Section 11.7; if the antenna signals for two quite different problems are very nearly the same under noise-free conditions, the small differences are soon masked as the noise level is increased, and the resolving power of the array is then inadequate to distinguish between the two solutions.

Practical measurements are of rather limited value in any comparison of methods, because the true values of the arrival angles (as distinct from the bearings of the signal sources) are not known; however, real data can be used to test the practicality of a processing scheme, and to compare the stability and plausibility of the outputs from different processing schemes. Controlled transmitters of known position in the near or far field of the array are sometimes used (see, for example, Hill (1989)).

12.4.2 Theoretical studies of accuracy

Theoretical comparisons of two or more methods have been made by some investigators.

For the special case of a single frame of data (snapshot), Hua and Sarkar (1987) compared the angle estimation performance of three methods, namely Prony, Pisarenko and a pencil-of-function approach suggested by Jain, Sarkar and Weiner (1983), all based on finding the roots of the Schelkunoff polynomial (see Section 12.3.3). Hua and Sarkar showed that the perturbation in angle is more sensitive to the noise components in side sensors (those at or near the end of the line) than in the middle sensors.

Loughrey (1989) examined the origin of the spurious peaks that occur in some solutions and showed that they are related to a defined spatial correlation between directions being searched and the incident signal directions. High resolution algorithms such as MLM and MUSIC obtain their resolution improvement through non-linear transformations on spatial correlations, whereas a conventional beamformer (Discrete Fourier Transform) transforms the correlation linearly and is more affected by spurious peaks. Simulations confirmed that the DFT algorithm has poor resolution that is independent of the SNR, while MLM and MUSIC show improved resolution with increasing SNR.

12.4.3 Simulated problems

We begin with an example of a simulated problem that throws light on the behaviour of MUSIC. R. L. Johnson (1986) examined the effects on MUSIC solutions of incorrect partitioning of the eigenvector matrix under various conditions of signal coherence, bearing separation, unequal signal strength and additive noise. The array was linear, with nine elements uniformly spaced at intervals of 0·4λ. Five rays arriving from zero elevation angle were modelled, with separations in azimuth of one beamwidth except in the study of bearing separation. Fig. 12.5, reproduced from Johnson's paper, shows the effect of solving for four, five and six rays.

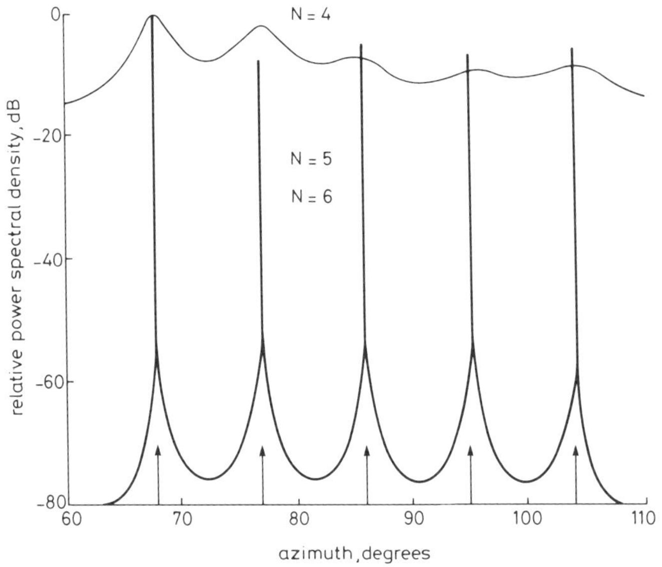

Fig. 12.5 Plot of spectral estimator for three MUSIC solutions

Five incoherent rays arriving from azimuths indicated by arrows. Solutions for N = 4, 5, and 6, where N is the assumed number of rays.

Based on Fig. 1 in a paper by Johnson, IEEE. Trans., **AP-34,** *August 1986, pp. 985–991*

This simple example, for incoherent rays and a favourable SNR (50 dB for each ray), demonstrates that in such cases it is easy to infer the correct number of rays and that overestimation (solving for six) is much less serious than underestimation (solving for four).

The effect of coherence between two of the five rays was found to be as follows: only four significant eigenvalues and only three dominant spectral peaks were found. However, when one of the pair of coherent rays was much stronger than the other, it could be resolved in spite of an underestimation of the number of rays. For small bearing separations of incoherent signals, it was observed that solutions for reduced estimates of the number of rays (referred to as ray order) tended to resolve the widely-separated rays and to converge to only one of the closely-spaced bearings. For rays with unequal amplitudes, the reduced ray order estimates tended to produce a reduction in the number of spectral peaks, but it was not easy to predict which peaks would dominate as the ray order was successively reduced.

For low SNR and low time–bandwidth product (i.e. number of frames), ray order is difficult to determine from the size of the eigenvalues. It was observed that averaging spectra for four subsets of 500 frames was not nearly as effective as forming one solution from all 2000 frames.

The literature on superresolution contains many examples of comparisons of two or more methods against a small set of simulated problems, but the problems show a wide diversity in terms of the array aperture and type, number of rays, arrival angles, ray amplitudes, signal-to-noise ratio, number of frames, correlation and other parameters. It is difficult to relate the results from different investigations, except in the most general terms.

Of most interest here are papers containing the results of simulations on several methods, particularly if their respective capabilities have been thoroughly explored by extensive computer runs of the Monte Carlo type.

Burg (1972) showed that the resolution of MEM is never worse than that of Capon's MLM. In turn, Thorvaldsen (1981) compared Burg's method with a least-squares solution for the coefficients of an AR process, by means of a simulated two-ray problem for an 8-element linear array. The least-squares solution was clearly superior, with the Burg solution showing false peaks and greater bias.

Ferrara and Parks (1983) showed that antennas with diverse polarisations can offer a significant advantage over identically-polarised elements for the determination of angles of arrival. They compared the results from the MLM, AAR (see Section 12.3.1) and MUSIC methods, but excluded MEM, which cannot be readily extended for use with such an array. All of the methods approached the same performance as the SNR was increased, but the authors drew attention to the following unique property of MUSIC: asymptotically perfect results can be obtained, even for low SNR, provided the form of the noise covariance matrix is known or can be measured.

R. L. Johnson and Miner (1986) made a particularly thorough comparison of five different methods, on problems involving two closely-spaced rays with varying degrees of spatial/temporal correlation, various levels of additive white Gaussian noise, different time–bandwidth products and varying levels of cochannel interference. No tests were made with completely coherent signals, but simulations were run for correlations of 0·8 and 0·3. The array consisted of seven elements spaced at intervals of 0·4 λ in three dimensions.

The methods were selected with the aim of satisfying the following four criteria:

(*a*) The algorithm must have high resolution capability
(*b*) The technique must be able to deal with three-dimensional non-uniform antenna array deployments
(*c*) The algorithm must be able to estimate azimuth/elevation angle of arrival
(*d*) The method should not require a prior knowledge of the number of signals present.

MUSIC does require the number of rays to be inferred or estimated in some way, but was nevertheless included. The other four methods chosen were MEM as described by Lang and McClellan (1982) and versions of MLM, AAR and TNA (see Section 12.3.1). Fig. 12.6 shows an example of a two-ray problem solved with the TNA algorithm. Overall, MUSIC was shown to be clearly superior to the other methods and was the only one to resolve the correlated signals correctly.

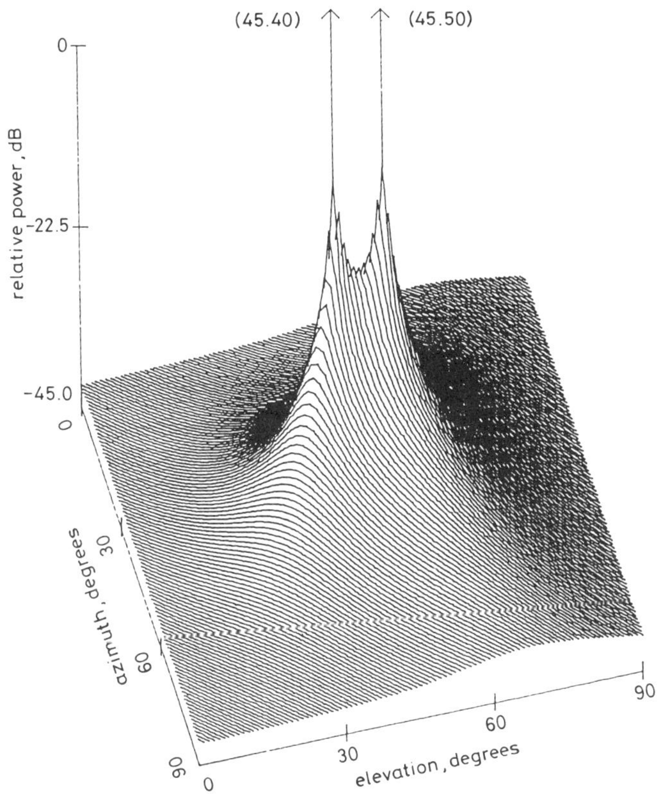

Fig. 12.6 Plot of TNA spectral estimator for two incoherent rays of equal amplitude
True angles: Ray 1, Az = 45·0° El = 40·0°
Ray 2, Az = 45·0° El = 50·0°
Based on Fig. 2 in a paper by Johnson and Miner, IEEE. Trans., **AES-22,** *July 1986, pp. 432–442*

The Alternating Projection (AP) version of MLM was tested by Ziskind and Wax (1988) against a number of two-ray problems, for a three-element linear array with inter-element spacing of 0·5 λ. The results were compared with the theoretical Cramer-Rao lower bound on variances (see Chapter 13) and with results obtained from MUSIC with the same input data. The decrease of variance as a function of increasing SNR and of increasing number of frames was investigated. In each case the AP method performed better than MUSIC, which failed completely against coherent signals. The number of iterations required for AP to converge to a stable solution was tested on a number of problems involving four rays and seven elements; the average requirement was for four to five iterations and never more than seven.

12.4.4. Empirical results

Based mainly on the results of simulations, but taking some account of theoretical considerations, a number of authors have put forward empirical results for the effects of various factors such as the number of frames and the SNR on resolving power.

For the maximum entropy method, the SNR required to separate two signals varies approximately as the inverse cube of the separation (Luthra and Steinberg, 1982). Friedlander and Porat (1989*b*) developed an expression for the asymptotic output-interference-to-signal ratio (OISR). For two signals arriving from angles θ_1 and θ_2 which are close together but not equal, the OISR behaves approximately as $(\theta_1 - \theta_2)^{-4}$.

12.4.5 Practical performance

As already noted, tests on live signals from distant radio transmitters are of rather limited value in comparing superresolution algorithms. However, when a local transmitting source is under the control of the experimenter and its true bearing is known, each solution from the same sample of data can be examined to see how closely the bearing is determined under conditions of noise, multipath and interference. Two or more local sources are needed for controlled experiments on resolution.

Various methods for the determination of elevation angles were used by Clarke and Tibble (1978) in their analysis of data collected on transmissions via the ionosphere from distant sites. Some difficulties were experienced with the maximum entropy method; for example in deciding the length of the optimum prediction filter.

Schmidt and Franks (1986) used an eight-element circular array with a diameter of 13λ and resolved rays down to $\theta_B/3$, where θ_B is the beamwidth. The limit could be reduced to $\theta_B/5$ or $\theta_B/6$ if the two rays were orthogonally polarised.

Starkey (1986) attempted to resolve two uncorrelated noise-like sources with the 24 elements of an S-band phased array. The sources were separated in azimuth by $0{\cdot}35\theta_B$. A pitched roof between the sources and the array gave a multipath environment characterised as varying between fair and severe as the positions of the sources were varied. The methods compared were Capon's MLM, a sidelobe canceller method (SLC) equivalent to an AR method, a Marple solution (Kay and Marple, 1981) and four different forms of eigenanalysis, referred to by the author as Schmidt, modified Schmidt (MS), Johnson and de Graaf (JG) and modified JG (MJG). The SNR in each channel was 32 dB and 2048 frames of data were available for analysis.

Simulations showed that, with this SNR and in the absence of multipath, the sources should be resolved by all algorithms. However, the presence of multipath provided a more severe test in practice. The sources were never resolved by the MLM, MS and JG methods. MJG and Schmidt did not resolve the sources properly, but gave indications sometimes of more than one source. The greatest success was obtained with two combinations of methods:

(*a*) MLM with Marple for AR order 16
(*b*) MLM with SLC for AR order 16.

Starkey commented that a guaranteed improvement in performance will require improved estimation of the AR order and the number of signals present, together with the use of several methods of signal processing.

Scott (1987) collected samples of 100 frames with an HF array in southern England. The array consisted of eight elements in an irregular heptagon, with one element at the centre; the aperture was approximately 1λ. Most of the results presented relate to experiments performed with a mobile signal source taken to a number of sites at a typical distance of 2 km from the array, but the final series of measurements was made on a broadcast signal from Spain. The two processing algorithms used were a version of MLM called scanned adaptive beam with triangular factorisation (SABTF) and Brandwood's version of MUSIC without eigenvectors (Brandwood, 1987*b*).

The practicality of using the algorithms was demonstrated, but the experiments were not designed as a critical comparison of their resolution performance. Theoretically, the two methods (and other algorithms) should give identical results for a single dominant signal; in the experimental conditions there appears to have been little to choose between them.

Results obtained at VHF by the same research group were reported by Hill (1989). The array consisted of five elements in a pentagon. The wavefields were generated with one fixed source and other mobile sources. Five algorithms were compared:

(*a*) A digital implementation of Adcock processing, using only three of the elements
(*b*) Scanned Fixed Beam (SFB)
(*c*) Scanned Adaptive Beam (SAB), equivalent to MLM
(*d*) MUSIC
(*e*) Noise–Space Projection (NSP).

The Adcock (single) bearing was determined from a phase comparison between orthogonal pairs of equispaced elements. NSP mimics MUSIC, but without the need for eigenanalysis.

MUSIC, NSP and SAB displayed distinct advantages in a two-signal environment, with NSP and SAB more reliable than MUSIC in the determination of the number of signals. The same three algorithms were then applied to the data from three elements used for the Adcock-type algorithm. All three indicated the presence of two signals, but with large errors in the angle of the smaller signal.

Martin (1989) applied versions of eigenanalysis mainly to HF radio data collected on a path from San Antonio to Ottawa, but with some acoustic data. He found that the noise eigenvalues did not cluster around a common value of one, but tended to increase in geometrical progression from the smallest upwards. A test based on eigenvalue ratios was therefore used to estimate the number of arriving rays. With this form of analysis, MUSIC was shown to have 'very favourable directional characteristics' compared with other methods.

References

BARABELL, A.J. (1983): 'Improving the resolution performance of eigenstructure-based direction-finding algorithms'. ICASSP 83, Boston, 14–16 April, **1,** pp. 336–339 (IEEE, New York)

BENJAMIN, R. (1980): 'Generalisations of maximum-entropy pattern analysis', *Proc. IEE Pt. F,* **127,** pp. 341–353

BIENVENU, G., and KOPP, L. (1980): 'Adaptivity to background noise spatial coherence for high resolution passive methods', ICASSP 80, Denver, 9–11 April, pp. 307–310 (IEEE, New York)

BIENVENU, G., and KOPP, L. (1981): 'Source power estimation method associated with high resolution bearing estimator'. ICASSP 81, Atlanta, 30 March–1 April, pp. 153–156 (IEEE, New York)

BORGIOTTO, G.V., and KAPLAN, L.J. (1979): 'Superresolution of uncorrelated interference sources by using adaptive array techniques', *IEEE Trans.,* **AP-27,** pp. 842–845

BRANDWOOD, D.H. (1987*a*): 'Noise space projection and economical methods for obtaining the MUSIC DF function without eigenanalysis'. ICAP 87, IEE. Conf. Publn. 274, Vol. **1,** pp. 415–418

BRANDWOOD, D.H. (1987*b*): 'Noise-space projection: MUSIC without eigenvectors', *Proc. IEE Pt. H,* **134,** pp. 303–309

BRANDWOOD, D.H. (1989): 'Multiple signal DF with coherent signals using a general array'. Passive DF Colloquium, IEE Digest No. 1989/16, pp. 5/1–4

BURG, J.P. (1972): 'The relationship between maximum entropy spectra and maximum likelihood spectra', *Geophys.,* **37,** pp. 375–376

CAPON, J. (1969): 'High-resolution frequency-wavenumber spectrum analysis', *Proc. IEEE,* **57,** pp. 1408–1418

CAPON, J. (1970): 'Applications of detection and estimation theory to large array seismology', *Proc. IEEE,* **58,** pp. 760–770

CHILDERS, D.G. (Ed) (1978): 'Modern spectral analysis', (IEEE, New York)

CLARKE, R.H., and TIBBLE, D.V. (1978): 'Measurement of the elevation angles of arrival of multicomponent HF skywaves', *Proc. IEE,* **125,** pp. 17–24

DU FORT, E.C. (1983): 'High resolution emitter direction finding using a phased array antenna', *IEEE Trans.,* **ASSP-31,** pp. 1409–1416

ELLIS, J.H. (1980): 'Fundamental limitations of wave-front analysis by least squares', *Proc. IEE Pt. H,* **127,** pp. 219–227

FARRIER, D.R. (1989): 'High-performance signal subspace beamformer', *Proc. IEE Pt. F,* **136,** pp. 283–288

FERRARA, E.R., and PARKS, T.M. (1983): 'Direction finding with an array of antennas having diverse polarizations', *IEEE Trans.,* **AP-31,** pp. 231–236

FRIEDLANDER, B., and PORAT, B. (1989*a*): 'The exact Cramer–Rao bound for Gaussian autoregressive processes', *IEEE Trans.,* **AES-25,** pp. 3–7

FRIEDLANDER, B., and PORAT, B. (1989*b*)): 'Performance analysis of a null-steering algorithm based on direction-of-arrival estimation', *IEEE Trans.*, **ASSP-37,** pp. 461–466

GABRIEL, W.F. (1976): 'Adaptive arrays – an introduction', *Proc. IEEE*, **64,** pp. 239–272

GABRIEL, W.F. (1980): 'Spectral analysis and adaptive array superresolution techniques', *Proc. IEEE*, **68,** pp. 654–666

GABRIEL, W.F. (1989): 'Large-aperture sparse array antenna systems of moderate bandwidth for multiple emitter location', *IEEE Trans.*, **AP-37,** pp. 16–29

GETHING, P.J.D. (1978): 'Radio direction finding and the resolution of multicomponent wave-fields', (Peter Peregrinus) chap. 11

HABERMAN, E., and GRIFFITHS, L.J. (1989): 'On combining the eigenstructure-based and modeled-based approaches in direction of arrival estimation problems'. ICASSP 89, Glasgow, 25–28 April, **4,** pp. 2621–2624 (IEEE, New York)

HILL, D.S. (1989): 'A practical comparison of multiple element DF techniques'. Passive DF Colloqium, IEE Digest No. 1989/16, pp. 6/1–5

HUA, Y., and SARKAR, T. (1987): 'Analysis of three high resolution techniques for radio direction estimation', ICASSP 87, Dallas 6–9 April, **3,** pp. 1728–1731 (IEEE, New York)

HUDSON, J.E. (1981): 'Adaptive array principles', (Peter Peregrinus), p. 27

HUDSON, J.E. (1985*a*): 'Fast near-optimum resolution of coherent sources by line arrays'. ICAP 85, IEE Conf. Publn. 248, pp. 237–240

HUDSON, J.E. (1985*b*): 'Decomposition of antenna signals into plane waves', *Proc. IEE Pt. H*, **132,** pp. 53–57

HUDSON, J.E. (1985*c*): 'Antenna adaptivity by direction finding and null steering', *Proc. IEE Pt. H*, **132,** pp. 307–311

JAIN, V.K., SARKAR, T.K., and WEINER, D.D. (1983): 'Rational modelling by pencil-of-functions method', IEEE Trans., **ASSP-31,** pp. 564–573

JEFFRIES, D.J., and FARRIER, D.R. (1985): 'Asymptotic results for eigenvector methods', *Proc. IEE Pt. F*, **132,** pp. 589–594

JOHNSON, D.H. (1982): 'The application of spectral estimation methods to bearing estimation problems', *Proc. IEEE*, **70,** pp. 1018–1028

JOHNSON, D.H., and de GRAAF, S.R. (1982): 'Improving the resolution of bearing in passive sonar arrays by eigenvalue analysis', *IEEE Trans.*, **ASSP-30,** pp. 638–647

JOHNSON, R.L. (1986): 'Eigenvector matrix partition and radio direction finding performance', *IEEE Trans.*, **AP-34,** pp. 985–991

JOHNSON, R.L. (1988): 'Antenna array design for covariance-based direction-finding methods', *IEEE Trans.*, **AP-36,** pp. 1537–1544

JOHNSON, R.L., and MINER, G.E. (1986): 'Comparison of superresolution algorithms for radio direction finding', *IEEE Trans.*, **AES-22,** pp. 432–442

JOHNSON, R.L., and SHERRILL, W.M. (1982): 'Modern signal processing applied to radio direction finding'. Final Report for SWRI Project 16–9284

KAY, S., and DEMEURE, C. (1984): 'The high resolution estimator – a subjective entity', *Proc. IEEE*, **72,** pp. 1815–1816

KAY, S.M. and MARPLE, S.L. (1981): 'Spectral analysis — a modern perspective', *Proc. IEEE*, **69,** pp. 1380–1419

KIM, J.W., and UN, C.K. (1989): 'Signal subspace method for beam-steered adaptive arrays', *Electronics Letters*, **25,** pp. 1076–1077

KUMARESAN, R., and TUFTS, D.W. (1983): 'Estimating the angles of arrival of multiple plane waves', *IEEE Trans.*, **AES-19,** pp. 134–139

LAGUNAS-HERNANDEZ, M.A. and GASULL-LLAMPALLAS, A. (1984): 'An improved maximum likelihood method for power spectral density estimation', *IEEE Trans.*, **ASSP-32,** pp. 170–173

LAI, H.H., and DYSON, J.D. (1981): 'The determination of the direction of arrival of an interference field', *Radio Sci.*, **16,** pp. 365–376

LANG, S.W., and McCLELLAN, J.H. (1982): 'Multidimensional MEM spectral estimation', *IEEE Trans.*, **ASSP-30,** pp. 880–887

LE CADRE, J.P. (1989): 'Parametric methods for spatial signal processing in the presence of unknown colored noise fields', *IEEE Trans.*, **ASSP-37,** pp. 965–983

LEE, C.S., ER, M.H., and EVANS, A.J. (1989): 'Simple and fast method for direction-finding of multiple wideband sources', *Elecronics Letters*, **25,** pp. 1522–1523

LIU, Q.G., and ZOU, L.H. (1988): 'Radio direction finding by source covariance rotation'. Proc. AP–S 88, Internat. Symposium Digest: Antennas and Propagation, Syracuse, 6–10 June, Vol. **1,** pp. 270–273 (IEEE, New York)

LOUGHREY, M. (1989): 'Spatial correlation effects in direction finding algorithms'. Passive DF Colloquium, IEE. Digest No. 1989/16, pp. 4/1–5

LUTHRA, A.K., and STEINBERG, B.D. (1982): 'The maximum entropy image of two targets', *Proc. IEEE*, **70,** pp. 98–99

McCLELLAN, J.H. (1982): 'Multidimensional spectral estimation', *Proc. IEEE*, **70,** pp. 1029–1039

MARTIN, G.E. (1989): 'Signal subspace processing of experimental radio data', *Proc. Soc. Photo-optical Instrumentation Engineers*, **975,** pp. 101–107

MAYHAN, J.T., and NIRO, L. (1987): 'Spatial spectral estimation using multiple beam antennas', *IEEE Trans.*, **AP-35,** pp. 897–906

NICKEL, U. (1987): 'Angular superresolution with phased array radar: a review of algorithms and operational constraints', *Proc. IEE Pt. F*, **134,** pp. 53–59

NICKEL, U. (1988): 'Algebraic formulation of Kumaresan-Tufts superresolution method, showing relation to ME and MUSIC methods', *Proc. IEE Pt. F*, **135,** pp. 7–10

OH, S. K., and UN, C.K. (1989*a*): 'Computationally efficient maximum likelihood algorithm for localising multiple sources', *Electronics Letters*, **25,** pp. 1540–1542

OH, S. K., and UN, C.K. (1989*b*): 'Improved MUSIC algorithm for high-resolution array processing', *Electronics Letters*, **25,** pp. 1523–1525

OUIBRAHIM, H. (1989): 'Prony, Pisarenko and the matrix pencil: a unified presentation', *IEEE Trans.*, **ASSP-37,** pp. 133–134

OUIBRAHIM, H., WEINER, D.D., and SARKAR, T.K. (1988): 'A generalized approach to direction finding', *IEEE Trans.*, **ASSP-36,** pp. 610–612

OWSLEY, N.L. (1971): 'Source location with an adaptive antenna array'. Naval Underwater Systems Center Report NL–3015

PAULRAJ, A., ROY, R., and KAILATH, T. (1986): 'Estimation of signal parameters via rotational invariance techniques – ESPRIT' 19th Asilomar Conference on Circuits, Systems and Computers, Pacific Grove, 6 Nov 1985, pp. 83–89 (IEE, New York)

PISARENKO, V.F. (1972): 'On the estimation of spectra by means of non-linear functions of the covariance matrix', *Geophys. J.R. Astr. Soc.*, **28,** pp. 511–531

PISARENKO, V.F. (1973): 'The retrieval of harmonics from a covariance function', *Geophys. J.R. Astr. Soc.*, **33,** pp. 347–366

REDDI, S.S. (1979): 'Multiple source location — a digital approach', *IEEE Trans.*, **AES-15,** pp. 95–105

REED, I.S., MALLET, J.D., and BRENNAN, L.E. (1974): 'Rapid convergence rate in adaptive arrays', *IEEE Trans.*, **AES-10,** pp. 853–863

REILLY, J.P. (1987): 'A real-time high-resolution technique for angle of arrival estimation', *Proc. IEEE*, **75,** pp. 1692–1694

REILLY, J.P., WONG, K.M., and REILLY, P.M. (1989): 'Direction of arrival estimation in the presence of noise with unknown, arbitrary covariance matrices', ICASSP 89, Glasgow, 25–28 April, **4,** pp. 2609–2611 (IEEE, New York)

ROY, R., and KAILATH, T. (1989): 'ESPRIT – estimation of signal parameters via rotational invariance techniques', *IEEE Trans.*, **ASSP-37,** pp. 984–995

SCHELKUNOFF, S.A. (1943): 'A mathematical theory of linear arrays', *Bell Syst. Tech. J.*, **22,** pp. 80–107

SCHMIDT, R.O. (1979): 'Multiple emitter location and signal parameter estimation', Proc. RADC Spectrum Estimation Workshop, Griffiths AFB, Rome, NY, pp. 243–258; republished in *IEEE Trans.*, **AP-34**, pp. 276–280

SCHMIDT, R.O. (1981): 'Signal subspace approach to multiple emitter location and spectral estimation'. PhD Thesis, Stanford University

SCHMIDT, R.O. (1983): 'New mathematical tools in direction finding and spectral analysis', *Proc. Soc. Photo-optical and Instrumentation Engineers*, **431,** pp. 7–19

SCHMIDT, R.O. (1986): 'Multiple emitter location and signal parameter estimation', *IEEE Trans.*, **AP-34,** pp. 276–280; republication of SCHMIDT (1979)

SCHMIDT, R.O., and FRANKS, R.E. (1986): 'Multiple source DF signal processing: an experimental system', *IEEE Trans.*, **AP-34,** pp. 281–290

SCHREIBER, R. (1986): 'Implementation of adaptive array algorithms', *IEEE Trans.*, **ASSP-34,** pp. 1038–1045

SCOTT, K.K. (1987): 'A practical implementation of superresolution algorithms for direction finding'. ICAP 87, IEE Conf. Publn. 274, Vol. **1,** pp. 411–414

SHARMAN, K.C., and DURRANI, T.S. (1986): 'A comparative study of modern eigenstructure methods for bearing estimation — a new high performance approach', Proc. 25th Conf. on Decision and Control, Athens, December, pp. 1737–1742 (IEEE, New York)

SPIELMAN, D., PAULRAJ, A., and KAILATH, T. (1986): 'A high resolution algorithm for combined time-of-arrival and direction-of-arrival estimation'. 19th Asilomar Conference on Circuits, Systems and Computers, Pacific Grove, 6 Nov. 1985, pp. 90–93 (IEEE, New York)

STARKEY, P.G. (1986): 'Experimental performance of phased array direction finding algorithms in severe multipath conditions'. Colloquium on direction finding systems in their operating environment, IEE Digest no. 1986/10, pp. 5/1–5

STOICA, P., and NEHORAI, A. (1988): 'Study of the statistical performance of the Pisarenko harmonic decomposition method', *Proc. IEE Pt. F*, **135,** pp. 161–168

STOICA, P., and SHARMAN, K.C. (1990): 'Novel eigenanalysis method for direction estimation', *Proc. IEE Pt. F*, **137,** pp. 19–26

THORVALDSEN, T. (1981): 'A comparison of the least squares method and the Burg method for autoregressive spectral analysis', *IEEE Trans.*, **AP-29,** pp. 675–679

TUFTS, D.W., KUMARESAN, R., and KIRSTEINS, I. (1982): 'Data adaptive signal estimation by singular value decomposition of a data matrix', *Proc. IEEE*, **70,** pp. 684–685

VAN BLARICUM, M.L., and MITTRA, R. (1978): 'Problems and solutions associated with Prony's method for processing transient data', *IEEE Trans.*, **AP-26,** pp. 174–182

VAN DEN BOS, A. (1971): 'Alternative interpretation of maximum entropy spectral analysis', *IEEE Trans.*, **IT-17,** pp. 493–494

WANG, H., and KAVEH, M. (1985): 'Coherent signal-subspace processing for the detection and estimation of angles of arrival of multiple wide-band sources', *IEEE Trans.*, **ASSP-33,** pp. 823–831

WAX, M., SHAN, T.J., and KAILATH, T. (1984): 'Spatio-temporal spectral analysis by eigenstructure methods', *IEEE Trans.*, **ASSP-32,** pp. 817–827

ZISKIND, I., and WAX, M. (1988): Maximum likelihood localization of multiple sources by alternating projection', *IEEE Trans.*, **ASSP-36,** pp. 1553–1560

ZOLTOWSKI, M.D., and STAVRINIDES, D. (1989): 'Sensor array signal processing via a Procrustes rotations based eigenanalysis of the ESPRIT data pencil', *IEEE Trans.*, **ASSP-37,** pp. 832–861

Chapter 13

Additional tools for superresolution

13.1 Introduction

A spectral estimator such as one of those described in Chapter 12 is likely to be at the heart of any processing scheme to achieve superresolution in angle measurements. However, the total processing scheme will involve other steps, prior and subsequent to the use of the main algorithm, in order to improve overall performance. Many published papers deal with refinements to the basic procedures, or with approximations that shorten the calculations with very little loss of performance. Now that the main categories of algorithms have been described in Chapter 12 it is possible to review some of these developments.

13.2 Coherent signals

13.2.1 How the problem arises

A definition of coherence was given in Section 12.1.3. We now consider the physical causes, the impact of the problem on radio direction finding and ways in which it can be overcome.

In the HF band, each antenna element is likely to receive a ground-reflected version of each arriving ionospheric mode. For a horizontally-deployed array of identical sensors, whether horizontally or vertically polarised, above a flat, uniform ground plane this reflected ray simply modifies the apparent amplitude and phase of the sky wave and causes no difficulty. The effect on a vertically-stacked set of sensors needs special treatment, but this subject has already been extensively discussed in earlier chapters.

Apart from this special case, it seems unlikely that perfect coherence would persist over (say) several seconds, for ionospherically-propagated rays that have travelled by different paths. With brief transmissions, however, lasting for a few seconds or less, there may be a difficulty that can be better described as a lack of relative fading. Some measure of partial correlation (partial coherence) might be expected between one ray received via the ionosphere and a version of it arising from a local reflecting object near the DF site.

The problem of perfect coherence may be more serious in other contexts. For groundwave or line-of-sight signals, whether at HF or higher frequencies, the correlation between a direct arrival and reflected or re-radiated components from site obstacles, or re-radiated signals from an intelligent jammer, could approach unity.

As noted in Section 12.1.3, partial coherence between (say) two ionospherically-propagated rays tends to increase the sampling time required to achieve the same resolution performance as with two incoherent rays; further information on the magnitude of the increase may be found in a paper by R. L. Johnson (1988).

13.2.2 Effects on angle solutions

Simulated scenarios show that the arrival of two or more coherent rays can cause problems with some standard methods of superresolution. Zoltowski and Haber (1986) modelled direct and specularly-reflected arrivals at a linear array of seven elements. For an angular separation of greater than λ/L, where L is the length of the array in wavelengths, no significant peak was found in the MUSIC angular spectrum, that is both rays were undetected. For a separation of less than λ/L a single peak occurred, approximately midway between the true direct and specular arrival angles. Other investigators have also found that the coherent rays tend to be merged in the solution into one apparent arrival from some intermediate direction, which may be considerably in error compared with the true directions.

Any wavefront analysis method capable of obtaining a solution from a single frame of data is able to deal with either coherent or incoherent rays; the distinction has no meaning at one instant in time. Hence single-frame methods such as the Prony method can be used to overcome the problem. Single-frame solutions can of course be averaged over many frames to improve accuracy. Other methods of solution, requiring more than one frame of data, are described below. For a mixture of coherent and incoherent rays, multi-frame solutions have the advantage of being able to exploit the relative fading between the incoherent arrivals.

It is perfect coherence over all the available frames that causes difficulty. No real problem exists with correlated signals, provided the frames available cover a reasonable part of the fading cycle; for example, Barton (1983) and Brandwood (1989) demonstrated successful resolution of two-ray problems with correlation coefficients of 0·985 and 0·99 respectively.

With a small sample of frames, however, resolution may be sharply degraded; the wavefield must be considered to be composed of mutually coherent signals. R. L. Johnson (1988) discussed the rate of convergence of the $(r \times r)$ elements of the signal correlation (coherence) matrix $\boldsymbol{S}$ to their asymptotic values.

13.2.3 Spatial smoothing

Spatial smoothing has frequently been advocated as a means of dealing with the problem of coherent signals. It is very closely related to the concept of space stepping in a distorted wavefield introduced in Section 10.4.2, where it was shown that a particular wavefield could be resolved from a single frame of data. Clearly, no use was made of the relative fading between rays in this example and the method is therefore immune to the problem of coherency.

The concept of spatial smoothing is best explained with a specific example. Consider a uniformly-spaced linear array of 12 elements. It can be mentally divided into two uniformly-spaced sub-arrays of 11 elements, namely 1–11 and 2–12. These sub-arrays are identical apart from a displacement in absolute position, which is of no significance when determining angles of arrival from distant sources. We can also form three sub-arrays of 10 elements, four sub-arrays of 9 elements, and so on, preserving the uniform spacing in each case.

Instead of forming one (12 × 12) signal covariance matrix for the whole sensor array we can form an average of two (11 × 11) matrices, or three (10 × 10) matrices, and so on, choosing whatever seem to be the best tactics. To avoid giving undue weight to certain sensors, combinations such as 1–6 with 7–12, or 1–4 with 5–8 and 9–12, can be used. Whatever the choice, the resulting covariance matrix is said to have been spatially smoothed. When f frames of data are available we can also form a time average of the f spatially-smoothed matrices.

In passing, it may be noted that a third form of smoothing, in the frequency domain, can be used when dealing with wideband sources and noise (Shan, Wax and Kailath, 1984; Wang and Kaveh, 1985; Zhu and Wang 1988*a*, 1989; Cadzow, Kim and Shiue, 1989; Kim, Cadzow and Park, 1989). In effect, data sets collected at two or more frequencies are transformed or 'focused' to a single central frequency. Lee, Er and Evans (1989) used an alternative two-stage approach for wideband sources, based on the use of the minimum eigenvector (rather than the full noise subspace eigenvectors) as a weight vector.

The effect of spatial smoothing on the signal coherence matrix $\boldsymbol{S}$ is that the elements $s_{ij} \to 0$ for $i \neq j$. Thus by spatial smoothing the covariance matrix $\boldsymbol{R}$ averages to the result expected for a completely incoherent wavefield. A mathematical proof of this result may be found in a paper by Shan, Wax and Kailath (1985).

Spatial smoothing is usually illustrated with simulated problems for uniformly-spaced linear arrays. Such arrays are restricted to the determination of cone angle rather than azimuth and elevation angles separately. However, the same technique is applicable also to two-dimensional and three-dimensional sensor arrays (see Fig. 13.1). The requirement is to be able to perform space steps between identical sub-arrays. In principle it is not necessary for the sub-arrays, whether linear, planar or three-dimensional, to be uniformly spaced, although such a condition will normally greatly increase the number of sub-arrays that can be used and may also simplify the calculations required. The appropriate theory for

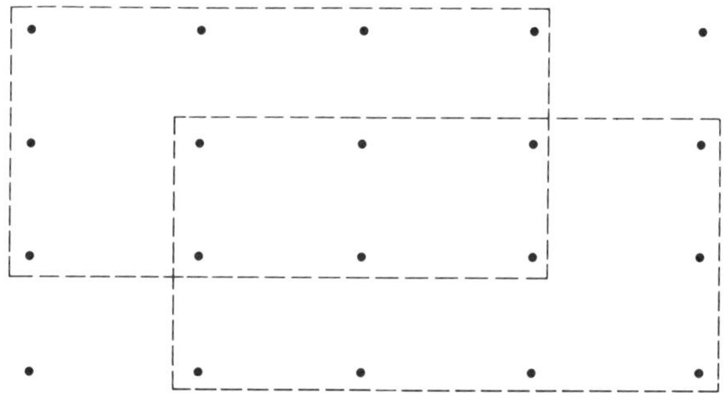

Fig. 13.1 Two-dimensional array of 20 sensors
● sensors
– – – – – boundaries of two identical sub-arrays of 12 sensors

uniformly-spaced arrays in two and three dimensions may be found in a paper by Yeh, Lee and Chen (1989).

The signals in all elements are usually used in the solution, although some versions of the smoothing over sub-arrays give increased weighting to the more central elements. Cadzow, Kim and Shiue (1989) drew attention to the undesirable feature of a loss of effective aperture. However, some other authors recommend that spatial smoothing be incorporated as a matter of routine into all processing methods such as MUSIC that break down for coherent signals. The best choice of sub-array size for use with MUSIC was discussed by Attia (1987), who pointed out that spatial smoothing also tends to reduce the processing time required, and should be used even if signal coherence is of no concern. Reddi (1987*a*) showed that the number of sub-arrays required to ensure a non-singular covariance matrix is $(r + 1)$, where r is the nullity of the source covariance matrix. A corrected proof was given by Reddi (1987*b*).

Coherent interference causes problems with adaptive arrays. Cantoni and Godara (1980) proposed an adaptive algorithm for resolving the directions of all sources, whether correlated or not. The use of spatial smoothing in this context has been analysed by Yeh, Chiu and Pei (1989).

13.2.4 Other methods

Certain processing methods have been specially developed to resolve coherent signals by means other that spatial smoothing. We have already seen in Section 12.3.1 that certain variants of the Maximum Likelihood Method can cope with coherent arrivals (Hudson, 1985; Ziskind and Wax, 1988*a*; Brandwood 1989; Oh and Un 1989). The effects of diverse polarisations can be included in the analysis (Ziskind and Wax 1988*b*). For any method that makes use of the signal covariance matrix it may still be helpful to use an appropriate form of spatial averaging to smooth ragged data and to reduce the processing load.

Other methods described in the literature are as follows. A Toeplitz approximation method (TAM), designed for robustness in an arbitrary ambient noise environment, was suggested by Kung, Lo and Foka (1986). Takao and Kikuma (1987) suggested an adaptive spatial averaging technique (ASAT) in which the covariance matrices from various subarrays were adaptively averaged to produce a Toeplitz matrix; the aim was to achieve satisfactory operation of an adaptive array when coherent signals are present. Tarres and Fernandez (1989) refined the Takao and Kikuma theory by linearly transforming the matrices so that each contains only one ray.

Zoltowski and Haber (1986) replaced the usual DOA search vector by a linear combination of such vectors. The procedure described for finding the proper linear combination leads to the disadvantage of a multidimensional search, but in angle only. The spectral estimator is plotted as a vertical co-ordinate above the θ_d, θ_s plane, where θ_d and θ_s are the arrival angles of the direct and specularly-reflected rays; a single ray (no specular component) gives rise to a ridge, whereas each pair of rays gives rise to a peak, as illustrated in Fig. 13.2. Bugnon and Mohler (1986) suggested a non-linear alternative to spatial smoothing. The shifted array correlation technique of Struckman (1988) solves for r angles of arrival by correlating r shifted array vectors with testing vectors.

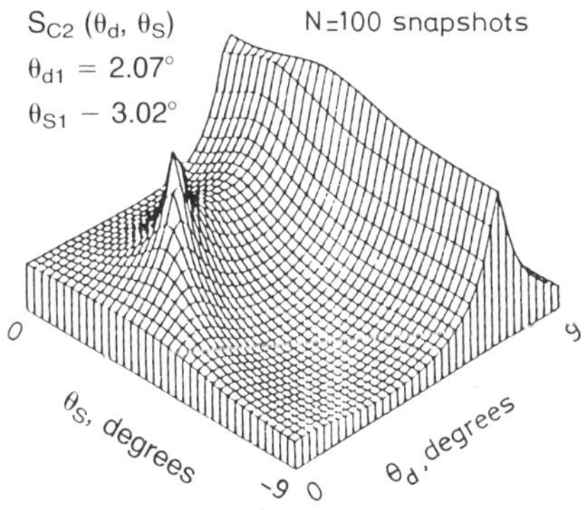

Fig. 13.2 Plot of spectral estimator S_{c2} for direct and specularly-reflected rays

True angles: coherent pair of rays from $\theta_d = 2°$, $\theta_s = -3°$
single (incoherent) ray from $\theta_d = 7°$

Based on Fig. 5c in a paper by Zoltowski and Haber, IEEE. Trans., ***AP-34,*** *pp. 1069–1079*

In the method investigated by Shan, Paulraj and Kailath (1986), the smoothed rank profile of a telescoping series of matrices obtained by averaging smaller and smaller principal diagonal submatrices was used to deduce the source coherency structure. The authors proposed a statistical procedure to verify problem solvability in the presence of coherent signals. Prasad and Chandna (1989) suggested that the chances of success with this method can be augmented by the incorporation of forward–backward averaging techniques.

Pillai and Kwon (1989*a*, *b*) also advocated a forward–backward smoothing scheme. In addition to the forward sub-arrays, this scheme makes use of complex-conjugated backward sub-arrays. The authors showed that the minimum number of sensors required to resolve r arrivals is the integer part of $3r/2$, as compared with $2r$ for conventional spatial smoothing.

Chung and Robertson (1988) described a method in which the signal subspace rank is determined; the method works even when sensor positions are not known precisely.

Symmetry considerations for a uniformly-spaced array of n elements underlie the method proposed by Pillai and Lee (1989). The array is effectively extended in aperture by considering what signals would be received on a further $(n - 1)$ elements added beyond one end element; this element becomes the centre of symmetry of the new configuration. It is then possible to process the array output correlations to generate a new set of quantities, functionally similar to the actual correlations in an incoherent environment. From these quantities a Toeplitz matrix is formed which leads on to the angle solution.

13.2.5 Summary on coherent signals

There is an extensive literature on the problem of coherent signals and a number of solution methods are available. Broadly, space-stepping and spatial smoothing methods as described to date have been restricted to uniformly-spaced arrays, although we have noted that the real requirement is for an array that contains two or more identical sub-arrays.

A number of papers contain lucid surveys and comparisons of more than one method of dealing with coherent signals; examples are Hudson (1985), Zoltowski and Haber (1986), Chung and Robertson (1988) and Cadzow, Kim and Shiue (1989). However, the individual papers referred to in Sections 13.2.3 and 13.2.4 should be consulted for a fuller appreciation of the assumptions, restrictions, underlying models and range of problems that can be solved with each of the methods: the finer details tend to be context-specific and have been omitted here.

It seems prudent to use some form of spatial smoothing in the initial analysis of data, in case coherent signals are present. If no evidence is found of the presence of coherent rays, the option of re-processing the data by some other method becomes available.

13.3 Adjusting and smoothing the initial data

13.3.1 Objectives

In addition to spatial smoothing, other forms of data smoothing are commonly used. A number of investigators have considered ways of preprocessing the raw data and/or the signal covariance matrix, in order to achieve one or more of the following objectives:

(*a*) To improve conformance to an assumed model

(*b*) To overcome the problem of the near-singularity of the covariance matrix, e.g. by deliberately adding simulated noise
(*c*) To correct for missing or clearly-corrupted data from a sensor (e.g. when using an algorithm that depends on uniform spacing)
(*d*) To reduce the computational load, for example by simplifying or avoiding the calculation of eigenvalues
(*e*) To improve resolution performance.

The motives for forcing $\boldsymbol{R}$ into an exact Toeplitz form, for example, may be to obtain a supposed statistical improvement by ensuring conformance to a model (assumed from prior knowledge to be a realistic model) and/or to simplify the calculation of eigenvalues.

13.3.2 Noise model

The data to be analysed will be affected by external noise (sometimes called convolutive noise) received at the senor elements, internal noise from the receiver system and, as part of the internal noise, measurement errors arising from such causes as channel imbalance and measurement quantisation. External noise arriving from a single well-defined direction may be classified as an interfering signal, so it is normal to assume that noise energy does not arrive from discrete directions; however, there is no reason to suppose that the noise is uniformly distributed in azimuth or elevation.

There will be a noise power contribution to each element of the covariance matrix $\boldsymbol{R}$. This contribution will not be zero, as with noiseless data, and cannot be assumed to conform exactly to any assumed noise model. For a single snapshot we certainly cannot assume that the noise component e_{ij}, $i \neq j$, has been averaged over all possible phase differences between sensors i and j; it is only for large samples of frames and distributed noise with no predominant direction of arrival that we can begin to have any confidence in the model.

It is usual to assume that the noise power probability distribution is of Gaussian form. The term Wishart noise is sometimes used for Gaussian noise from a finite sample. In narrow-band DF applications, noise is normally also assumed to be white, i.e. of uniform power across the pass-band of the receiving system. Some authors, e.g. D. H. Johnson (1982), refer to a similar concept of spatial whiteness, meaning that the noise is of uniform power across the elements of the array and uncorrelated from element to element. Spatial coherence of the background noise was considered by Bienvenu (1979) and Bienvenu and Kopp (1980).

Schmidt (1981) stressed the significance of the exact noise model adopted. It is, of course, possible to solve for noise parameters as well as ray parameters, perhaps by a method of successive refinement in which the ray solution is found first, then the noise solution, then an improved ray solution and so on.

13.3.3 Pre-whitening

The covariance matrix $\boldsymbol{R}$ can be expressed in the form

$$\boldsymbol{R} = \boldsymbol{R}_S + \boldsymbol{R}_N$$

where $\boldsymbol{R}_S$ is the matrix in the absence of noise and $\boldsymbol{R}_N$ represents the contribution of noise components to **R**. The diagonal elements of **R** from measurements made over a large sample of frames should be equal; if not, a (spatial) whitening process to force equality is sometimes used (Brandwood, 1987*a*).

The near-singularity of an ill-conditioned matrix from a small number of frames is sometimes stabilised by effectively increasing the contribution due to noise; a fixed value is added to all the principal diagonal elements of the matrix (Brandwood, 1985). Conversely, the effects of noise can usefully be subtracted when methods other than eigenanalysis are to be used (Brandwood, 1987*b*).

Farrier, Jeffries and Mardani (1988) recommended pre-whitening of correlated noise fields and showed how to predict the performance of the MUSIC algorithm when this action has been taken. They considered the possibility of measuring $\boldsymbol{R}_N$ as a 'background' covariance matrix at moments when the signals are absent, or on a slightly different frequency. However, they pointed out that the number of such data samples is likely to be restricted and that whitening is not as simple as might appear at first sight; performance can be degraded if the background statistics are poor.

The representation of noise fields by a small number of parameters was investigated by Le Cadre (1989), who suggested a method for estimating the noise correlation matrix from the overall signal-plus-noise correlation matrix; the noise parameters can then be used to whiten the observations.

13.3.4 Corrections for instrumental effects

Corrections for any known errors arising from instrumental imperfections should be made where possible.

The effects of mutual coupling on radiation patterns were considered in Section 5.6. Shau, Adams and Sarkar (1988) stated that the performance of a foward–backward linear predictor could be improved by correcting for the effects of mutual coupling between sensor elements. Yeh, Leou and Ucci (1989) considered the distorting effect of coupling on the eigenstructure of the covariance matrix and proposed methods for overcoming the problem; for coherent sources, the distortion of the signal components of the covariance matrix is eliminated before spatial smoothing is applied. Computer simulations showed that the effects of mutual coupling could be counteracted successfully.

The exact position of sensor elements used on a moving, flexing platform will be subject to small fluctuations and to the uncertainties introduced by navigational errors. A sonar array towed behind a ship will not maintain an absolutely constant configuration. In circumstances such as these there is an interest in the effects of small errors in the assumed sensor positions on the measured angles. It was shown by Seymour, Cowan and Grant (1987) that high resolution techniques such as MUSIC are very sensitive to such errors. A calibration procedure, in which bearings from known transmitters are anlaysed to determine the true sensor positions, was shown by means of simulations to lead to an improvement. Seymour, Grant and Cowan (1989) adopted a normal distribution for the positional error of each sensor, with mean zero and variance σ_p^2; they showed that, for small errors, the variance of the errors of angles determined with MUSIC is directly proportional to σ_p^2.

The effects in a sonar environment of random sensor motions were investigated by Wong, Walker and Niezgoda (1988). The Cramer–Rao lower bound on variances (see Section 13.5) under the influence of this source of error was determined and it was shown that angle estimates obtained with the MUSIC algorithm conformed fairly closely to the bound for small sensor motions. Sensor position errors that remain stationary over the period of observation have a bigger effect than random movements without correlation from frame to frame. Luo (1990) has drawn attention to the need for corrections to Section 3.1 of this paper.

Weiss and Friedlander (1988) described an exact maximum-likelihood method for simultaneous estimation of arrival angles and sensor locations; the array must be non-linear, with the location of one sensor and the direction of another known. Two-dimensional positioning errors of Gaussian distribution in a linear array intended to be uniformly spaced were investigated by Chen, Lee and Yeh (1988), who adjusted the covariance matrix to Toeplitz form; the effects of the errors are thereby reduced without the use of a calibration procedure. Position perturbations were stated by Zhu and Wang (1988*b*) to be more serious than pattern perturbations, in the sense that raising the SNR offers only limited compensation in the former case.

Friedlander and Weiss (1988) considered the effects on eigenanalysis methods of gain and phase errors in the measuring equipment and showed how to solve for calibration parameters as well as the signal parameters. They used an iterative procedure, solving first for the angles of arrival with no calibration corrections, then minimising an expression representing the scatter of the observations interpreted in terms of these angles; with comparatively small balance errors a typical problem should lead to a stable solution after about 2–6 iterations. A technique of dynamic phase correction was used by Cadzow, Arnold and Kocsis (1989) to determine corrections that 'sharpen up' the peaks of the spectral estimator.

13.3.5 Short cuts

Because the computer calculations required for repeated matrix manipulations can be time-consuming, many research teams have found short cuts or simplifications to reduce the computing load with little or no loss of resolving power. A few papers describing interesting approaches are briefly mentioned here, but the relative merits of various mathematical library routines are not discussed.

Brandwood's method of computing the MUSIC estimator without the use of eigenanalysis, and Schreiber's recursive procedure for updating the covariance matrix have already been referred to in Section 12.3.6.

A standard method of finding the inverse of the covariance matrix is to factorise it into lower triangular–diagonal form. For an $(n \times n)$ matrix, the number of multiplications is reduced from about n^3 to about $n^3/6$, according to Brandwood (1986). An alternative factorisation, trapezoidal–diagonal, is said by the same author to be more stable, and to require fewer multiplications if the number of rays is less than $n/3$.

Inversion and eigenanalysis are easier for a Toeplitz matrix than for the non-Toeplitz form. Toeplitz approximation methods (TAM), which exploit this feature by finding a Toeplitz form that approximates to the actual covariance matrix, have been widely used. For example, Beex and Fargues (1989) suggested a highly-parallel recursive/iterative procedure for the efficient decomposition of the matrix into its eigencomponents; their paper also contains many references to earlier work. A useful survey of various conjugate-gradient algorithms for finding the smallest eigenvalue and associated eigenvector of a symmetrical matrix may be found in a paper by Yang, Sarkar and Arvas (1989).

13.4 Number of rays

13.4.1 Need for objective criteria

Some of the superresolution algorithms require, as input, an estimate of the number of arriving rays within the passband of the receiver; the number is sometimes referred to as the ray (or signal) order. It is not normally known and a hypothesis, or series of hypotheses, must be made in order to proceed.

The problem of arriving at the correct estimate is by no means straightforward. For example, the difficulties of partitioning the eigenvalues of the covariance matrix into large and small categories were illustrated in various simulations examined by Johnson (1986). In many conditions of signal coherence, bearing separation, unequal signal strength and SNR, the effects of noise were seen in the merging of eigenvalues, making the separation of signal and noise subspaces difficult.

Whereas imprecise terms and subjective judgments can be used by a human analyst in deciding whether eigenvalues are 'large' or 'small', or whether a 'very low' ray amplitude represents a spurious arrival, or whether the fit of the solution to the original data is 'significantly' improved by solving for an additional ray, more objective criteria are desirable in any form of analysis and are essential in completely automated methods. We now consider statistical criteria that can be used in the analysis of imperfect data.

Liu and Nolte (1980) examined the problem of detecting a wideband acoustic signal from a known direction in the presence of noise. An optimum array detector is defined in the paper as one which points a beam at the wanted signal and a null at the main noise source. The authors obtained closed-form expressions for the probability of detection P_D and the probability of false alarms P_{FA} for the performance of several optimum and sub-optimum array detectors.

Although the context of Liu and Nolte's work was rather different from the radio direction finding problem with multiple arrivals from unknown directions, it is of interest that optimum detection performance was shown to depend on the eigenvectors of the covariance matrices of the received data vectors. Appropriate expressions for P_D and P_{FA} in the case of angle determinations may be found in Kaveh, Wang and Hung (1987).

13.4.2 Information-theoretic criteria

The problem of determining the number of signals present in a time series has been extensively studied and has led to the formulation of various criteria for model selection. Two approaches that have attracted particular attention in the context of superresolution are an information criterion (AIC) suggested by Akaike (1974) and the minimum description length (MDL) criterion of Schwarz (1978) and Rissanen (1983). The MDL principle asserts that for a given data set and a class of probabilistic models, one should choose the model that yields the shortest data description length as measured by the number of bits required to encode the data.

A treatment covering both criteria may be found in a paper by Zhang, Wong, Yip and Reilly (1989). The estimated number of arrivals m is chosen to minimise a function $J(m)$ that is the sum of a log-likelihood function $L(m)$ and a penalty function $p(m)$:

$$J(m) = L(m) + p(m)$$

The log-likelihood function for descending eigenvalues μ_1 to μ_n, where n is the number of elements, is

$$L(m) = f \ln \left\{ \left(\frac{1}{(n-m)} \sum_{i=m+1}^{n} \mu_i \right)^{n-m} \Bigg/ \prod_{i=m+1}^{n} \mu_i \right\}$$

The value of m is chosen to minimise $L(m)$. The penalty function for f observations and n elements is of the form

$$p(m) = \alpha(f)\,[m\,(2n - m)]$$

where $\alpha(f)$ is a penalty coefficient that is a constant function of the number of observations:

$$\alpha(f) = 1 \qquad \text{for AIC}$$

$$\alpha(f) = (1/2)\ln(f) \qquad \text{for MDL}$$

Hannan and Quinn (1979) suggested another form of the penalty coefficient:

$$\alpha(f) = (1/2)\ln\{\ln(f)\} \qquad \text{for HQ criterion}$$

The conclusions reached by Zhang, Wong, Yip and Reilly were based on asymptotic analysis for $f \to \infty$ and were confirmed to be in very close agreement with computer simulation results. The MDL criterion is more efficient than AIC in reducing the probability of false alarms (overestimation), which approaches zero for MDL as f increases. However, AIC is better than MDL in reducing the probability of missing a signal (underestimation). The performance of both can be improved with a better choice of $L(m)$, as suggested in a paper by Wong, Zhang, Reilly and Yip (1988).

Martin (1989) described an instance in which both AIC and MDL gave misleading results with actual noise. A further problem is that the basic information needed to calculate the criteria may not be readily available in a practical situation.

13.5 The Cramer–Rao Lower Bound on variances

13.5.1 Background

The Cramer–Rao Lower Bound is a standard statistical parameter calculated from the Fisher information matrix (Melsa and Cohn, 1978). We suppose that a set of parameters denoted by the vector θ are to be estimated from a set of consecutive measurements denoted by the vector **y**. Let the probability density function of **y** be denoted by f_θ (**y**). Then the Fisher information matrix J with elements

$$J_{k,1}(\chi) = E\left\{ \frac{\partial \log f_\theta(y)}{\partial \theta_k} \frac{\partial \log f_\theta(y)}{\partial \theta_1} \right\}$$

is calculated and the lower bound on the variance of the unknown parameters is given by

$$\text{var}(\theta) \geqslant J^{-1}(\theta)$$

The right-hand side of this equation is the CRLB. It sets a minimum theoretical value on the variances determined from a given set of input data. Any method that produces variances on or only just above this bound is judged to be good, in that the accuracy of the output is close to the limit of what can be achieved from the input.

13.5.2 Accuracy studies

To calculate the so-called exact value of the CRLB in the context of angle measurements from f frames of data, the symbols in these definitions must be replaced by appropriate analytic expressions. The terminology is initially confusing, in that the exact value must be calculated from the imperfect data and is therefore itself subject to uncertainty. Furthermore, the computational load is often reduced by using an approximation to the full expression for the exact CRLB. For small samples, the uncertainty in the CRLB means that its estimated value tends to be set conservatively low. The convergence of the calculated CRLB towards its asymptotic value as the sample size is increased is clearly of interest and is a major topic in the literature.

Details of the procedure for calculating the CRLB may be found in a number of papers. The appendix to a paper by Wang and Kaveh (1985) provides a helpful introduction. Jeffries and Farrier (1985) considered the asymptotic performance of MUSIC and of the method proposed by

Kumaresan and Tufts (1983); neither was clearly superior to the other. A general expression for the variance of angles determined by eigenvector methods, including MUSIC, was derived by Tressens *et al.* (1988) and compared with the CRLB. Optimality, i.e. variance very close to the lower bound, is achieved only if all the signal eigenvalues are much greater than the noise level.

A paper by Stoica and Nehorai (1989*a*) provided a comprehensive treatment of the expected behaviour of the CRLB for angles determined with MUSIC. The authors, by assuming the true number of arriving rays to be known or correctly determined, avoided one cause of wild solutions. The relationship of MUSIC to maximum likelihood estimators (MLE) was also considered. The MLE results do not achieve the CRLB for large samples, except for the impossible case of an infinite number of sensors. The MUSIC error variance does asymptotically approach zero provided the rays are uncorrelated and the number of sensors exceeds the number of rays; for correlated signals the variance may occasionally increase as the number of sensors is increased, but the general tendency is in the opposite direction. The scope of the investigation was extended in Stoica and Nehorai (1989*b*).

Pillai and Kwon (1989*b*) have also considered the performance of MUSIC-type estimators for correlated and coherent signals. They considered forward–backward and forward-only smoothing schemes, with MUSIC as a special case. The variance of angles from MUSIC is shown to approach zero for very large samples. Not surprisingly, the resolution limit in the coherent case can be substantially larger than in the uncorrelated case. To achieve the same limit, the number of snapshots (frames) must be increased by a factor of approximately

$$\left\{\left(\frac{1}{n\,\omega_d}\right)^2 - 1\right\}$$

where

$$\omega_d = \pi(\cos\theta_1 - \cos\theta_2)/2$$

$$n = \text{the number of sensors}$$

$$\theta_1, \theta_2 = \text{the arrival angles of the two rays}$$

The theoretical results in this paper were shown by the authors to be in close agreement with the results of simulations.

Some research papers concerned with more general problems than the specific case of angle determinations are also worth consulting. Several techniques for the calculation of the CRLB for ARMA models were described in a paper by Friedlander (1984). Porat and Friedlander (1986) used, for the calculation of the exact Fisher matrix, an efficient recursive method based on an algorithm for computing orthogonal polynomials of a Toeplitz matrix. The results were again applied to ARMA models.

The exact bound for autoregressive (AR) processes was derived by the same authors (Friedlander and Porat 1989). Their main conclusions were as follows. The exact Fisher matrix is the sum of a constant term plus the matrix appearing in a standard asymptotic expression for large samples. The exact CRLB can be larger or smaller than the value calculated from this standard approximation; for narrowband processes it tends to be considerably smaller than the asymptotic approximation, whereas the opposite is true for wideband processes. The results do not carry over into moving average and ARMA processes.

In principle, the CRLB theory provides a useful lower bound on accuracy, but the limit may not be easy to calculate.

13.6 State of development of superresolution

13.6.1 Resolving power

We now attempt an assessment of what has been achieved to date with superresolution processing techniques, which may include any of the additional tools mentioned in this chapter.

The first factor to be considered is resolution, if possible in absolute terms. The resolution limit for two incoherent rays depends on many factors, particularly array aperture, SNR and number of frames. The effects of aperture can be partly removed by expressing results in terms of beamwidth θ_B. Because a small improvement in resolution requires a comparatively large improvement in SNR, if other factors are held constant, it might be thought that, conversely, small changes of SNR within a practical range would have rather a small effect on results achieved. However, Gabriel (1980) found an approximately linear relationship between resolving power and SNR in dB, from about $0{\cdot}1\theta_B$ at 33 dB to $0{\cdot}97\theta_B$ at 0 dB.

Results from different papers show condiserable scatter because of differences of SNR and other factors, and it is difficult to state any 'rule of thumb'. A few examples are therefore quoted for illustrative purposes. Broadly, there seems to be no great difficulty in resolving at about $0{\cdot}5\theta_B$ from a single frame and at about half this limit from many frames, both under favourable conditions of SNR. However, it is possible to do better: Gabriel (1980) and Mayhan and Niro (1987) achieved $0{\cdot}1\theta_B$, while Du Fort (1983) reached $0{\cdot}030\theta_B$ and Alsup (1984) achieved $0{\cdot}025\theta_B$ in simulations for a two-dimensional array and was optimistic that $0{\cdot}01\theta_B$ could be reached. For coherent rays, Hudson (1985) suggested that about $0{\cdot}5\theta_B$ can be resolved, but not $0{\cdot}25\theta_B$. Struckman (1988) found that SNRs in excess of 50 dB would be required to resolve two coherent signals at $0{\cdot}1\theta_B$.

The resolution limit is often subconsciously visualised in terms of the azimuthal separation at zero elevation angle; however, rays very close or coincident in azimuth but widely separated in elevation can be resolved with an appropriate form of array.

13.6.2 Automated analysis

Although the subject of superresolution is a very lively research topic, it is by no means clear that the subject has developed to the point at which a fully-automated analysis program can be used to cope with any form of problem, including those arising in routine operational DF. The alternative is for a human analyst to make intelligent use of a variety of lines of attack, often using more than one method on the same data.

Even the most automatic of analysis programs, designed to cope with any type of problem, requires inputs. Clarke (1986) emphasised that a unique solution or target model cannot be inferred from observed data without some additional assumptions or background knowledge. He stated that such knowledge may take many forms and it is often not easy to know how to use it, or to appreciate what assumptions underlie the use of a particular algorithm. Prior information, for example about the receiver noise level or the likely number of rays, should be incorporated into the solution method if possible.

It may be questioned whether it is realistic to expect the same algorithm to be the best in all contexts; for example wideband sonar with poor SNRs, radar with rapidly moving targets and jammers but known transmitted waveforms, mapping of many discrete sources in radioastronomy, and for a whole range of problems within a given context.

There may, for example, be particular processing tricks that we cannot do without in special circumstances but which are a disadvantage in other circumstances. Some form of spatial smoothing may be required to resolve two or more completely coherent signals. A matrix inversion that can be carried out relatively easily for an HF array of (say) 10 elements may be totally impracticable for a 10 000-element radar array. For an adaptive array feeding a single-channel receiver, a robust algorithm leading to rapid convergence to a desirable radiation pattern is more important than an unstable optimal solution.

It must also be appreciated that no algorithm, however powerful or sophisticated, can overcome the problem of masquerading (Section 11.7), in which two quite distinct problems can generate virtually identical data on every frame of a sample. While some algorithms are better than others, superresolution provides no magic solution to the problem of inadequate resolving power arising from inadequate aperture. The following quotation from a paper by D. H. Johnson (1982) is apposite: 'The enhanced resolution algorithms provide incrementally improved performance but do not implement new principles which defeat the fundamental limitations of aperture sampling'.

13.6.3 Choice of estimator

Subject to the above reservations, the best choice of estimator for some specified range of problems is clearly a question that arouses considerable interest. In recent years MUSIC has been widely regarded as the best of the superresolution algorithms for use in normal circumstances against non-coherent rays and the norm by which other algorithms are judged; see, for example, Paulraj, Roy and Kailath (1986) and R. L. Johnson and Miner (1986).

The author sees no reason to disagree with this conclusion, provided the term MUSIC is taken to include the most recent improvements and refinements to the basic method. However, some other recent developments such as the Alternating Projection and MODE algorithms described in Chapter 12 appear to provide a superior performance in particular circumstances. In addition, methods have been found for simplifying the MUSIC calculations with only marginal loss of performance.

MUSIC can also be used, with spatial smoothing, to resolve coherent arrivals. Some of the more recent methods specially developed for this situation, for example the ASRP method suggested by Prasad and Chandna (1989), appear to extend the range of problems that can be solved.

13.6.4 Design of arrays

One of the important tools for angle determinations is the sensor array. In the context of HF direction finding, there is a natural desire to incorporate superresolution techniques into existing wide aperture systems, rather than make a major new investment in hardware. Nevertheless, the question of how a new array should be designed to exploit superresolution effectively deserves consideration.

Certain general principles apply to any type of direction finder, whether or not superresolution techniques are to be used. Aperture is clearly of prime importance, and the elements must be deployed in two or three dimensions if the total angle of arrival of each ray is to be determined. Wide inter-element spacings have the desirable features of reducing total costs for a given aperture, reducing unwanted mutual coupling effects and, for spacings greater than about $0{\cdot}5\lambda$, producing more independent samples of the wavefield than elements at narrower spacings. However, a disadvantage of wide spacings is that ambiguities appear for inter-element spacings greater than $0{\cdot}5\lambda$. The requirements are therefore contradictory, as pointed out by Mayhan and Niro (1987). The problem becomes worse for a wide-band array, or for a narrow-band array intended to operate over a substantial frequency band.

For wavefront analysis methods in general, including the modern methods of superresolution, the digital techniques normally employed require the use of multi-channel receiving systems. A further advantage of wide inter-element spacings for a given aperture is then the reduction in the number of channels required. However, the number of sensor elements helps to determine the theoretical upper limit on the number of arriving rays that can be resolved. Above (say) 20 elements this may not be a significant practical limitation in the HF band, but arrays of less than (say) six elements have more limited capabilities.

Certain methods of solution, such as the space-stepping methods of the Prony type (see Section 10.4.1) and the Kumaresan–Tufts method (Section 12.3.6) can be used only with uniformly-spaced arrays. For others, such as MUSIC, simplifications that shorten the calculations apply when the spacings are equal. Spatial smoothing, the most common line of attack on the problem of coherent rays, also requires uniform spacing. These considerations may well outweigh the economies that would be achieved with an array of the minimum-redundancy type, with non-uniform spacing.

The design of superresolution direction finders is considered in only a few published papers. An early paper by Bain (1956) on wide-aperture radio direction finders for the HF band contains a brief general discussion of 'systems in which the equations of the wave interference field are solved'. Burtnyk and Wolfe (1967) considered the balance to be struck between temporal and spatial diversity, but for a conventional beam-forming system.

Lang, Duckworth and McClellan (1981) studied the design of minimum redundancy arrays to resolve incoherent wavefields with MEM and MLM methods. Schmidt (1981) considered the form of a planar array, occupying a given circular area, that gives the best Cramer–Rao accuracy; he showed that elements spaced uniformly on the circular boundary achieve this result. The effects of array design on the performance of the MUSIC algorithm were also considered by Spielman, Paulraj and Kailath (1986).

The most systematic approach to the design of systems for use with covariance-based direction-finding methods is to be found in a paper by R. L. Johnson (1988). After considering a number of techniques for dealing with coherent signals, Johnson restricted his attention to uniformly-spaced planar arrays in order to exploit spatial smoothing. Subject to this imposed constraint, he developed a procedure for minimising the number of elements and the dimensionality of the spatial covariance matrix. The design procedure begins with a specification of the worst-case signal environment likely to be encountered and leads to a variational calculus problem of constrained optimisation. The results of a number of simulations for two coherent rays impinging on an array are presented in graphical form in order to delimit the area of acceptable design. One solution in terms of an array of finite length always exists for non-zero ray separation, no matter how small the ray separation becomes, but the existence of other local minima has not been disproved: the author stated that the form of the design contours is not yet fully understood.

Wax and Ziskind (1989) considered the number of elements required in a general array (i.e. one not necessarily uniformly spaced) for a specified problem to have a unique solution. A sufficient condition is

$$r < (n + \eta)/2$$

where η is the rank of the covariance matrix. A weaker condition that is nearly always satisfied by a sample of data is

$$r < 2\eta n/(2\eta + 1)$$

Gabriel (1989) presented the design considerations for the special case of a sparse array with restricted degrees of freedom to cover a restricted angular sector.

13.6.5 Cross-fertilisation

No assessment of the current state of superresolution would be complete without a mention of the vigour with which research is being pursued in a number of related fields. New papers of high quality are continuing to appear regularly.

The quest for an optimum spectral estimator has already generated a great deal of fruitful research on signal processing methods. Many papers on spectrum analysis, time series analysis, maximum entropy methods and adaptive antenna arrays contain information, theory and results that are relevant to the direction-finding problem and there has been a cross-fertilisation of ideas between research groups working in different fields.

For readers who wish to study the extent to which the different fields are connected, the following papers are recommended: In a tutorial paper, Gabriel (1980) discussed the application of maximum entropy spectrum analysis to superresolution techniques for adaptive arrays. Reviews by Kay and Marple (1981), McClellan (1982) and Marple (1989) provide insights into the relationship with modern spectrum analysis. A textbook by Hudson (1981) and a tutorial paper by Griffiths (1983) provide good surveys of work on adaptive antennas.

References

AKAIKE, H. (1974): 'A new look at the statistical model identification', *IEEE Trans.*, **AC-19,** pp. 716–723

ALSUP, J.M. (1984): 'High resolution techniques for two-dimensional estimation of angle-of-arrival for planar arrays'. ICASSP 84, San Diego 19–21 Mar 1984, vol. 3, pp. 39B.4/1–4 (IEEE, New York)

ATTIA, E.H. (1987): 'Efficient computation of the MUSIC algorithm as applied to a low-angle elevation estimation problem in a severe multipath environment'. ICASSP 87, Dallas, 6–9 April, vol. 3, pp. 1760–1763 (IEEE, New York)

BAIN, W.C. (1956): 'The theoretical design of direction-finding systems for high frequencies', *Proc. IEE Pt. B*, **103,** pp. 113–119

BARTON, P. (1983): 'Direction finding using an adaptive null tracker', *Proc. IEE Pts. F and H*, **130,** pp. 78–83

BEEX, A.A., and FARGUES, M.P. (1989): 'Highly parallel recursive/iterative Toeplitz eigenspace decomposition', *IEEE Trans.*, **ASSP-37,** pp. 1765–1768

BIENVENU, G. (1979): 'Influence of the spatial coherence of the background noise on high resolution passive methods'. ICASSP 79, Washington, 2–4 April, pp. 306–309 (IEEE, New York)

BIENVENU, G., and KOPP, L. (1980): 'Adaptivity to background noise spatial coherence for high resolution passive methods', ICASSP 80, Denver, 9–11 April, pp. 307–310 (IEEE, New York)

BRANDWOOD, D.H. (1985): 'Adaptive array techniques for direction finding'. Proc. ICAP 85, IEE. Conf. Publn. 248, pp. 246–250

BRANDWOOD, D.H. (1986): 'Stable matrix inverse method for array DF using trapezoidal factorisation', *Electronics Letters*, **22,** p. 177

BRANDWOOD, D.H. (1987*a*): 'Noise–space projection: MUSIC without eigenvectors', *Proc. IEE Pt. H*, **134,** pp. 303–309

BRANDWOOD, D.H. (1987*b*): 'Noise space projection and economical methods for obtaining the MUSIC DF function without eigenanalysis'. ICAP 87, IEE Conf. Publn. 274, pp. 415–418

BRANDWOOD, D.H. (1989): 'Multiple signal DF with coherent signals using a general array'. Passive DF Colloquium, IEE Digest No. 1989/16, pp. 5/1–5/4

BUGNON, F.J., and MOHLER, R.R. (1986): 'Nonlinear signal processing for direction finding', *Proc. IEEE* Conf. on Decision and Control, Athens, 10–12 December, Vol. 2, pp. 1054–1055 (IEEE, New York)

BURTNYK, N., and WOLFE, J.L. (1967): 'Direction-finding performance of a wide-aperture sampling linear array', *Proc. IEE*, **114,** pp. 581–588

CADZOW, J.A., ARNOLD, H., and KOCSIS, S. (1989): 'Direction finding using dynamic phase correction', *Proc. Soc. Photo-optical Instrumentation Engineers*, **975,** pp. 116–125

CADZOW, J.A., KIM, Y.S., and SHIUE, D.C. (1989): 'General direction-of-arrival estimation: a signal subspace approach', *IEEE Trans.*, **AES-25,** pp. 31–47

CANTONI, A., and GODARA, L.C. (1980): 'Resolving the directions of sources in a correlated field incident on an array', *J. Acoust. Soc. Am.*, **67,** pp. 1247–1255

CHEN, Y.M., LEE, J.H., and YEH, C.C. (1988): 'Calibration-free bearing estimation for arrays with randomly perturbed sensor locations'. ICASSP 88, New York, 11–14 April, **5,** pp. 2917–2920 (IEEE, New York)

CHUNG, J.C., and ROBERTSON, A.W. (1988): 'Resolution of coherent signals using random arrays', ICASSP 88, New York, 11–14 April, **5**, pp. 2678–2680 (IEEE, New York)

CLARKE, I.J. (1986): 'Comparison of advanced signal processing algorithms'. Colloquium on direction finding systems in their operating environment, IEE Digest No. 1986/10, pp. 9/1–9/9

DU FORT, E.C. (1983): 'High resolution emitter direction finding using a phased array antenna', *IEEE Trans.*, **ASSP-31,** pp. 1409–1416

FARRIER, D.R., JEFFRIES, D.J., and MARDANI, R. (1988): 'Theoretical performance prediction of the MUSIC algorithm', *Proc. IEE Pt. F*, **135,** pp. 216–224

FRIEDLANDER, B. (1984): 'On the computation of the Cramer–Rao bound for ARMA parameter estimation', *IEEE Trans.*, **ASSP-32,** pp. 721–727

FRIEDLANDER, B., and PORAT, B. (1989): 'The exact Cramer–Rao bound for Gaussian autoregressive processes', *IEEE Trans.*, **AES-25,** pp. 3–7

FRIEDLANDER, B., and WEISS, A.J. (1988): 'Eigenstructure methods for direction finding with sensor gain and phase uncertainties'. ICASSP 88, New York, 11–14 April, **5,** pp. 2681–2684 (IEEE, New York)

GABRIEL, W.F. (1980): 'Spectral analysis and adaptive array superresolution techniques', *Proc. IEEE*, **68,** pp. 654–666

GABRIEL, W.F. (1989): 'Large-aperture sparse array antenna systems of moderate bandwidth for multiple emitter location', *IEEE Trans.*, **AP-37,** pp. 16–29

GRIFFITHS, J.W.R. (1983): 'Adaptive array processing – a tutorial', *Proc. IEE Pts. F and H*, **130,** pp. 3–16

HANNAN, E.J., and QUINN, B.G. (1979): 'The determination of the order of an autoregression', *J. Roy. Stat. Soc.* Series B, **41,** pp. 190–195

HUDSON, J.E. (1981): 'Adaptive array principles' (Peter Peregrinus)

HUDSON, J.E. (1985): 'Fast near-optimum resolution of coherent sources by line arrays'. ICAP 85, IEE. Conf. Publn. 248, pp. 237–240

JEFFRIES, D.J., and FARRIER, D.R. (1985): 'Asymptotic results for eigenvector methods', *Proc. IEE Pt. F*, **132,** pp. 589–594

JOHNSON, D.H. (1982): 'The application of spectral estimation methods to bearing estimation problems', *Proc. IEEE*, **70,** pp. 1018–1028

JOHNSON, R.L. (1986): 'Eigenvector matrix partition and radio direction finding performance', *IEEE Trans.*, **AP-34,** pp. 985–991

JOHNSON, R.L. (1988): 'Antenna array design for covariance-based direction-finding methods', *IEEE Trans.*, **AP-36,** pp. 1537–1544

JOHNSON, R.L., and MINER, G.E. (1986): 'Comparison of superresolution algorithms for radio direction finding', *IEEE Trans.*, **AES-22,** pp. 432–442

KAVEH, M., WANG, H. and HUNG, H. (1987): 'On the theoretical performance of a class of estimators of the number of narrow-band sources', *IEEE Trans.*, **ASSP-35,** pp. 1350–1352

KAY, S.M., and MARPLE, S.L. (1981): 'Spectral analysis – a modern perspective', *Proc. IEEE*, **69,** pp. 1380–1419

KIM, Y.S., CADZOW, J.A., and PARK, H.K. (1989): 'Signal enhancement approach for high resolution of multiple broadband incoherent sources'. ICASSP 89, Glasgow, 25–28 April, Vol. 4, pp. 2617–2620 (IEEE, New York)

KUMARESAN, R., and TUFTS, D.W. (1983): 'Estimating the angles of arrival of multiple plane waves', *IEEE Trans.*, **AES-19,** pp. 134–139

KUNG, S.Y., LO, C.K., and FOKA, R. (1986): 'A Toeplitz approximation approach to coherent source direction finding'. ICASSP 86, Tokyo, 7–11 April, Vol. 1. pp. 193–196 (IEEE, New York)

LANG, S.W., DUCKWORTH, G.L., and McCLELLAN, J.H. (1981): 'Array design for MEM and MLM processing'. ICASSP 81, Atlanta, 30 March—1 April, Vol. 1, pp. 145–148 (IEEE, New York)

LE CADRE, J.P. (1989): 'Parametric methods for spatial signal processing in the presence of unknown colored noise fields', *IEEE Trans.*, **ASSP-37,** pp. 965–983

LEE, C.S., ER, M.H., and EVANS, A.J. (1989): 'Simple and fast method for direction-finding of multiple wideband sources', *Electronics Letters*, **25,** pp. 1522–1523

LIU, S.C.C., and NOLTE, L.W. (1980): 'Performance evaluation of array processors for detecting Gaussian acoustic signals', *IEEE Trans.*, **ASSP-28,** pp. 328–333

LUO FA-LONG (1990): 'Comments on the effects of random sensor motion on bearing estimation by the MUSIC algorithm', *Proc IEE Pt. F*, **137**, pp. 289–290

McCLELLAN, J.H. (1982): 'Multidimensional spectral estimation', *Proc. IEEE*, **70,** pp. 1029–1039

MARPLE, S.L. (1989): 'A tutorial overview of modern spectral estimation'. ICASSP 89, Glasgow, 25–28 April, Vol. 4, pp. 2152–2157 (IEEE, New York)

MARTIN, G.E. (1989): 'Signal subspace processing of experimental radio data', *Proc. Soc. Photo-optical Instrumentation Engineers*, **975,** pp. 101–107

MAYHAN, J.T., and NIRO, L. (1987): 'Spatial spectral estimation using multiple beam antennas', *IEEE Trans.*, **AP-35,** pp. 897–906

MELSA, M.A., and COHN, D.L. (1978): 'Decision and estimation theory', (McGraw Hill), pp. 231–235

OH, S.K., and UN, C.K. (1989): 'Computationally efficient maximum likelihood algorithm for localising multiple sources', *Electronics Letters*, **25,** pp. 1540–1542

PAULRAJ, A., ROY, R., and KAILATH, T. (1986): 'Estimation of signal parameters via rotational invariance techniques – ESPRIT'. 19th Asilomar Conference, 6 Nov. 1985, (IEEE. Comput. Soc. Press) pp. 83–89

PILLAI, S.U., and KWON, B.H. (1989*a*): 'Forward/backward spatial smoothing techniques for coherent signal identification', *IEEE Trans.*, **ASSP-37,** pp. 8–15

PILLAI, S.U., and KWON, B.H. (1989*b*): 'Performance analysis of MUSIC-type high resolution estimators for direction finding in correlated and coherent scenes', *IEEE Trans.*, **ASSP-37,** pp. 1176–1189

PILLAI, S.U., and LEE, Y. (1989): 'Coherent signal classification using symmetry considerations', *IEEE Trans.*, **ASSP-37,** pp. 135–138

PORAT, B., and FRIEDLANDER, B. (1986): 'Computation of the exact information matrix of Gaussian time series with stationary random components', *IEEE Trans.*, **ASSP-34,** pp. 118–130

PRASAD, S., and CHANDNA, B. (1989): 'An augmented smoothed rank profile algorithm for determination of source coherency structure', *IEEE Trans.*, **ASSP-37,** pp. 1144–1146

REDDI, S.S. (1987*a*): 'On a spatial smoothing technique for multiple source location', *IEEE Trans.*, **ASSP-35,** p. 709

REDDI, S.S. (1987*b*): 'Corrections and modifications to "On a spatial smoothing technique for multiple source location"', *IEEE Trans.*, **ASSP-35,** pp. 1352–1353

RISSANEN, J. (1983): 'A universal prior for the integers and estimation by minimum descriptive length', *Ann. Stat.*, **11,** pp. 416–431

SCHMIDT, R.O. (1981): 'A signal subspace approach to multiple emitter location and spectral estimation'. PhD Thesis, Stanford University, CA

SCHWARZ, G. (1978): 'Estimating the dimension of a model', *Ann. Stat.*, **6,** pp. 461–464

SEYMOUR, L.P.H.K., COWAN, C.F.N., and GRANT, P.M. (1987): 'A new calibration algorithm to detect sensor positioning errors in direction finding arrays'. Colloquium on Digital Signal Processing, IEE Digest 14, pp.10/1–4

SEYMOUR, L.P.H.K., GRANT, P.M., and COWAN, C.F.N. (1989): 'The effects of sensor positioning errors on bearing estimation'. Passive DF Colloquium, IEE Digest No. 89/16, pp. 9/1–8

SHAN, T.J., PAULRAJ, A., and KAILATH, T. (1986): 'On smoothed rank profile tests in eigenstructure approach to directions-of-arrival estimation'. ICASSP 86, Tokyo, 7–11 April, pp. 1905–1908 (IEEE, New York)

SHAN, T.J., WAX, M., and KAILATH, T., (1984): 'Spatial smoothing approach for location estimation of coherent sources.' Conf. Record, 17th Asilomar Conf. on Circuits, Systems and Computers, Santa Clara, 31 October–2 November 1983, pp. 367–371 (IEEE, New York)

SHAN, T.J., WAX, M., and KAILATH, T. (1985): 'On spatial smoothing for direction-of-arrival estimation of coherent signals', *IEEE Trans.*, **ASSP-33,** pp. 806–811

SHAU, D.H., ADAMS, A.T., and SARKAR, T.K. (1988): 'A study of the effects of mutual coupling on the direction-finding performance of a linear array using the method of moments'. Proc AP–S 88, Internat. symposium digest: antennas and propagation, Syracuse, 6–10 June, **3,** p. 1384 (IEEE, New York)

SPIELMAN, D., PAULRAJ, A., and KAILATH, T. (1986): 'Performance analysis of the MUSIC algorithm'. ICASSP 86, Tokyo, 7–11 April, Vol. 3, pp. 1909–1912 (IEEE, New York)

STOICA, P., and NEHORAI, A. (1989*a*): 'MUSIC, maximum likelihood, and Cramer–Rao bound. *IEEE Trans.*, **ASSP-37,** pp. 720–741

STOICA, P., and NEHORAI, A., (1989*b*): 'MUSIC, maximum likelihood and Cramer–Rao bound: further results and comparisons'. ICASSP 89, Glasgow, 25–28 April, Vol. 4, pp. 2605–2608 (IEEE, New York)

STRUCKMAN, K.A. (1988): 'Angle of arrival estimation of coherent signals using a shifted array correlation technique'. Proc. AP–S 88, Internat. symposium digest: antennas and propagation, Syracuse, NY, 6–10 June. Vol. 1, pp. 266–269 (IEEE, New York)

TAKAO, K., and KIKUMA, N. (1987): 'An adaptive array utilizing an Adaptive Spatial Averaging Technique for multipath environments', *IEEE Trans.*, **AP-35,** pp. 1389–1396

TARRES, F., and FERNANDEZ, J.A. (1989): 'A new method for cancelling coherence in adaptive beamforming or angle of arrival estimators'. Proc. ICAP 89, IEE Conf. Publn. 301, pp. 336–340

TRESSENS, S., CLERGEOT, H., OUAMRI, A., and BOUVET, M. (1988): 'More on optimality of EVD methods for bearing estimation'. ICASSP 88, New York,

11–14 April, **5,** pp. 2881–2884 (IEEE, New York)

WANG, H., and KAVEH, M. (1985): 'Coherent signal-subspace processing for the detection and estimation of angles of arrival of multiple wide-band sources', *IEEE Trans.*, **ASSP-33,** pp. 823–831

WAX, M., and ZISKIND, I. (1989): 'On unique localization of multiple sources by passive sensor arrays', *IEEE Trans.*, **ASSP-37,** pp. 996–1000

WEISS, A.J., and FRIEDLANDER, B. (1988): 'Array shape calibration using sources in unknown locations – a maximum likelihood approach'. ICASSP 88, New York, 11–14 April, Vol. 5, pp. 2670–2673 (IEEE, New York)

WONG, K.M., WALKER, R.S., and NIEZGODA, G. (1988): 'Effects of random sensor motion on bearing estimation by the MUSIC algorithm', *Proc. IEE Pt. F*, **135,** pp. 233–250

WONG, K.M., ZHANG, Q.T., REILLY, J.P., and YIP, P. (1988): 'A new criterion for the determination of the number of signals in high-resolution array processing'. ICASSP–88, New York, 11–14 April, pp. 2769–2772 (IEEE, New York)

YANG, X., SARKAR, T.K., and ARVAS, E. (1989): 'A survey of conjugate gradient algorithms for solution of extreme eigen-problems of a symmetric matrix', *IEEE Trans.*, **ASSP-37,** pp. 1550–1556

YEH, C.C., CHIU, S.C., and PEI, S.C. (1989): 'On the coherent interference suppression using a spatially smoothing adaptive array', *IEEE Trans.*, **AP-37,** pp. 851–857

YEH, C.C., LEE, J.H., and CHEN, Y.M. (1989): 'Estimating two-dimensional angles of arrival in coherent source environment', *IEEE Trans.*, **ASSP-37,** pp. 153–155

YEH, C.C., LEOU, M.W., and UCCI, D.R. (1989): 'Bearing estimation with mutual coupling present', *IEEE Trans.*, **AP-37,** pp. 1332–1335

ZHANG, Q.T., WONG, K.M., YIP, P.C., and REILLY, J.P. (1989): 'Statistical analysis of the performance of information theoretic criteria in the detection of the number of signals in array processing', *IEEE Trans.*, **ASSP-37,** pp. 1557–1567

ZHU, J.X., and WANG, H. (1988*a*): 'A performance comparison of smoothing approaches for high-resolution active direction-finding of completely-correlated targets'. ICASSP 88, New York, 11–14 April, **4,** pp. 2428–2431 (IEEE, New York)

ZHU, J.X., and WANG, H. (1988*b*): 'On fundamental performance limitations of direction finding of multiple narrow-band sources under sensor position and pattern perturbations'. ICASSP 88, New York, 11–14 April, **4,** pp. 2452–2455 (IEEE, New York)

ZHU, J.X., and WANG, H. (1989): 'A comparison of smoothing approaches for angle measurement', *IEEE Trans.*, **AES-25,** pp. 529–535

ZISKIND, I., and WAX, M. (1988*a*): 'Maximum likelihood localization of multiple sources by alternating projection', *IEEE Trans.*, **ASSP-36,** pp. 1553–1560

ZISKIND, I., and WAX, M. (1988*b*): 'Optimum localization of coherent signals having diverse polarizations'. Proc. AP–S 88, Internat. Symposium Digest: Antennas and Propagation, Syracuse, 6–10 June, **1,** pp. 250–253 (IEEE, New York)

ZOLTOWSKI, M., and HABER, F. (1986): 'A vector space approach to direction finding in a coherent multipath environment', *IEEE Trans.*, **AP-34,** pp. 1069–1079

Chapter 14

Ray paths

14.1 Ray re-tracing: the objective

Suppose that the DOAs of several rays from the target transmitter have been determined. Suppose further that the state of the ionosphere in the general area between the transmitter and the DF is known in some detail. Then it is possible in principle to reverse the ray directions at the point of reception and re-trace the ray paths back to the transmitter, using a suitable computer program to find the paths in the ionosphere. Ideally, all ray paths would ultimately be found to intersect at one point on the earth's surface, possibly after multiple hops. Not only would the DOAs at the DF site be correctly explained; the relative time delays between different paths would also be correctly predicted. That is to say, the versions of a given signal element received over various paths should trace back to a common time-origin as well as a common space-origin.

In practice, of course, exact intersections in time and space are not obtained because of measurement errors and of incomplete or imperfect knowledge about the state of the ionosphere. However, the objective of ray re-tracing procedures is to make the best possible use of the available DOAs and ionospheric data. Note that an estimate of the position of the transmitter is now obtained from a single DF site; in effect, a range is determined from elevation angles and ionospheric data, while a direction is determined from the measured azimuth angle or angles.

The aim in this chapter is to summarise those parts of ray-tracing theory that are relevant to DF. For further details, textbooks such as those by Budden (1985) and Kelso (1964) should be consulted.

14.2 Definitions

14.2.1 The ray path

The ray path can be briefly defined as the locus of energy flow from a transmitter to a point of reception. In practice, we cannot launch a single ray from a transmitting-antenna array of finite size. Suppose that each element of a finite array launches a spherical wave. These waves produce a wave-interference field, in which the surfaces of constant phase and constant amplitude can be plotted at any moment. We might then define the ray (in the far field) as the locus of maximum amplitude; this may be thought of, in free space, as the central axis of the main lobe of the array radiation pattern. An infinitely sharp lobe, and hence a single ray, can be produced by an infinitely large two-dimensional filled array; a single ray can therefore be defined mathematically by a function of the form

$$\int_{-\infty}^{\infty} \int_{-\infty}^{\infty} f(x, y)dxdy$$

This type of function is used, for example, by Budden (1985) as the foundation from which he develops ray theory.

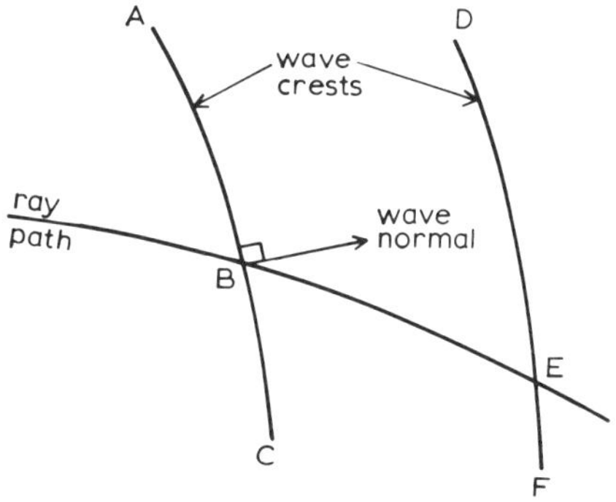

Fig. 14.1 Wave crests and ray path

14.2.2 Surfaces of constant phase

If we now imagine a transmitting array designed to produce two or more lobes of equal amplitude, we can launch two or more rays simultaneously but in different directions. The surfaces of constant phase referred to above now link points of equal phase on these rays. In free space the rays are, of course, straight lines radially outwards from the transmitter, and the surfaces of constant phase are spheres centred on the transmitter; the rays intersect the surfaces perpendicularly. In an anisotropic refracting medium, such as the ionosphere with magnetic field, this condition no longer holds; the 'wave-normal', perpendicular to the surfaces of constant phase, does not necessarily coincide with the ray direction.

14.2.3 Phase velocity

We can define a phase velocity along the ray by following the intersections of a particular phase surface with the ray path. In Fig. 14.1, suppose that the surfaces *ABC*, *DEF* represent particular phase features such as wave crests. Suppose that the crest at *ABC* moves to the position *DEF* in a time interval δt. Then the phase velocity along the ray is $BE/\delta t$.

14.2.4 Phase refractive index

The phase refractive index, n, is a complex quantity when absorption and collisions are taken into account and is a function of direction when an imposed magnetic field is present. The real part μ determines the separation of surfaces of constant phase, i.e. the wavelength λ in the medium: thus,

$$\mu\lambda = \lambda_0$$

where λ_0 is the wavelength in free space.

14.2.5 Wave packets

A wave of specified frequency may be pictured as moving from the transmitter to the receiver with the appropriate phase velocity along the ray path. However, modulation or signal intelligence is always associated with a spread of frequencies, and must be thought of as travelling in a 'wave packet' with a velocity called the group velocity.

In general, pulses become distorted in the ionosphere because of the variation of phase velocity with frequency, and there is no entirely satisfactory definition of group velocity. However, the definition usually adopted for the group velocity V of a wave packet with frequency spread $d\omega$ centred on frequency ω is

$$V = \frac{d\omega}{dk}$$

where the carrier wave is

$$E = E_0 \exp \{j(\omega t - kx)\}$$

Here E_0 is field strength, ω is the angular wave frequency in rads/s, k is the angular wave number ($= 2\pi/\lambda$) and x is a spatial co-ordinate along the direction of motion of the wave.

If we write

$$\mu' = c/V$$

$$\mu = ck/\omega$$

where c is the velocity of light in free space, we obtain

$$\mu' = \mu + \omega \left(\frac{d\mu}{d\omega}\right) \qquad (1)$$

It may be shown that the path of a wave packet is identical with the ray path.

14.2.6 Phase path

Suppose a surface of constant phase takes time t to travel between two arbitrary points A and B. It is sometimes convenient to change the units from seconds to kilometres by writing $P = ct$, where P is referred to as the phase path. Phase path may be thought of as the distance the wave would have travelled in free space in the same time interval. Alternatively, it is perhaps simpler to think of it always as a time interval, expressed in units of distance. Phase path determines the shape of surfaces of constant phase in the composite wave-field at the DF and is of considerable importance in any analysis of the effects of ionospheric tilts. This parameter is therefore discussed in more detail in Section 14.5.

14.2.7 Group path

Suppose a wave packet takes a time T to travel from A to B. Then the group path, sometimes called the effective path, is given by $Q = cT$. This is again a time interval converted to units of kilometres and it is worth noting that there is no difference in the form of the definitions of phase path and group path. Group path is also discussed in more detail in Section 14.5.

14.3 Ray tracing in the presence of a magnetic field

14.3.1 The Haselgrove equations

In 1837, Hamilton published, in a treatise on geometrical optics, a set of differential equations governing the path of the ray in an anisotropic medium. These equations were expressed in canonical form, very similar to the more familiar Hamiltonian equations of dynamics.

Dr Jennifer Haselgrove (1954) pointed out that these differential equations applied to radio propagation as well as to optics and were suitable for numerical integration with a digital computer. Since then, Haselgrove's equations (as they are now called) have been extensively used in ray-tracing computer programs in which the effects of the earth's magnetic field are to be taken into account. Simpler methods can be used if the earth's field is to be neglected.

For practical applications it is convenient to express the Haselgrove equations in spherical co-ordinates (r, θ, φ), where r is radial distance from the earth's centre and θ, φ can be conveniently defined as magnetic colatitude and magnetic longitude (relative to a meridian through the magnetic and geographic poles), respectively. Both Haselgrove and Budden used phase path as the variable of integration. Equal increments of phase path correspond to larger and larger steps in geometrical path length as the reflection point is approached, whereas it would be better to reduce the geometrical step length as this critical condition is approached. In many computer programs group path Q is therefore used as the variable of integration, instead of phase path.

Suppose $\boldsymbol{p}$ represents a vector of magnitude μ in the direction of the wave-normal, with components $(p_r, p_\theta, p_\varphi)$, so that

$$p^2{}_r + p^2{}_\theta + p^2{}_\varphi = \mu^2$$

Then the six Haselgrove equations can be written in the form adopted by Jones (1966)

$$\frac{dr}{dQ} = \frac{1}{\mu\mu'}\left(p_r - \mu\frac{\partial\mu}{\partial p_r}\right)$$

$$\frac{d\theta}{dQ} = \frac{1}{\mu\mu' r}\left(p_\theta - \mu\frac{\partial\mu}{\partial p_\theta}\right)$$

$$\frac{d\varphi}{dQ} = \frac{1}{\mu\mu' r\sin\theta}\left(p_\varphi - \mu\frac{\partial\mu}{\partial p_\varphi}\right)$$

$$\frac{dp_r}{dQ} = \frac{1}{\mu'}\frac{\partial\mu}{\partial r} + p_\theta\frac{d\theta}{dQ} + p_\varphi\sin\theta\frac{d\varphi}{dQ}$$

$$\frac{dp_\theta}{dQ} = \frac{1}{r}\left(\frac{1}{\mu'}\frac{\partial\mu}{\partial\theta} - p_\theta\frac{dr}{dQ} + rp_\varphi\cos\theta\frac{d\varphi}{dQ}\right)$$

$$\frac{dp_\varphi}{dQ} = \frac{1}{r\sin\theta}\left(\frac{1}{\mu'}\frac{\partial\mu}{\partial\varphi} - p_\varphi\sin\theta\frac{dr}{dQ} - rp_\varphi\cos\theta\frac{d\theta}{dQ}\right)$$

When suitable expressions for μ and its derivatives are available, integration of these equations (usually numerically) gives values of the six quantities (r, θ, φ, p_r, p_θ, p_φ) at successive points on the ray path. The trivial integration of dQ gives group path. Phase path is obtained by integrating the equation

$$\frac{dP}{dQ} = \frac{\mu}{\mu'}$$

14.3.2 The Appleton–Hartree expression for refractive index

We may write

$$\mu = \mu(X, Y, Z, \psi)$$

where X, Y, Z, are standard parameters in magneto-ionic (MI) theory, defined as follows:

$$X = f_N^2/f^2$$

$$Y = f_H/f$$

$$Z = 2\pi v/f$$

Here f_N is the plasma frequency at a particular point in the ionized medium, f_H is the gyrofrequency and v is the collision frequency. A full discussion of the basic definitions and equations encountered in MI theory can be found in a textbook by Ratcliffe (1962). ψ is the angle between the wave normal and the imposed magnetic field. Without loss of generality we can therefore write

$$\frac{\partial \mu}{\partial r} = \frac{\partial \mu}{\partial X}\frac{\partial X}{\partial r} + \frac{\partial \mu}{\partial Y}\frac{\partial Y}{\partial r} + \frac{\partial \mu}{\partial Z}\frac{\partial Z}{\partial r} + \frac{\partial \mu}{\partial \psi}\frac{\partial \psi}{\partial r}$$

and similarly for $\partial\mu/\partial\theta$ etc.

Up to this point, no particular expression has been assumed for μ, nor has any particular model of electron density etc. been adopted. The Haselgrove differential equations are valid for any magneto-ionic medium. In order to apply them to a practical problem it is of course necessary to define

(*a*) the refractive-index function, e.g. the Appleton–Hartree expression (see below) in order to calculate

$$\mu, \mu', \frac{\partial \mu}{\partial X} \quad \text{etc} \quad \text{and} \quad \frac{\partial \mu}{\partial \psi},$$

(*b*) the electron-density model in order to calculate X,

$$\frac{\partial X}{\partial r} \quad \text{etc},$$

(*c*) the magnetic-field model, in order to calculate Y,

$$\frac{\partial Y}{\partial r} \quad \text{etc},$$

(*d*) the collision-frequency model, in order to calculate Z,

$$\frac{\partial Z}{\partial r} \quad \text{etc}.$$

Suitable models may be defined either analytically or numerically, provided the various differentials can be calculated.

The complex refractive index n at a point in the ionosphere, in the presence of a magnetic field, is given by the Appleton–Hartree expression

$$n^2 = 1 - \frac{X}{(1 - jZ) - \dfrac{Y_T^2}{2(1 - X - jZ)} \pm \left\{ \dfrac{Y_T^4}{4(1 - X - jZ)^2} + Y_L^2 \right\}^{1/2}}$$

where Y_T and Y_L are the transverse and longitudinal components of Y with respect to the direction of propagation. The derivation of this result from Maxwell's equations can be found in standard textbooks, e.g. Kelso (1964) Section 4.3.

When collisions between electrons and heavy particles can be neglected, Z can be set to zero; n is then real and is equal to the phase refractive index μ. Then

$$\mu^2 = 1 - \frac{X}{1 - \lambda} \tag{2}$$

where

$$\lambda = \frac{Y_T^2}{2(1 - X)} \pm \left\{\frac{Y_T^4}{4(1 - X)^2} + Y_L^2\right\}^{1/2} \tag{3}$$

In the absence of a magnetic field, $\lambda = 0$ and $\mu^2 = 1 - f^2{}_N/f^2$. The plasma frequency f_N is proportional to $\sqrt{N}$, where N is the electron density per unit volume.

14.3.3 The Millington quadratic

The path of a ray launched at a specified elevation angle and incident obliquely on the ionosphere can be determined by numerical integration of the Haselgrove equations, with the Appleton–Hartree expressions used for μ. If full account is to be taken of curvature of the ionosphere and of the earth's magnetic field, a three-dimensional ray-tracing computer program such as that described by Jones (1966) is required. Before such programs became available, simplified methods for use with desk calculating machines were devised.

For example, on comparatively short paths it may be legitimate to neglect curvature effects. If we assume horizontal stratification within the ionosphere, Snell's law for any point on the ray path can be written

$$\mu \cos \Delta = \cos \Delta_0$$

where Δ is elevation angle and Δ_0 is the initial elevation angle at which the ray is launched. We now use the symbol q for $\mu \sin \Delta$, which may be regarded as the vertical component of a vector of magnitude μ; the horizontal component is, from the above equation, constant.

Similarly, let Y_x and Y_z be the horizontal and vertical components respectively (for propagation in the xz plane) of the magnetic field Y.
Thus

$$Y_L = Y_x \cos \Delta + Y_z \sin \Delta$$

i.e.

$$\mu Y_L = Y_x \cos \Delta_0 + Y_z q$$

Also,

$$Y^2 = Y_L^2 + Y_T^2$$

By squaring a version of eqn. 3 and eliminating Y_L and Y_T, using the above equations, we obtain after some manipulation

$$(Y_x \cos \Delta_0 + Y_z q)^2 = \lambda^2 (\cos^2 \Delta_0 + q^2) - \frac{\lambda(\lambda^2 - Y^2)}{(\lambda - 1)} \tag{4}$$

This is essentially a quadratic equation in q first derived by Millington (1951). Clemmow and Mullaly (1955) summarised his procedure in the standard notation adopted here, but quoted the last term in the above equation incorrectly as $-\lambda(\lambda - Y^2)/(\lambda - 1)$.

If the earth's magnetic field can be regarded as constant in magnitude and direction over the whole of the propagation path, Y_x, Y_z and Y are (known) constants. We can then find pairs of (q, λ) values that satisfy the equation. For each (q, λ), X can be calculated from eqn. 2, rewritten as

$$q^2 - \sin^2 \Delta_0 = \frac{-X}{(1 - \lambda)} \tag{5}$$

Thus for a specified profile (X known as a function of height) the height corresponding to an assumed value of q can be calculated and the path is defined by a sequence of such points.

14.3.4 The Booker quartic

The well-known quartic equation first derived by Booker (1938, 1949) can be obtained from the Millington quadratic by substitution for λ from eqn. 5 into eqn. 4. The result, expressed as a cubic in X, is

$$X^3 + AX^2 + BX + \Gamma = 0$$

where

$$A = -\{1 + 2(C^2 - q^2)\}$$

$$B = (C^2 - q^2)\{(2 - Y^2) + (SY_x + qY_z)^2 + (C^2 - q^2)\}$$

$$\Gamma = -(1 - Y^2)(C^2 - q^2)^2$$

and

$$C = \sin \Delta_0$$

It should be noted that A is given incorrectly as $\{1 - 2(C^2 - q^2)\}$ by Booker when first quoted in eqn. 37 of his second paper, but appears correctly in the rest of the paper. The cubic in X can also be regarded as a quartic in q

$$\alpha q^4 + \beta q^3 + \gamma q^2 + \delta q + \epsilon = 0$$

where the coefficients can be deduced from the cubic equation.

The quartic gives four values of q for a specified value of X; these roots correspond to the up-going and down-going ordinary (O) and extraordinary (X) modes. At the base of a horizontally-stratified ionosphere, $X = 0$ and hence

$$\Gamma = 0$$

and

$$q = \pm \sin \Delta_0$$

Thus the angle of exit from the ionosphere is the same as the angle of incidence for both O and X rays, even though the ray paths within the ionosphere are different and are asymmetric about the apex in general (see Fig. 14.2). This property might be used as a check on the accuracy of a numerical integration for the path within the ionosphere.

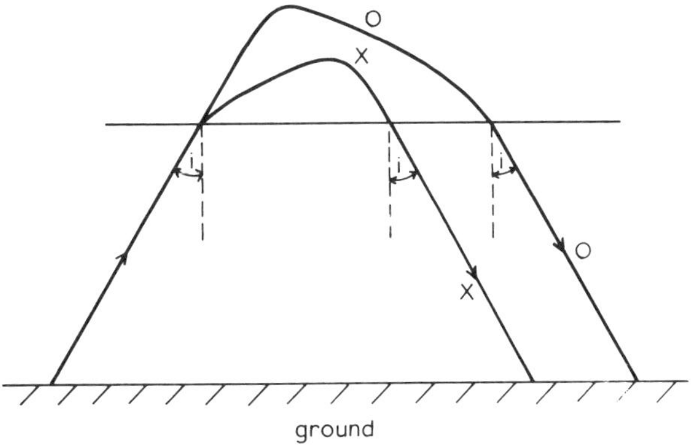

Fig. 14.2 Sketch of ray paths for ordinary (*o*) and extraordinary (*x*) rays

Booker's quartic is usually applied to a horizontally stratified ionosphere, for which the horizontal component of the 'vector' μ can be regarded as a constant, i.e.

$$S = \mu \cos \Delta = \cos \Delta_0$$

where S is the sine of the angle of incidence on the base of the ionosphere. For a spherically-stratified ionosphere S becomes a variable along the ray path and the use of the Haselgrove equations is indicated.

14.3.5 Magneto-ionic deviations

A wave entering the ionosphere is split into O and X components which then follow different paths. Under the assumptions of a plane earth and a horizontally-stratified ionosphere it can be shown that the component wave packets emerge from the layer moving in planes parallel to, but not necessarily coincident with, the vertical plane of launch. There is no deviation when propagation is in the magnetic meridian and no *net* deviation for propagation in the plane perpendicular to the meridian.

Table 14.1 Elghozi's results for a parabolic layer

Distance between transmitter and receiver, km	Ratio of signal frequency to MUF	Azimuth plane from mag *N*	Bearing error, *O* wave
45	0·96	135°	−16°
45	0·996	135°	−27°
200	0·94	135°	− 1°
200	0·94	160°	−0·6°

The question arises whether the net deviations for paths on intermediate azimuths give rise to significant DF errors. Since the deviations are of opposite sign for the *O* and *X* paths between two terminals, their effects tend to cancel when both rays are received by a DF. On the other hand, the deviations are systematic for a given path/frequency/mode combination. Moreover, the *X* ray will often be so much weaker than the *O* ray that the cancellation argument cannot be safely applied. Systematic errors of 0·1° or greater would certainly be worth removing from averaged-azimuth measurements made with a WADF.

Early papers on the magnitudes of magneto-ionic deviations (MIDs) were based on laborious hand calculations for a few rays. Elghozi (1953) calculated ray paths for a 'linear' ionospheric layer, in which electron density increases linearly with height, and found bearing errors of only a few hundredths of a degree on certain oblique paths with ground ranges of 200, 600 and 1400 km. However, with a parabolic layer he found that rather larger errors can occur when the signal frequency is close to the maximum usable frequency (MUF) for the path; these results are shown in Table 14.1 above. Millington (1954) found for near-vertical paths that, when the signal frequency is well below the MUF, the bearing error is likely to be less than 5°.

Now that large samples of rays can be traced with computer programs, it has become clear that much smaller errors than 5° arise on low-angle long-range paths. For example, Davies (1968) stated that 'on frequencies well below the penetration frequency (the so-called junction frequency) the deviations are of the order of 1 km or so for low-angle propagation over distances of the order of 2000 km.' Rao (1969) used the Jones 3-D program to trace a large number of rays from a transmitter position in Texas and studied MIDs as a function of frequency for (*a*) fixed ray paths, (*b*) fixed ground ranges and (*c*) fixed elevation angles and for two types of profile. Fig. 14.3, based on Fig. 3 in his paper, shows MID as a function of azimuth and elevation angles for a quasiparabolic layer. Errors of a few tenths of a degree are seen to occur at the higher-elevation angles. The largest error tabulated by Rao is −4·965° for an *O* ray at an elevation of 85°, with a quasiparabolic layer; the ground range in this case was 43 km. The combined effects of MIDs and ionospheric tilts are also discussed in his paper.

The author has traced a sample of 70 *O* rays and 70 *X* rays launched from a fixed transmitter site in England. Results obtained with the aid of the Jones 3-D program are shown in Appendix 6 and are consistent with Rao's conclusions. For ground ranges of 700 km or more, none of the bearing errors exceeded 0·05° for either the *O* or *X* rays. Errors of the order of 0·1° occurred only on comparatively short, high-angle paths. The product of bearing error and ground range decreases with increasing range. The positional error (in km) arising from an uncorrected MID in a single bearing is therefore more serious on short paths than on long paths. Systematic corrections for MIDs are probably worth applying when the path length is less than about 500 km, or if the signal frequency is within a few per cent of the MUF. MIDs decrease as magnetic latitude increases.

For *O* and *X* rays launched from the same point on the same azimuth and frequency and at the same elevation angle, the one-hop ground range reached is always smaller for the *X* rays than the *O* rays. Differences of the order of 5 or 10% in range can easily occur and in the extreme case the *O* ray might penetrate the layer while the *X* ray is reflected. A simple ray-

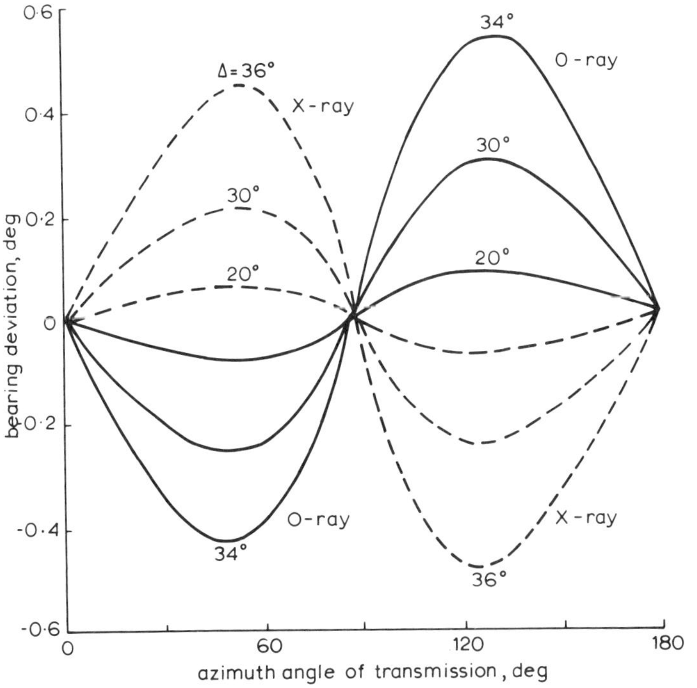

Fig. 14.3 Azimuth-angle dependence of *O*-, and *X*-ray bearing deviations for different elevation angles of transmission Δ
QP layer: critical frequency 5 MHz, semithickness 150 km, base height 100 km

Based on Fig. 3 in a paper by Rao, Radio Sci., **4**, *pp. 153–161, Feb 1969, copyrighted by American Geophysical Union.*

tracing program in which the earth's magnetic field is neglected gives 'no field' results intermediate between the O and X values.

14.3.6 Relationship between ray-path deviations in azimuth and elevation

The idealised model of the ionosphere and the earth's magnetic field used for the above MID calculations was as follows: a spherically-stratified ionosphere concentric with the earth's surface, plus a dipole magnetic field, the axis of the dipole passing through the centre of the earth. This model, which is often adopted in ray-tracing programs, may be regarded as one member of a class of axially-symmetric models in which neither the electron density nor the magnetic field varies with the magnetic longitude φ.

It may be shown (Gething, 1972) that the final Haselgrove equation as quoted in Section 14.3.1 can be integrated, when axial symmetry applies, to give

$$rp_\varphi \sin\theta = Ka$$

where K is a constant and a is the radius of the earth.

We now use the subscript T for the point of transmission and the subscript R for the point of reception, both on the earth's surface. The sign convention adopted for the angles of reception is that α_R is approximately equal to α_T on a short path and that Δ_R is positive; in effect, α_R and Δ_R are the azimuth and elevation of the ray after ground reflection at R. Then

$$\sin\alpha_T \cos\Delta_T \sin\theta_T = \sin\alpha_R \cos\Delta_R \sin\theta_R \qquad (6)$$

This equation is a very general one and can be applied to magneto-ionic deviations or to deviations due to ionospheric tilts, or to a combination of both, subject only to the requirement of axial symmetry. No assumption has been made that the deviations are small.

For a common point of transmission below the ionosphere the value of K is the same for all rays launched in the same direction (or, more generally, with the same value of the direction cosine p_φ). Under these conditions, the value of K is the same for O and X rays of all frequencies, including penetrating rays. It therefore represents an interesting common characteristic of such a family of rays, and can provide a useful check on the accuracy of a numerical integration along a selected ray path. Various deductions can be made from eqn. 6 about the relationships between small angular deviations in azimuth and elevation arising, for example, from MID effects.

14.4 Ray tracing when the earth's field is neglected

14.4.1 Parabolic and quasiparabolic layers

Analytic expressions for ground range, phase path and group path can be obtained for certain idealised models of the electron density distribution in the ionosphere, when the effects of the earth's magnetic field and of collisions are neglected. Neglect of the former, i.e. magneto-ionic effects, should not cause significant errors when the signal frequency is large compared with the gyrofrequency (1·2 MHz in the UK). For example, exact path integrals are available (Budden, 1985) for a linear, exponential or parabolic variation of density as a function of height above the base of the layer, for a flat earth and horizontally-stratified ionosphere. Of these models, only the parabolic exhibits the important practial phenomena of skip distance, MUF, and ray penetration; this model has therefore been extensively adopted for simple calculations, for example by Appleton and Beynon (1948), Kift (1960) and Bradley and Dudeney (1973).

When earth curvature effects cannot be neglected (e.g. on comparatively long paths), the path integrals for the parabolic layer must be modified and become approximate rather than exact. Expressions for ground range D, phase path P and group path P', correct to first-order curvature terms, are quoted in Appendix 7. Expressions are also given for penetration through one layer, with reflection at a higher layer. De Voogt (1953) showed that the exact integrals can be obtained in the curved earth case if the squared plasma frequency f^2_N can be expressed in the form

$$f_N^2 = a_1 + \frac{a_2}{\rho} + \frac{a_3}{\rho^2}$$

where a_1, a_2, a_3 are constants; ρ is the normalised radius r/R, where r is the distance from the centre of the earth to the thin spherical shell in which the plasma frequency is f_N and R is the radius of the earth. Muldrew (1959) pointed out that a solution could also be obtained if

$$f_N^2 = a_4\rho^2 + a_5$$

and Maliphant (1968) has used

$$f_N^2 = \frac{a_6}{\rho^2} + a_7 + a_8\,\rho^2$$

A close approximation to a parabolic distribution can be obtained by writing de Voogt's expression in the form

$$f_N^2 = f_c^2\left\{1 - \left(\frac{h - h_m}{y_m}\right)^2 \left(\frac{R + h_0}{R + h}\right)^2\right\} \qquad (7)$$

in the height range $h_0 < h < (h_0 + 2y_m)$, and $f_N = 0$ elsewhere. Here f_c is the critical frequency, i.e. the frequency that just penetrates at vertical incidence, h is height above the surface of the earth, h_0 is the height of the base of the layer, h_m is the height of maximum ionisation and y_m is the semithickness of the layer. For a 'true' parabolic layer, the factor $\{(R + h_0)/(R + h)\}^2$ would be omitted.

Path integrals for the distribution defined by eqn. 7, often referred to as a quasiparabolic (QP) layer, are available (Croft and Hoogasian, 1968) and are quoted in Appendix 7. Because these analytic expressions are exact the QP-layer provides, in principle, a useful method of checking the accuracy of ray-tracing computer programs based on numerical integration. Unfortunately, it is not always possible in practice to evaluate the analytic expressions to adequate accuracy, particularly near penetration conditions; phase path, for example, depends on a comparatively small difference between two very large numbers.

14.4.2 Other distributions

It is sometimes desirable to use, as the input to a ray-tracing program, an experimentally-determined profile of electron density against height. This may be of irregular form, and not easily represented by a single mathematical function. A common procedure in such a problem is to divide the profile into small height intervals (say) 1 km thick, to assume that the electron density is constant within each spherical shell thus defined, and to apply Snell's Law at each boundary between successive shells. In a curved ionosphere

$$\mu r \sin i = \text{constant}$$

where r is the radius vector from the centre of the earth to a shell in which the refractive index is μ and i is the angle between the radius vector and the ray path within the shell; this version of Snell's Law is sometimes called Bouger's rule. An alternative is to use thicker shells, the plasma frequency within each shell being approximated by an expression of the type

$$f_N^2 = \frac{a_6}{\rho^2} + a_7 + a_8\rho^2$$

Maliphant (1968) found that typical profiles can be adequately approximated by a few segments of this type; the values of a_6, a_7 and a_8 differ from one segment to another, but are chosen to give a 'best fit' to the profile within each segment and to give continuity in f_N at the segment boundaries. The contributions to D, P and P' for each segment are calculated from analytic expressions and all contributions are then added to obtain total-path lengths.

14.5 Phase path and group path

14.5.1 Importance in DF

DF instruments are basically phase-sensitive devices, used to determine the orientation of lines of constant phase in the wave-field at the reception point. These lines are the intersections of surfaces of constant phase with the horizontal plane, and a surface of constant phase links points on various ray paths with the same value of the phase-path parameter P. Thus, phase path is an important parameter in theoretical discussions of a number of aspects of DF theory, particularly in the consideration of the effect of ionospheric tilts on the orientation of lines of constant phase. It is also a key parameter in any treatment of Doppler frequency shifts arising, for example, from travelling ionospheric disturbances; the frequency shift is

$$-(1/\lambda)\left(\frac{dP}{dt}\right)$$

Group path P' is the quantity normally determined experimentally in vertical, oblique bistatic and oblique backscatter sounding of the ionosphere; from a knowledge of the values of P' at various frequencies an ionospheric profile can be derived in suitable circumstances. The differences in P' values for two or more ionospheric modes give rise to multipath distortion; the relative time delay between two modes arriving at known elevation angles can be used to obtain a crude estimate of the distance of an unknown transmitter from the receiver (see Section 14.6.4).

Some of the main characteristics of phase path and group path, and of the relationships between them, will be summarised in the remainder of this chapter.

14.5.2 Group refractive index

From eqn. 1 in Section 14.2.5, the group refractive index μ' can be written

$$\mu' = \mu + f\frac{\partial \mu}{\partial f}$$

$$= \frac{\partial}{\partial f}(\mu f)$$

When the earth's magnetic field and collisions are neglected, μ is given by

$$\mu = \left(1 - \frac{f_N^2}{f^2}\right)^{1/2}$$

and hence

$$\mu' = \left(1 - \frac{f_N^2}{f^2}\right)^{-1/2}$$

$$= \frac{1}{\mu}$$

The group velocity U and phase velocity V are then related by the equation

$$UV = (c\mu)\,(c/\mu)$$

$$= c^2$$

where c is the velocity of light. Within the ionosphere, and in the presence of a magnetic field,

$$\mu\mu' - 1 \neq 0$$

in general; it may be shown that there are three cases in which $\mu\mu' = 1$, namely

(*a*) $X = 0$
(*b*) $\lambda = 0$
(*c*) $X = (\lambda/Y_L)^2$

$X = 0$ applies outside the layer, i.e.

$$\mu = \mu' = 1$$

below the ionosphere, with or without a magnetic field. The second condition can apply only to the *O*-ray (Gething, 1963), for which λ is generally negative but is equal to zero for any point on the ray path at which $Y_L = 0$. The third case, *c*, can apparently apply to either ray and leads to the condition

$$\lambda^2 - \lambda(1 - X)^2 - XY^2 = 0$$

which is satisfied if $Y = \pm(1 - X)$.

14.6 Breit and Tuve's theorem

14.6.1 Flat earth, no field

For a flat earth and a horizontally-stratified ionosphere with no magnetic field we can write, for a ray launched at elevation angle Δ_0

$$P' \cos \Delta_0 - D = \int \mu' \cos \Delta_0 \, ds - \int dx$$

where ds is an element of ray path, dx is the corresponding horizontal distance and the integrals are taken over the whole path. From Snell's Law

$$ds \cos \Delta_0 = \mu dx$$

Hence

$$P' \cos \Delta_0 - D = \int (\mu\mu' - 1) dx$$

$$= 0$$

i.e.

$$P' \cos \Delta_0 = D$$

This is Breit and Tuve's theorem (Davies, 1965) from which one of the quantities P', Δ_0 and D can be calculated if the other two are known. In the simplified conditions assumed, the horizontal component U_x of the group velocity has a constant value of $c \cos \Delta_0$ at all points on the ray path and the time taken by a wave packet to travel over the path is

$$\frac{P'}{c} = \frac{D}{c \cos \Delta_0}$$

14.6.2 Curved earth, no field

With a curved earth and ionosphere the relation between P' and D is no longer independent of the profile, and the simple form of Breit and Tuve's theorem is approximate rather than exact. The parts of the path below the ionosphere, denoted by a subscript 2, are related by the exact equation

$$P'_2 = 2R \sin\left(\frac{D_2}{2R}\right) \sec\left(\Delta_0 + \frac{D_2}{2R}\right)$$

Fairly close limits can be set on P'_1, the segment of group path within the ionosphere. Let r represent the distance of a point on the ray path from the centre of the earth and suppose that we can put definite limits on r within the layer, such that it lies between r_{min} and r_{max}. Then it can be shown (Gething, 1962) that

$$\frac{r_{min}^2}{R^2} D_1 \sec \Delta_0 \leqslant P'_1 \leqslant \frac{r_{max}^2}{R^2} D_1 \sec \Delta_0$$

Thus, Breit and Tuve's theorem for this segment of the path may be written in the form

$$P'_1 = \alpha D_1 \sec \Delta_0$$

where $\alpha = 1$ for a plane earth and is slightly greater than 1 for a curved earth.

Close limits can be placed on α with only very limited information about the ionosphere. Suppose, for example, that we know the height h_0 of the lower boundary of a layer and the height h_m of the level of maximum ionisation. Then we can put

$$r_{min} = R + h_0$$

and, for reflection

$$r_{max} \leqslant R + h_m$$

We now put $h_0 = 240\,\text{km}$ and $h_m = 300\,\text{km}$ to obtain some idea of magnitudes. The values of r_{min} and r_{max} then differ by less than 1% and the corresponding limits on α by less than 2%:

$$1{\cdot}077 \leqslant \alpha \leqslant 1{\cdot}096$$

If we use the mean of these limits as our estimate of α, the corresponding value of P'_1 cannot be more than 1% in error in this example. The percentage error in P', the total path, will be considerably smaller.

In the absence of any ionospheric information, we can use the exact equation given above for P'_2 as an approximate equation for the total group path P'. Thus

$$P' = 2R \sin\left(\frac{D}{2R}\right) \sec\left(\Delta_0 + \frac{D}{2R}\right) \tag{8}$$

This equation may be regarded as one in which full account is taken of the curvature of the earth's surface but no correction is made for curvature of the ionosphere. It can provide an excellent approximation when P'_1 is much smaller than P'_2. If P' and Δ_0 are known, the value of D can be found by iteration in the more approximate equation

$$D_{(r+1)} = P' \cos(\Delta_0 + D_r/2R) \tag{9}$$

Here D_r is the rth approximation to the ground range. The process starts with $D_0 = 0$ and usually converges to a stable value of D (to, say, the nearest 0·1 km) after about four iterations.

Table 14.2 Ground ranges D_c calculated from Breit and Tuve's theorem

Ray ref. no.	Launch conditions			Ordinary ray			Extraordinary ray		
	f	Δ_0	φ	D_t	D_c	% error	D_t	D_c	% error
11	4·0	5·0°	45°	2230·4	2219·8	−0·5	2221·8	2211·8	−0·5
15		15·0	45	1249·9	1248·1	−0·1	1247·6	1245·7	−0·2
19		25·0	45	818·8	818·1	−0·1	811·1	811·2	0·0
23		35·0	45	588·8	588·1	−0·1	575·5	577·1	0·3
27		45·0	45	441·2	439·9	−0·3	421·4	425·4	1·0
31	6·0	5·0	45	2260·9	2250·0	−0·5	2269·5	2257·8	−0·5
35		15·0	45	1298·3	1296·0	−0·2	1290·9	1289·1	−0·1
39		25·0	45	886·8	885·6	−0·1	872·8	873·1	0·0
43		35·0	45	682·5	680·5	−0·3	655·8	658·4	0·4
47		45·0	45	574·6	570·0	−0·8	522·1	529·0	1·3
51	8·0	5·0	45	2323·0	2310·5	−0·5	2317·2	2305·0	−0·5
55		15·0	45	1374·4	1371·6	−0·2	1362·3	1360·1	−0·2
59		25·0	45	1014·6	1012·0	−0·3	988·3	988·1	0·0

14.6.3 Curved earth, with field

Eqn. 9 takes no account of magneto-ionic effects but, because both P' and D for a given launch angle Δ_0 are affected by the field, the errors resulting from the use of this equation appear to be small. Table 14.2 shows the values D_c of ground range calculated by the iterative method compared with the range D_t obtained by ray-tracing, for some of the rays specified in Appendix 6.

Results are shown only for an azimuth angle of $\varphi = 45°$; similar percentage errors occur on other azimuths.

The following tentative conclusions can be drawn from the very limited sample of rays shown in the table:

(*a*) D_c is usually within 1% of the true value D_t (ray 47 *X* is an exception).

(*b*) For *O* rays, D_c is always less than D_t. (In effect, D_c is measured along the chord linking transmitter and receiver).

(*c*) For *X* rays, D_c is less than D_t at long ranges (> 1000 km) but D_c is greater than D_t at short ranges (< 700 km). At long ranges, the dominant error is presumably 'geometrical' in origin, i.e. due to earth curvature. At short ranges, the magneto-ionic effects appear to dominate. At intermediate ranges, the cancellation of the two effects can lead to near-zero errors.

14.6.4 Calculation of approximate ground range

Suppose that the time taken for a wave packet to travel from a transmitter to a receiver is t_1 via one mode and t_2 via another mode, and that the corresponding elevation angles are Δ_1 and Δ_2. The time at which a given signal element is transmitted is not generally known, but the difference in time of arrival, $(t_2 - t_1)$, is a measurable quantity. From this time difference and with measured values of Δ_1 and Δ_2 an approximate value of the ground range *D* of a transmitter of unknown position can be deduced. From the simplest version of Breit and Tuve's theorem,

$$ct_1 \simeq D \sec \Delta_1$$

$$ct_2 \simeq D \sec \Delta_2$$

Hence

$$D \simeq \frac{c(t_2 - t_1)}{(\sec \Delta_2 - \sec \Delta_1)}$$

A distance estimate obtained in this way will be in error partly because of the neglect of earth curvature and of magneto-ionic effects and partly because of errors in the measurement of $(t_2 - t_1)$, Δ_1 and Δ_2. Numerical simulations suggest that a calculated range will often be within about 10% of the true range. This first estimate can be refined if the modes of propagation are known, or can be inferred from a knowledge of Δ_1, Δ_2 and the approximate range.

14.7 Relationship between phase path and group path

Appleton (1928) showed that, for a flat earth and horizontally-stratified ionosphere

$$\frac{dP}{df} = \frac{1}{f}(P' - P) \tag{10}$$

for a path between two fixed points. Here dP is the total change in P arising from a change df in frequency *and* a related change in the elevation angle Δ such that the ground range D remains constant. The author has shown (Gething, 1965) that the same relationship is still rigorously satisfied on oblique paths for which earth-curvature effects are significant, subject only to the common assumptions that the magnetic field can be neglected and that the ionospheric region concerned is spherically stratified. We write

$$\frac{dP}{df} = \frac{\partial P}{\partial f} + \frac{\partial P}{\partial \Delta}\frac{d\Delta}{df}$$

subject to the condition that $$\frac{dD}{df} = 0$$

i.e. that $$\frac{\partial D}{\partial f} + \frac{\partial D}{\partial \Delta}\frac{d\Delta}{df} = 0$$

It can also be shown that, for a constant frequency, the changes in phase path and ground range resulting from a small change in elevation angle are related by the equation

$$\frac{\partial P}{\partial \Delta} = \cos\Delta \frac{\partial D}{\partial \Delta}$$

Hence $$\frac{dP}{df} = \frac{\partial P}{\partial f} - \cos\Delta \frac{\partial D}{\partial f}$$

from which it can be shown that

$$\frac{dP}{df} = \frac{1}{f}(P' - P)$$

This equation is valid for any electron density profile and for any (fixed) positions of the path terminals, which need not lie on the surface of the earth.

14.8 Variation of phase path and group path with frequency

A plot of P' (as ordinate) against f (abscissa) for a fixed ground range constitutes an oblique ionogram for that path; the plot can be obtained experimentally, or simulated with a suitable ray-tracing program. Similarly, a plot of P against f can be obtained and such a plot can be described as a phase path oblique ionogram. The computer program should include a 'homing' option, for the selection of pairs of values of f and Δ corresponding to a specified ground range.

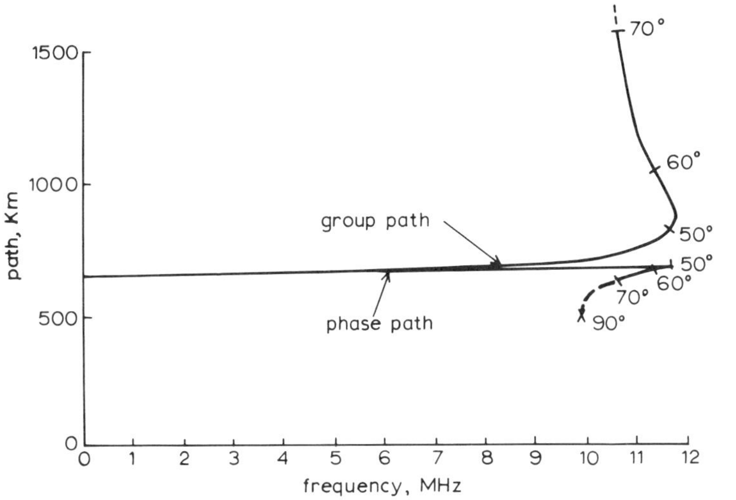

Fig. 14.4 Phase-path and group-path oblique ionograms for a ground range of 500 km
parabolic layer: f_c = 10 MHz, h_m = 280 km, y_m = 80 km

Plots of P' against f and P against f, computed for a single parabolic layer and a ground range of 500 km, are shown in Fig. 14.4. The broken lines are used to indicate that the values plotted become very sensitive to small changes in elevation angle or frequency at high-elevation angles.

For a finite ground range, P is always finite and approaches a finite limiting value P_c for the penetration condition $\Delta = 90°$, at which P' becomes infinite. It can be shown (Gething, 1974) that, for any profile,

$$P_c = 2h_c$$

where h_c is the limiting value of phase height at vertical incidence, i.e. the phase height for $f = f_c$ where f_c is the critical frequency. For a parabolic layer,

$$P_c = (2h_m - y_m)$$

and the appropriate value (480 km) is shown by a cross in Fig. 14.4. Since dP/df becomes very large as P' becomes very large, a vertical line through f_c must be tangential to the (f, P) curve; this result applies for any ground range. P' is always greater than P for a given frequency; hence, from eqn. 10, the slope of the (f, P) curve must be everywhere positive.

An estimate of the variation across a finite bandwidth of the phase difference between two modes of propagation is sometimes required in discussions of selective fading. Suppose P_1 is the phase path for (say) a one-hop mode and P_2 is the total phase path for (say) a two-hop mode between the same terminals. The phase difference φ between signals received over these two paths is

$$\varphi = \frac{2\pi f}{c}(P_2 - P_1)$$

Hence

$$\frac{d\varphi}{df} = \frac{2\pi}{c}(P_2 - P_1) + \frac{2\pi f}{c}\left(\frac{dP_2}{df} - \frac{dP_1}{df}\right)$$

$$= \frac{2\pi}{c}(P_2' - P_1')$$

The change $\delta\varphi$ of phase difference, across a small but finite bandwidth δf, is therefore given approximately by

$$\delta\varphi = \frac{2\pi\delta f\,(P_2' - P_1')}{c}$$

where P'_1 and P'_2 are computed at the centre frequency.

14.9 Applications to transmitter location

We noted in Section 1.6 that an ionospheric height error occurs in time difference direction finding (TDDF) if measured time differences (expressed in units of distance) for ionospherically-propagated signals are interpreted directly as differences of ground range. Estimates of ionospheric parameters, such as those given in routine predictions, can be used to calculate a first-order correction to each line of position, leading to a significant increase in accuracy. Based on data collected over two pairs of long oblique paths, Rao and Beckwith (1975) obtained a reduction in position errors by a factor of between three and four by such means. However, corrections based on predictions cannot be other than approximate. Rose (1988) found that the time stability of a number of HF ionospheric paths was affected by an unexpectedly high degree of ionospheric movement.

In single station location (SSL) techniques, the range coordinate is determined from measured elevation angles. Simplified calculations of ray paths tend to be used, because the detailed knowledge of the state of the ionosphere needed for more elaborate ray-tracing methods is not normally available. This lack of knowledge is, of course, one source of error; see, for example, Baker (1983).

The true profile of electron density against height is not uniquely determined by the standard methods of interpreting vertical ionograms; in particular, the depth of any valley in the profile is not normally determined. The resulting error component in calculated ranges is,

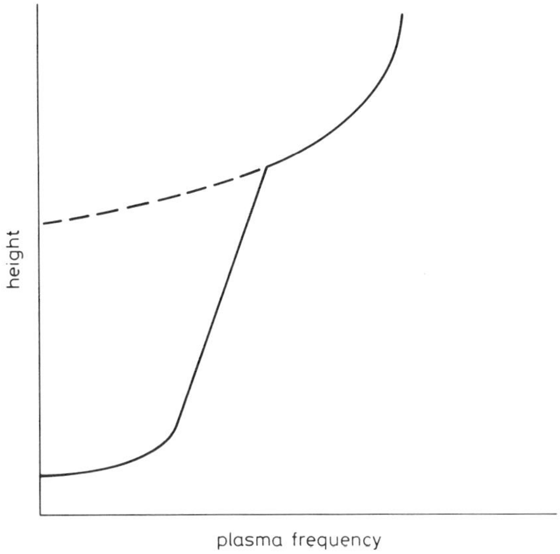

Fig. 14.5 Bradley–Dudeney ionospheric profile
Based on Fig. 1 in a paper by Milsom, Proc. IEE., ***132H,*** *pp. 33–38*

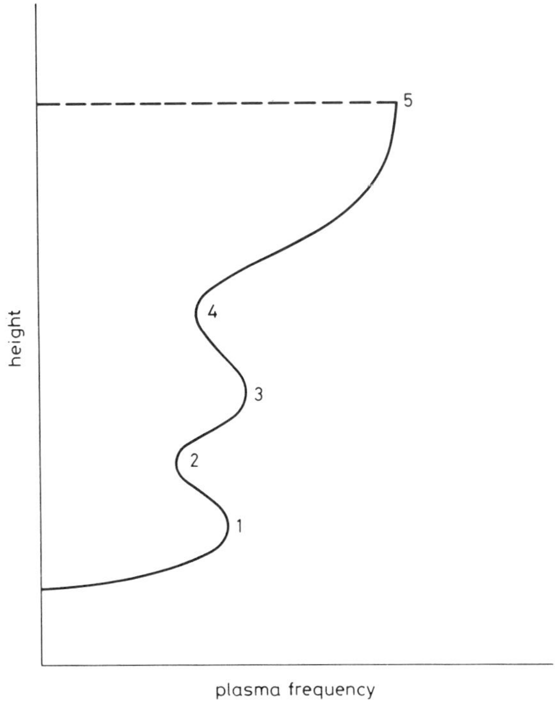

Fig. 14.6 Ionospheric profile consisting of five parabolic or quasi-parabolic layers

however, comparatively small; Barabashov, Vertogradov and Khondu (1987) found errors of less than 2·5% of range at ranges up to 350 km.

An alternative to calculating a complete electron density profile is to measure a few key parameters from ionograms taken at vertical incidence and use them to define a simplified model ionosphere, such as that suggested by Bradley and Dudeney (1973). The Bradley–Dudeney model consists of a parabolic E-layer, a parabolic F-layer and a linear profile joining them (Fig. 14.5). Milsom (1985) showed that, if a slightly modified version of this profile is adopted, exact expressions can be derived for the ground range and other characteristics of the oblique path for a specified elevation angle.

Baker and Lambert (1988) used instead a model consisting of three parabolic layers linked by two inverted parabolic layers to represent respectively the E, F1 and F2 regions of the ionosphere and the two transition regions between them; see Fig. 14.6. The same authors showed (1989) that a similar model composed of quasiparabolic layers can also be used, and gives results that compare favourably with numerical ray-tracing through a Bradley–Dudeney model.

McNamara (1988) obtained median miss distances of about 7% of range for both E and F2 propagation, from measurements selected on the basis of a reliable identification of propagation mode.

References

APPLETON, E.V. (1928): 'Some notes on wireless methods of investigating the electrical structure of the upper atmosphere', *Proc. Phys. Soc.*, **41,** pp. 43–59

APPLETON, E.V., and BEYNON, W.J.G. (1948): 'The application of ionospheric data to radio communication', DSIR Radio Research Special Report No. 18.

BAKER, D.C. (1983): 'Accuracy limitations of HFDF and SSL systems as the result of ionospheric propagation', S. Afr. Inst. of Electr. Eng., Symposium on Ant. and Propn., pp. CC/1–10

BAKER, D.C., and LAMBERT, S. (1988): 'Multiparabolic ionospheric model for SSL application', *Electronics Letters*, **24,** pp. 425–426

BAKER, D.C., and LAMBERT, S. (1989): 'Range estimation for SSL HFDF systems by means of a multi-quasiparabolic ionospheric model', *Proc. IEE.*, **136H,** pp. 120–125

BARABASHOV, B.G., VERTOGRADOV, G.G., and KHONDU, A.A. (1987): 'Assessment of vertical sounding utilization efficiency in decameter-wave angle-and-range measuring systems', *Radiotekhnika* **42,** no. 9, pp. 3–7. English translation in *Telecom. Radio Eng.*, **42,** no. 10, pp. 27–31

BOOKER, H.G. (1938): 'Propagation of wave packets incident obliquely upon a stratified doubly refracting ionosphere'. *Philos. Trans.*, **237,** pp. 411–451

BOOKER, H.G. (1949): 'Application of the magneto-ionic theory to radio waves incident obliquely upon a horizontally-stratified ionosphere'. *J. Geophys. Res.*, **54,** pp. 243–274

BRADLEY, P.A., and DUDENEY, J.R. (1973): 'A simple model of the vertical distribution of the electron concentration of the ionosphere'. *J. Atmos & Terr Phys.* **35,** pp. 2131–2146

BUDDEN, K.G. (1985): 'The propagation of radio waves' (Cambridge University Press)

CLEMMOW, P.C., and MULLALY, R.F. (1955): 'The dependence of the refractive index in magneto-ionic theory on the direction of the wave normal'. Report on the Conference on the Physics of the Ionosphere, Cambridge 1954 (The Physical Society, London)

CROFT, T.A., and HOOGASIAN, H. (1968): 'Exact ray calculations in a quasiparabolic ionosphere with no magnetic field', *Radio Sci.*, **3,** pp. 69–74

DAVIES, K. (1965): 'Ionospheric radio propagation', NBS Monograph 80, (US Govt. Printing Office), pp.161–162

DAVIES, K. (1968): Magneto-ionic theory of radio propagation', Agard lecture series XXIX, July 1968: 'Radio Wave Propagation'.

ELGHOZI, G. (1953): 'Un problème de radiogoniometrie: l'anisotropie ionosphérique', *Ann. Télécomm.*, **8,** pp. 78–92

GETHING, P.J.D. (1962): 'Relation between oblique and ground path-lengths in ionospheric propagation over a curved earth', *Nature*, **193,** pp. 260–261

GETHING, P.J.D. (1963): 'Lateral deviation and effective path length for oblique propagation through the ionosphere', Proceedings of the International conference on the ionosphere, London 1962 (Institute of Physics and the Physical Society), pp. 421–427

GETHING, P.J.D. (1965): 'Relationship between phase path and effective path for oblique ionospheric propagation', *J. Atmos. & Terr. Phys.*, **27,** pp. 57–66

GETHING, P.J.D. (1972): 'Relationship between ray-path deviations in azimuth and elevation for axially-symmetric ionospheric models', *J. Atmos. & Terr. Phys.*, **34,** pp. 1137–1140

GETHING, P.J.D. (1974): 'Phase path characteristics for oblique ray paths', *J. Atmos. & Terr. Phys.*, **36,** pp. 1–8

HASELGROVE, J. (1955): 'Ray theory and a new method for ray-tracing', Report on the conf. on the physics of the ionosphere, Cambridge 1954 (Physical Society, London), pp. 355–364

JONES, R.M. (1966): 'A three-dimensional ray tracing computer program', Essa Technical Report IER 17–ITSA 17

KELSO, J.M. (1964): 'Radio ray propagation in the ionosphere' (McGraw–Hill)

KIFT, F. (1960): 'The propagation of high frequency radio waves to long distances', *Proc. IEE.*, **107B,** pp. 127–140

McNAMARA, L.F. (1988): 'Ionospheric modelling in support of single station location of long range transmitters', *J. Atmos. & Terr. Phys.*, **50,** pp. 781–795

MALIPHANT, R.G. (1968): 'HF ground backscatter analysis with the aid of computer ray tracing', Agard Conf. Proc. no. 37, Part I, pp. 7–1 to 7–16

MILLINGTON, G. (1951): 'The effect of the earth's magnetic field on short-wave communication by the ionosphere', IEE. Monograph no. 1.

MILLINGTON, G. (1954): 'Ray path characteristics in the ionosphere', IEE. Monograph no. 94

MILSOM, J.D. (1985): 'Exact ray path calculations in a modified Bradley/Dudeney model ionosphere', *Proc. IEE Pt. H.*, **132,** pp. 33–38

MULDREW, D.B. (1959): 'An ionospheric ray-tracing technique and its application to a problem in long-distance radio propagation', *IRE. Trans.*, **AP-7,** pp. 393–396

RAO, N.N. (1969): 'Bearing deviation in high-frequency transionospheric propagation. 3. Ray tracing investigation of the magneto-ionic effect', *Radio Sci.*, **4,** pp. 153–161

RAO, N.N., and BECKWITH, R.I. (1975): 'Reduction of positional estimate errors in HF hyperbolic position-fixing systems', *IEEE. Trans.*, **AES-11,** pp. 113–114

RATCLIFFE, J.A. (1962): 'The magneto-ionic theory and its applications to the ionosphere' (Cambridge University Press)

ROSE, R.B. (1988): 'High-resolution HF time of arrival measurements (1981–1985)', *Radio Sci.*, **23,** pp. 257–264

de VOOGT, A.H. (1953): 'The calculation of the path of a radio-ray in a given ionosphere', *Proc. IRE.*, **41,** pp. 1183–1186

Chapter 15

The effects of ionospheric tilts

15.1 Introduction: thin tilted-mirror model

When the properties of an ionospheric layer vary with latitude and longitude, we can speak of gradients in the ionospheric parameters. The layer can be described, for brevity, as tilted, but it must be borne in mind that a thick layer of variable properties does not necessarily behave in the same way as a tilted mirror. With this proviso, we now consider the effects of tilts on DF measurements, possible ways in which tilt corrections might be applied and possible ways of acquiring the necessary information about the ionosphere.

The effective tilt of an ionospheric layer of finite thickness can vary with the depth of penetration of a ray into the layer and is therefore dependent on the frequency and elevation angle of the ray, as well as the gradients of various ionospheric parameters. In brief, the tilt depends on the ray as well as the layer and can vary from ray to ray. Nevertheless, it is helpful to consider first a simplified model in which the ionosphere is replaced by a thin reflecting sheet (sometimes called a 'copper sheet') that acts as a mirror. The tilt effects that we have to consider then become a matter of geometry alone.

The transverse component θ_T of the tilt deviates the ray path out of the great-circle plane and hence produces a difference between the azimuth of arrival α_R of the ray and true bearing β of the transmitter. For brevity, this difference $(\alpha_R - \beta)$ will be referred to as the bearing error $\delta\beta$. For a plane mirror at a height h (at the reflection point) above a plane earth, the bearing error on a one-hop path is approximately

$$\delta\beta_1 = \frac{2h\theta_T}{D}$$

for a ground path-length D, provided θ_T is small (see Fig. 15.1). If, for example, $h = 300$ km, $D = 600$ km, $\theta_T = 1°$ then $\delta\beta_1$ is also 1°. Clearly, the effect of a given tilt is greater on a short, high-angle path than on a long path.

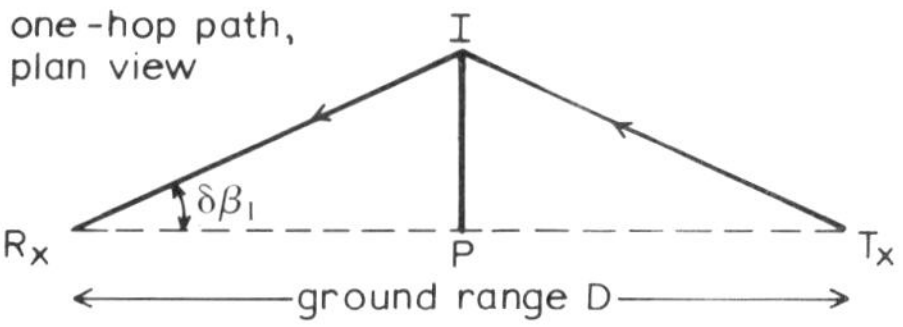

Fig. 15.1 Effect of a transverse-tilt component
θ_T = transverse tilt at I
h = height of mirror
$IP = h\theta_T$
$\delta\beta_1 = \dfrac{2h\theta_T}{D}$

On a two-hop path, under the same assumptions, each hop of length $D/2$ contributes $2h\theta_T/(D/2)$ bearing error and the total error is

$$\delta\beta_2 = \frac{8h\theta_T}{D}$$

This result can easily be verified by geometrical construction. Generalising to n hops, we add together n contributions each of $2hn\theta_T/D$ and obtain

$$\delta\beta_n = \frac{2hn^2\theta_T}{D}$$

Because

$$\tan\Delta_n = \frac{2hn}{D}$$

where Δ_n is the elevation angle on the n-hop path, $\delta\beta_n$ can also be written as

$$\delta\beta_n = n\theta_T \tan\Delta_n \tag{1}$$

The expected variance component of bearing error due to tilts clearly builds up very rapidly with an increasing number of hops. If all reflection points are at the same height, and with the same tilt at all points, the variances are proportional to 1, 16, 81,...for 1, 2, 3...hops. If the tilts are uncorrelated, each tilt may be assumed to have a statistical distribution with variance v and if all the vs are equal, the corresponding figures are proportional to 1, 10, 35...for 1, 2, 3...hops.

There is some experimental confirmation that much larger bearing errors arise in multihop propagation than in single-hop propagation. Bramley (1955) found that, on a 700 km path, the variance of two-hop F-layer bearings was about 12 times that of one-hop F-layer bearings. Hayden (1961) also found a striking difference in accuracy between one-hop and multihop bearings on a 452 km path.

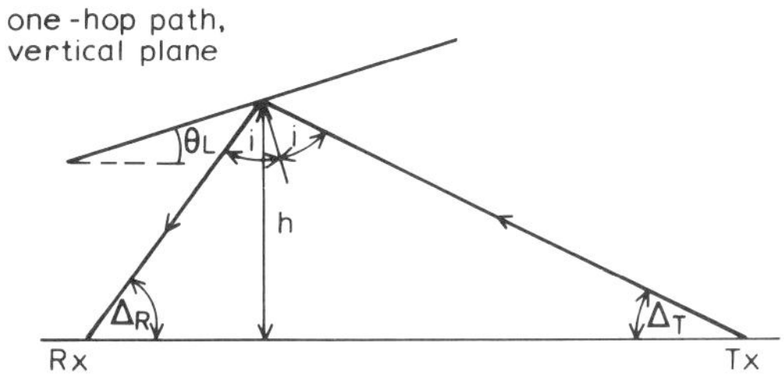

Fig. 15.2 Effect of a longitudinal-tilt component

The effect of a longitudinal tilt component θ_L is to produce a difference between the elevation angle of transmission Δ_T and the elevation angle of reception Δ_R for a given ray. From Fig. 15.2

$$\Delta_R - \Delta_T = 2\theta_L$$

The true ground-range of the transmitter is

$$D = h \cot \Delta_T + h \cot \Delta_R$$

whereas the 'no-tilt' range inferred from a measurement of Δ_R and a knowledge of *h at the reflection point* is

$$D_c = 2h \cot \Delta_R$$

The range error is

$$\begin{aligned} D_c - D &= h(\cot \Delta_R - \cot \Delta_T) \\ &\simeq -2h\theta_L \operatorname{cosec}^2 \Delta_R \end{aligned}$$

which becomes large as Δ_R becomes small, i.e. on long paths. The fractional error in range is

$$\frac{D_c - D}{D_c} \simeq \frac{-2\theta_L}{\sin 2\Delta_R}$$

and is therefore a minimum when $\Delta_R = 45°$.

On long paths, where earth-curvature effects become significant, the natural extension of the thin-plane-mirror model is to a thin spherical reflecting shell surrounding a spherical earth. The departure of the geoid from a true sphere appears to be quite negligible in this context. Tilt effects can be introduced, without loss of some of the simple geometrical properties of spheres, by displacing the centre of the reflecting shell relative to the centre of the earth. Such a model is occasionally used in theoretical studies of tilt effects, but a more common procedure is to use a three-dimensional ray-tracing program so that the effects of earth curvature and of finite layer thickness can both be taken into account. Croft and Fenwick (1963) considered a plane tilted mirror above a curved earth and their paper contains graphs showing the effects on range and bearing of a specified tilt.

15.2 Effective tilt of a thick layer

15.2.1 General theory

For a layer of finite thickness, with gradients in one or more parameters (e.g. h_0, h_m, f_c in the case of a parabolic layer), the thin-mirror model is not satisfactory and a more sophisticated approach, based on a study of phase path, must be adopted in order to determine the effective tilt.

We now consider the simplest possible direction-finder, consisting of two antennas mounted a distance a apart on a line perpendicular to the great-circle bearing of the transmitter. Ideally in the absence of a lateral tilt, there would be no phase difference between the signals in the two antennas. If the respective phase paths from the transmitter to the two elements differ by an amount δP when a lateral tilt is present, the apparent bearing error $\delta\beta$ in radians is

$$\delta\beta = \delta P/(a \cos \Delta)$$

For a one-hop path, the two reflection points are separated by a distance of $a/2$. Suppose now that the difference of phase path is due to a transverse gradient of some ionospheric parameter I. If this gradient is $dI/d\xi$, where $d\xi$ is an element of distance transverse to the path, the difference in the value of I at the two reflection points is $(a/2)(dI/d\xi)$ and the bearing error can be written

$$\delta\beta_1 = \frac{1}{2 \cos \Delta} \left(\frac{dI}{d\xi}\right) \left(\frac{dP}{dI}\right)$$

If the value of (dP/dI) can be found by differentiation of a suitable expression for P and if $(dI/d\xi)$ can be estimated from ionospheric measurements or predictions, it is possible in principle to correct an observed azimuth angle for the tilt error $\delta\beta_1$.

It should be noted that the total differential (dP/dI) must be found (as in Section 14.7) subject to the condition that the ground range D is constant.

This condition can only be satisfied if the elevation angle Δ differs slightly between the two paths. Thus, we make use of the equation

$$\frac{dP}{dI} = \frac{\partial P}{\partial I} + \frac{\partial P}{\partial \Delta}\frac{d\Delta}{dI}$$

A simple illustration of the theory is provided by the case previously considered, of a tilted-plane thin mirror. The gradient is then in height h and the transverse tilt component is

$$\theta_T = \frac{dh}{d\xi}$$

Also

$$\frac{dP}{dh} = \frac{\partial P}{\partial h} - \frac{\partial D}{\partial h}\left(\frac{\partial P}{\partial \Delta}\bigg/\frac{\partial D}{\partial \Delta}\right)$$

for D constant, and

$$P = 2h \text{ cosec } \Delta$$

$$D = 2h \cot \Delta$$

Therefore

$$\frac{dP}{dh} = 2 \text{ cosec } \Delta - 2 \cot \Delta \cos \Delta$$

and

$$\delta\beta_1 = \frac{1}{2 \cos \Delta}\theta_T (2 \text{ cosec } \Delta - 2 \cot \Delta \cos \Delta)$$

$$= \theta_T \tan \Delta$$

in agreement with eqn. 1.

The theory can be extended to a two-hop path with reflection heights h_1, h_2 and tilts θ_1, θ_2. The subscript 1 refers to the reflection point nearer the transmitter. The resulting bearing error is found to be

$$\delta\beta_2 = \frac{\tan \Delta}{(h_1 + h_2)}\{\theta_1 h_1 + \theta_2 (2h_1 + h_2)\}$$

If $h_1 = h_2 = h$ this becomes

$$\delta\beta_2 = \frac{2h}{D}(\theta_1 + 3\theta_2)$$

It is the tilt at the reflection point nearer the receiver that has the larger effect. With n hops from reflection points at heights $h_1, h_2, \ldots, h_n$ (numbered from transmitter to receiver) the appropriate equation is

$$\delta\beta_2 = \frac{2}{D}\{\theta_1 h_1 + \theta_2 (2h_1 + h_2) + \ldots$$
$$+ \theta_n (2h_1 + 2h_2 + \ldots + 2h_{n-1} + h_n)\}$$

15.2.2 Lateral tilt of a single parabolic layer

An examination of the expressions given in Appendix 7 for D, P and P' for a parabolic layer shows that the frequency f enters only through the ratio x of f to the critical frequency f_c. It follows that *for a parabolic layer*, eqn. 10 of Section 14.7 can be written in the form

$$\frac{dP}{df_c} = \frac{-1}{fc}(P' - P)$$

and hence

$$\delta\beta_1 = \frac{(P - P')}{2f_c \cos \Delta} \frac{df_c}{d\xi}$$
$$= \frac{y_m}{4f_c \cos \Delta}\left[\left\{x \sin^2 i_0 (1 - 2\epsilon) - \left(x + \frac{1}{x}\right)\right\} \ln F\right.$$
$$\left. + 2 \cos i_0 (1 + 3\epsilon x^2 \sin^2 i_0)\right] \frac{df_c}{d\xi}$$

For a flat-earth approximation, $\epsilon = 0$, this reduces to

$$\delta\beta_1 = \frac{y_m \cot i_0}{4f_c}\left\{2 - \left(z + \frac{1}{z}\right) \ln \left(\frac{1 + z}{1 - z}\right)\right\} \frac{df_c}{d\xi} \qquad (2)$$

where $z = x \cos i_0$. This expression agrees with one derived by Titheridge (1958) by geometrical arguments and is consistent with an equation quoted by Whale (1959) for transmission *through* a layer with a gradient in f_c. It should be noted that the effective tilt and bearing error can become very large as the penetration condition $z = 1$ is approached.

The effects of gradients in other ionospheric parameters of a parabolic layer, and the case of penetration of a lower layer and reflection at an upper layer, are considered in Appendix 8. The effect of a penetrated layer can perhaps best be understood by an optical analogy. There is no DF error (to a first approximation) if the layer is equivalent to a parallel-sided slab of glass, even if the slab is tilted. However, if the layer is equivalent (in cross-section) to a prism, it will produce a deflection of the ray path, and the deflections on the upward and downward paths are additive.

15.2.3 Longitudinal tilt of a single parabolic layer

The effects of a uniform longitudinal tilt in the plane of propagation on, for example, the shape of the leading edge of a swept-frequency backscatter display have been studied by a number of investigators (Croft, 1968; Tushentsova, 1969; Hatfield, 1970). The commonly-adopted model (Folkestad, 1968) is one in which the centre of a spherically-stratified parabolic layer does not coincide with the centre of the earth, but both lie in the plane of propagation, see Fig. 15.3. Increasing tilt values correspond to increasing separation of these two points. The movement of one centre relative to the other can be so arranged that the shortest distance from the sounding station *S* to the layer does not change; that is to say, the same 'normal-incidence' ionogram would be obtained whatever the tilt value. The magnitude of the tilt is defined by the gradient in layer height in the neighbourhood of the station; it should be noted that, in this model, the gradient decreases with increasing ground range and eventually changes sign, but for paths out to a few thousand kilometres it is legitimate to regard the tilt as approximately constant. A positive tilt is defined arbitrarily here as one in which the height of the layer increases with increasing distance along the sounding path, as in Fig. 15.2.

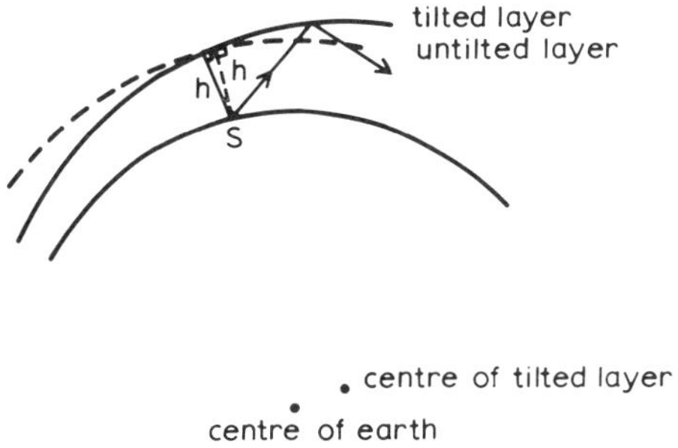

Fig. 15.3 Spherical layer with displaced centre gives simulated tilt

The geometrical steps needed to find the lengths of the ray-path segments below the ionosphere and the angles of incidence and emergence at the base of the ionosphere are set out in a paper by Folkestad (1968). Within the spherically-stratified layer, the ray can be traced in the normal way; for example, the standard results for the parabolic layer can be used when this form of distribution is appropriate. Ray-tracing studies with comparatively small tilts ($\sim 1°$ or $2°$) and short paths show that the elevation angles at the point of transmission and reception differ by approximately $2\theta_L$, as expected from simple tilted-mirror theory.

Longitudinal tilts produce errors in the determination of range with SSL systems, largely because of the effect on measured elevation angles at the point of reception. Lambert (1985) modelled the changes in the angles of arrival induced by a TID on a 300 km path.

15.2.4 Modelling of travelling ionospheric disturbances

Travelling ionospheric disturbances (TIDs) are known to occur in the ionosphere, particularly in the F-region. These disturbances produce wave-like corrugations in the iso-ionic contours, with quasiwavelengths of the order of 50–200 km and horizontal velocities in the region of 300 m/s. Disturbances in the E-region are generally of smaller amplitude.

As a TID moves across a propagation path, the effective tilt components vary with time. The measured azimuth angle and elevation angle of arrival of a ray at a DF will therefore fluctuate with time until a TID has moved away from the neighbourhood of the reflection zone. Some research workers such as Croft (1968), Georges and Stephenson (1969), Hunsucker (1971) and George (1972) have generated computer models of TIDs and have calculated the angle fluctuations for various path geometries. It is by no means clear that a time sequence of measured angles can be used to construct a unique and correct model of the TID.

15.3 Tilt-correction schemes

15.3.1 Introduction

Various possible methods of obtaining information about tilts and of applying corresponding corrections to observed azimuth angles will now be briefly considered. The main objective here is to classify the methods and to outline some of the difficulties. Various published papers, to which reference will be made, can be consulted for more information on the results obtained, which give a picture of the physical properties of tilts and TIDs, and the practical problems of analysing incomplete and imperfect ionospheric data.

15.3.2 Models and predictions

A few possible methods for the calculation of tilts that do not require real-time measurements are as follows.

Chapman model

The Chapman α model (see Davies, 1965) is often used as an approximation to the electron density in the E- and F1-regions of the ionosphere. The electron density N is given by

$$N = N_0 \exp \tfrac{1}{2} \{1 - Z - \sec \chi \exp(-Z)\}$$

where χ is solar zenith angle, Z is a normalised height and N_0 is the maximum electron density at $Z = 0$, $\chi = 0$. At any position, the height of maximum ionisation h_m and the critical frequency f_0 are functions of χ; h_m is a minimum, and f_0 is a maximum, at the subsolar point. The tilt of an iso-ionic contour can be calculated; it proves to be very small (0·03° on a typical path) except near the 'edges' of the layer at the sunrise/sunset boundary (terminator).

The F2-region does not behave like a Chapman layer and the model is therefore not recommended for this region. Observed systematic bearing errors for a mixture of propagation modes do not show a simple 1:1 dependence on zenith angle (Morgan, 1974) and appear to be larger than expected for a Chapman layer.

Predictions

Monthly median predictions of ionospheric parameters are issued in the form of 'numerical maps'. (Jones and Gallet, 1962; see also Jones and Stewart, 1970). Alternatively, the maps can be represented by numerical coefficients that are functions of sunspot number (Davies, 1981). From these, or the corresponding contour charts of f_0 and h_m, effective tilts can be calculated. The predictions provide a useful indication of the probable magnitudes of systematic sunrise and sunset tilts, *N–S* tilts near midday etc, but it must be appreciated that much smoothing of ionosonde measurements is involved in the preparation of numerical maps and that this smoothing process may lead to an under-estimate of the real tilts.

Regression analysis

Regression analysis is based on the assumption that the bearing correction required on a specified path is a function of many parameters such as path length, orientation, frequency, time of day, season etc. From an analysis of many observations on a variety of paths it might be possible to determine a suitable relationship between bearing correction and the most important factors.

Partial correlations between, for example, bearing error and time of day can undoubtedly be detected in good-quality WADF bearings. However, because of the existence of *E*s ionisation and of the considerable day-to-day variability of the ionosphere, the propagation modes on a given path are not simple functions of time of day and season. Regression analysis therefore appears to be a rather blunt instrument, even for the calculation of systematic tilts.

15.3.3 Use of real-time measurements

Some possible methods of obtaining real-time information about tilts will now be described. Apart from the first two below, all require the use of auxiliary equipment, in addition to the DF system.

Check transmitters

Bearings taken on transmitters of known position can provide information about tilts. In the simplest version, a check transmitter (CT) must be located close to a target transmitter (TT) and must be radiating at about the same time on about the same frequency. The bearing error on the CT is then subtracted from the observed bearing on the TT, on the assumption that the path geometries are so similar that the equation

$$\text{True bearing of TT} = \text{Observed bearing of TT} - (\text{Observed bearing of CT} - \text{true bearing of CT})$$

is approximately correct. An examination of the corresponding variance equation shows that the accuracy of the final bearing can be *degraded* by this procedure unless the observed bearings are of high quality: random errors should be sufficiently small and tilt-generated errors sufficiently large for the correlation coefficient between large samples of errors on the two targets to be at least 0·5.

The way in which the correlation of bearing errors on a group of collocated transmitters falls off with increasing frequency separation has been studied by Reynolds and Morgan (1975). Bearing fluctuations with periodicities of the order of 12 minutes were well correlated only when the frequencies differed by less than 2 MHz. Correlations in space and time depend on the quasiwavelength λ of the disturbance and its velocity. For good positive correlation between simultaneous bearing errors on two separated transmitters it is reasonable to suppose that the separation of the reflection points should not exceed $\frac{1}{8}\lambda$ and of the transmitters $\frac{1}{4}\lambda$. Thus, a ground separation of 25 km or less is required for TIDs with $\lambda = 100$ km. At greater separations, it might be possible to obtain satisfactory correlation between records of bearing errors against time if a time slip is applied to one of the records. The time slip required depends on the speed and direction of the TID, which might be determined from observations on several closely-spaced CTs.

A correction procedure based on the use of CTs is superficially very attractive because of its simplicity, but it is difficult in practice to find CTs satisfying the necessary criteria. For a 1:1 correction scheme, without time slip, we require that

(*a*) the CT should be close (and known to be close) to the TT,
(*b*) the true position of the CT should be accurately known,
(*c*) it should transmit at about the same time as the TT, on about the same frequency.

Even if the stringent requirements are satisfied, it cannot be guaranteed that the mode structures of the two signals are identical; relative mode strengths are affected by the radiation patterns of the transmitting arrays.

By searching over a wider area we should be able to find a greater number of suitable CTs, but at the price of two additional complications. Firstly, it is necessary to apply a time-slip, usually based on the assumptions that the TID moves in a fixed direction with a constant speed and without distortion. Secondly, because the path geometries may be very different for the different transmitters, it is necessary to convert from bearing error on each CT to tilt, then from tilt back to bearing correction for the TT; we therefore need to achieve successful mode resolution and identification on each path.

Partial success with the 1:1 correction method with time-slip has been reported by Jones and Reynolds (1975); a significant reduction of variance is obtained provided

(*a*) propagation is by one mode only,
(*b*) a long time sequence of bearings over a period of an hour or more is

obtained, from which the time-slip giving the best overall correlation is determined.

Extrapolation to zero-hop mode

For a tilted thin plane mirror about a flat earth, the bearing errors $\delta\beta_1$, $\delta\beta_2$ etc expected on 1, 2...hop paths (Section 15.1) can be written

$$\delta\beta_1 = m$$

$$\delta\beta_2 = 4m$$

$$\delta\beta_r = r^2 m$$

where

$$m = \frac{2h\theta_T}{D}$$

Suppose that the bearings β_1 and β_2 have been measured, and are known to correspond to one-hop and two-hop modes, respectively. Then

$$\begin{aligned} \beta_2 - \beta_1 &= \delta\beta_2 - \delta\beta_1 \\ &= 3m \end{aligned}$$

and the true bearing β is

$$\begin{aligned} \beta &= \beta_1 - m \\ &= \beta_1 - \tfrac{1}{3}(\beta_2 - \beta_1) \\ &= \tfrac{4}{3}\beta_1 - \tfrac{1}{3}\beta_2 \end{aligned}$$

Clearly, this procedure can be generalised to include other identified modes. In effect, the sequence of bearings β_1, β_2, β_3 etc is examined and an intelligent extrapolation is made to find β_0, the bearing for a zero-hop path, i.e. a hypothetical path not affected by the tilt.

This procedure does not appear to have been used, because of the difficulty of obtaining separate bearings on identified modes (from the same layer). It is also based on the assumption of a constant tilt at the two or more reflection points used in the analysis.

Direct measurement of local tilt

Given a pulsed sounder at the DF site and a DF system that determines the complete DOA of a returning echo from the ionosphere, i.e. both the elevation angle and azimuth angle, it is possible to measure the tilt of the layer above the site. A pulsed sounder is required in order to reject multiple echoes reflected by the earth and the ionosphere. The same tilt might then be assumed to apply over high-angle oblique paths, i.e. for short ground ranges. Treharne *et al* (1969) found good correlation between tilts measured on pulse echoes and the apparent tilt seen on very short-range paths.

Ionospheric profiles deduced from sounders

Measurements of ionospheric parameters by any of a variety of techniques of sounding, e.g. vertical incidence, oblique bistatic or oblique backscatter and topside sounding from a satellite, can provide information on the shapes of iso-ionic surfaces, from which effective tilts can be calculated.

For example, Rao (1968) made use of results from five vertical-incidence sounders near the midpoint of a path from Houston, Texas, to Urbana, Illinois and obtained excellent agreement between the bearing errors calculated from the sounder data and those observed with a DF at Urbana. The VI data are of great assistance in the identification of the propagation mode or modes on the oblique path. Rao (1969) also made use of measurements of total electron content in a layer, deduced from the Faraday rotation of the polarisation of signals from satellites, and showed that one correction computed in this way (when two satellite orbits happened to cross nearly above the path midpoint) lay very near the curve of bearing error against time observed on the Houston–Urbana path.

Provided suitable measurements are available for several sounding points in the desired area at the desired time, this method is clearly capable of providing valid measurements of the average tilts, over the area of the sounding network, of any selected iso-ionic surfaces. TIDs with sizes smaller than the network spacing could escape detection, yet cause significant DF errors.

The data-handling task of collection and analysis involved in this procedure should not be under-estimated; for example, true-height analysis of a time-sequence of VI ionograms from each of a network of established ionosondes might be required. The possible use of oblique and backscatter sounders to determine ionospheric profiles is discussed in Section 15.4.4.

Backscatter sounder as a tilt detector

If a backscatter (BS) sounder is available at or near the DF site, it can be used for at least three distinct purposes;

(*a*) as an aid to mode identification
(*b*) as a possible means of determining a synthetic electron-density profile for a remote area of the ionosphere (but elevation-angle measurements in the backscatter echoes are required – see Section 15.4.4),
(*c*) as an instrument for the detection and/or measurement of ionospheric tilts.

To date, the possibility of using a BS sounder for item (*c*) above has not received a great deal of attention and further developments, both in hardware and in methods of interpretation, would be required for implementation. For a typical BS system in the HF band, the ground cell defined by the pulse length and azimuthal beamwidth of the antenna array is large compared with an average TID and the resolving power of the system is therefore limited. Nevertheless, the effects of tilts and TIDs in BS records have been detected and described, e.g. by Hunsucker and Tveten (1967) and Ranzi and Giorgi (1972).

Doppler measurements
A change with time of the phase path P between a transmitter and a receiver leads to a Doppler shift of the received signal relative to a frequency standard at the receiver. The Doppler shift is proportional to rate of change of phase path and the total change of P over a time interval is therefore proportional to the integral, with respect to time, of the Doppler shift.

If the speed and direction of (say) a TID near the reflection point are to be determined, it is necessary to employ at least three spaced transmitters and a central receiver. Details of experimental techniques have been given by Davies (1962) and by Jones *et al* (1975). The velocity is found from the time-displacement of corresponding features in the Doppler records for the different transmitters and the known geometry of the transmitting and receiving locations.

Jones and Reynolds (1975) showed that this technique can be used successfully on an experimental path, provided a network of Doppler sounders working at near-vertical incidence can be established near the midpoint. Jones and Spracklen (1976) deduced the properties of TID's on an 850 km path by monitoring Doppler shifts at several separate receiving sites. The bearing deviation produced by each TID was then estimated. A very considerable improvement in accuracy was obtained when corresponding corrections were applied to observed bearings.

The possible extension of techniques of this kind to less favourable or less restrictive conditions deserves further study.

15.4 Determination of the state of the remote ionosphere

15.4.1 Objectives

In modern DF research it is very helpful to know as much as possible about the state of the ionosphere at the reflection point or points of an oblique path under study. The state of the local ionosphere above the DF site can be determined by VI sounding, but it may not be safe to adopt the same parameters (or parameters extrapolated geographically with the aid of monthly predictions) for the reflection points. It is convenient to refer to the reflection points collectively as the remote ionosphere, even when the path length is only a few hundred kilometres, and to say loosely that we wish to determine the state of the remote ionospere.

The following uses of ionospheric data, requiring increasing precision for the later items in the list, can be distinguished:

(*a*) Qualitative information about ionospheric storms, etc as an aid to classification of bearings, rejection of results obtained in abnormal conditions etc,
(*b*) Qualitative information about the presence of sporadic E ionisation, F1-layer etc, for the determination of possible modes of propagation,
(*c*) More precise information on critical frequencies etc for the determination of the probable modes of propagation, with greater statistical accuracy than in (*b*),

(*d*) Further information about layer heights, as well as critical frequencies as in (*c*), for the calculation of ranges from measured elevation angles,
(*e*) Information about lateral tilts, for the correction of bearing errors, and about longitudinal tilts, for the correction of range errors.

In principle, all this information could be obtained from a closely-spaced network of VI sounders. However, such sounders may be nonexistent (e.g. in an ocean region) or too widely spaced; and the output information may not be available sufficiently rapidly. The question then arises as to whether some or all of the required information can be obtained from oblique sounders, under the control of the experimenter. The two main techniques are

(*a*) oblique bistatic sounding, i.e. transmitter and receiver at different sites,
(*b*) oblique backscatter sounding, with the backscatter system near the DF site.

For brevity, these will be referred to as oblique and backscatter sounding respectively. In this Section the possibility of measuring the properties of the remote ionosphere using either of these techniques will be discussed briefly. It will be assumed that a sounder or sounders can be deployed on or near the path of interest and that the primary aim is to determine an electron-density profile for the remote ionosphere. Such a profile would be synthetic, that is to say it represents the average properties of the ionosphere over the area of interest rather than the true physical profile above a particular point on the earth's surface. For use as input to a ray-tracing computer program the synthetic profile is perhaps more appropriate, for an oblique path, than a true profile.

If the amount of information contained in a particular form of sounding record is insufficient for the determination of the synthetic profile, either because of the limitations of the sounding technique or because of the incomplete and imperfect nature of practical records, it may nevertheless be possible to determine a few parameters defining a simple approximate model of the ionosphere; for example, the heights and critical frequencies of one or more parabolic layers might be estimated.

15.4.2 General remarks on the analysis of oblique records

Gething and Maliphant (1967) pointed out that the mathematical process of analysing either form of oblique record can be regarded formally as consisting of two steps:

(*a*) conversion of the oblique data to an equivalent 'vertical' ionogram,
(*b*) performance of a true-height analysis on the 'vertical' ionogram.

In practice, the resulting synthetic profile may or may not be a useful representation of the state of the ionosphere. The problem is partly a mathematical one, depending on errors introduced in the geometrical transformation (step (*a*)) by approximations such as the neglect of earth curvature, and partly a physical one, depending on the variability of the ionosphere, the existence of tilts etc. The problem of calculating electron-density profiles from VI ionograms has received considerable theoretical attention; Wright and Smith (1967) have described several suitable methods for routine use, in which some account can be taken of magneto-ionic effects. Thus, step (*b*) should not introduce significant errors into the analysis.

For a specified profile, oblique and backscatter records can be readily simulated with the aid of a ray-tracing program. The converse problem of transforming from oblique incidence to vertical incidence is less straightforward, but it should be noted that a trial profile can be adjusted until satisfactory agreement is obtained between the computed and observed oblique records. Iterative methods therefore suggest themselves; the ultimate accuracy attainable depends on the sophistication of the ray-tracing program employed and on the quality of the practical records.

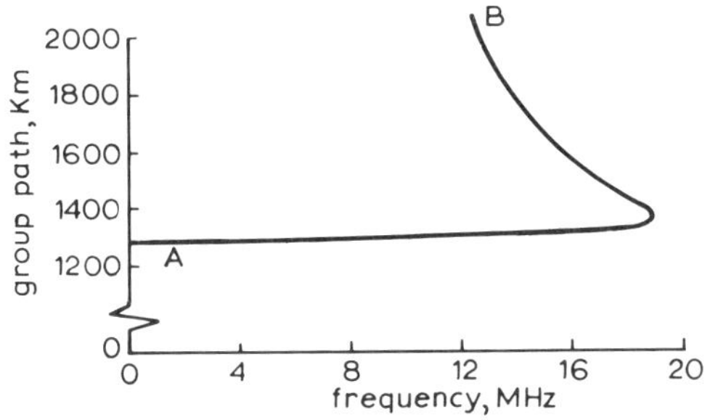

Fig. 15.4 Oblique ionogram for ground range of 1200 km
single parabolic layer: f_c = 10 MHz, h_m = 280 km, y_m = 80 km

Fig. 15.4 shows a calculated oblique ionogram for a path length of 1200 km. Practical records obtained by photography of an oscilloscope display depart from this ideal form for a variety of reasons. Portions of the low-frequency and high-angle segments of the curve, i.e. segments near *A* and *B* in Fig. 15.4, are lost because of inadequate signal strength. Bands of fixed-frequency interference may obscure some other parts of the curve, although chirp sounding techniques can be used to obtain greatly improved signal/noise ratios. The limited frequency bands of the transmitter/receiver combination and the antenna arrays employed, and the finite region of the (f, τ) domain covered by a single photograph, where $\tau = P'/c$, may further restrict the information available. The finite width of the transmitted pulse and the limited receiver bandwidth lead to a broadening of the curve. Various forms of ionospheric tilts and travelling disturbances can distort the curve from the shape expected for an untilted layer.

The same practical problems apply to backscatter ionograms. In addition, because of the comparatively broad beamwidth of a typical HF-antenna array and the existence of sidelobes in array patterns, the ground area illuminated cannot be regarded as a single small cell.

In practice, oblique sounding techniques including backscatter are of most potential value to the communicator when propagation conditions are difficult or abnormal and of most value in DF research when the properties of the remote ionosphere are likely to differ significantly from those of the overhead ionosphere. Records available for measurement and analysis are therefore likely to be imperfect and incomplete, and effects arising from ionospheric tilts will usually be present. The amount of information that can be extracted from such records naturally depends greatly on their quality and it is not possible to state conclusions of general validity on this subject. Clearly, however, robust methods of analysis that can be used with imperfect records are required.

15.4.3 Oblique ionograms

The following simplifying assumptions are usually made in analysis methods for oblique (bistatic) ionograms:

(*a*) The path length is known,

(*b*) The origin of the group path scale in, for example, Fig. 15.4 is known, i.e. absolute time delays rather than relative time delays have been determined,

(*c*) The record is complete from its origin at *A* (in Fig. 15.4) to some point *B* on the high-angle portion or, if not complete, missing portions can be sketched,

(*d*) Magneto-ionic effects can be neglected,

(*e*) The ionosphere over the region of interest is stratified into shells of constant electron density; these shells are spherical and centred on the centre of the earth,

(*f*) Hence, tilt effects can be neglected at least in the first analysis.

The assumption of spherical stratification is implicit in any analysis designed to calculate one synthetic profile from an oblique record. When tilts are negligible, curves for one-hop, two-hop etc propagation should yield the same profile; it is therefore sufficient to discuss the analysis of one-hop records.

The author has described (1969) an iterative method for the calculation of the synthetic profile. A first approximation to the profile is calculated from the observed ionogram, using 'flat earth' approximations. The oblique ionogram corresponding to this profile is then calculated, using curved-earth geometry in a suitable ray-tracing program and is compared with the observed record. The differences in frequency between the curves are measured at suitable values of τ and these residuals are then used to improve the accuracy of the profile, by a process of successive approximations. This procedure is based on the proportionality between the frequency scale of the profile and of the ionogram; that is to say, if the plasma frequency scale were multiplied by a constant factor, the frequency scale on the oblique ionogram would need to be multiplied by the same factor.

The synthetic profile can be defined either by the electron density in each of a large number of thin shells, or by a much smaller number of segments, as described in Section 14.4.2. M. S. Smith (1970) showed how the segment constants could be calculated to give a good fit between the corresponding oblique ionogram and the observed ionogram.

George (1970) assumed a linear variation of electron density with height within each of a series of slabs and showed how the profile could be built up by direct integration, starting from the base of the layer and the low-frequency end of the ionogram. A number of examples of the application of this method are given in his paper.

We now consider briefly whether some of the initial assumptions can be relaxed. If only relative time delays have been measured at the receiving station, the time origin might be estimated by comparing time delays for one-hop and two-hop modes reflected from the base of the lowest layer e.g. E-layer modes at the low frequency end of the record. Examination of theoretical oblique ionograms calculated by Kopka and Möller (1968), with a ray-tracing program that included the effects of the earth's magnetic field, suggests that the no-field ionogram lies approximately mid-way between the O and X traces. If this result is general, a good approximation to the electron-density profile should be obtained from practical records exhibiting magneto-ionic splitting by analysing a mean curve with a no-field program; certainly such an approximation should give a reliable solution at least up to the reflection height at which splitting becomes apparent. To obtain a more accurate solution or to extend the solution to greater heights might require the use of a more sophisticated ray-tracing program incorporating the effects of the earth's magnetic field.

When tilts along the sounding path are significant, the curves for various numbers of hops are not likely to lead to identical synthetic profiles. Mather, Wilson and Schneible (1965) showed that the separation between the one-hop and two-hop 'noses' of the curves gives an indication of the magnitude of the tilt.

In summary, a simple computer program applied to the routine analysis of imperfect records cannot always give the right answer; such a program should be regarded as an aid in the early stages of intelligent analysis. The principle of checking each approximation to the profile in order to find out what record it would produce is an effective safeguard against serious errors of interpretation.

15.4.4 Backscatter ionograms

We now consider a backscatter ionogram, i.e. a record of time delay against frequency for returned backscatter echoes, obtained with a swept-frequency or stepped-frequency sounder in a comparatively narrow azimuth sector defined by the radiation patterns of the transmitting and receiving arrays. For a detailed discussion of various features of such records, a review paper by Croft (1972) should be consulted.

It is known that the leading edge of a BS ionogram, i.e. the curve of minimum group time delay τ_{min} (two-way) against frequency f, depends on the electron density distribution of the area of the ionosphere involved. For a specified profile it is comparatively easy to generate the (f, τ_{min}) curve with the aid of a ray-tracing program. The transformation from a profile to a corresponding BS ionogram is therefore well understood; a paper by Maliphant (1968) shows several instructive examples. The leading edge for a very simple model of a single ionospheric layer is coincident with the two-hop vertical ionogram for all frequencies up to a 'branching' frequency f_b. For frequencies greater than f_b, a smaller value of travel time occurs at oblique incidence than at vertical incidence. Fig. 15.5 shows an example of a leading edge constructed for a single parabolic layer with the parameters shown on the diagram; in this simulation the effects of the earth's magnetic field have been neglected.

It is convenient to restrict the term 'leading edge' to the part of the curve for which the transmitted frequency f is greater than f_b. For the determination of the properties of the remote ionosphere in a specified distance range, the leading edge over the corresponding frequency range provides probably the only measurable feature on a BS ionogram. The amount of information obtainable from the shape and position of a segment of leading edge appears to be very limited. For example, it is clear from papers by Shearman (1956) and Hatfield (1970) that, for the very simple case of an untilted parabolic layer, it is difficult to determine both the critical frequency f_0 and the height of maximum ionization h_m, although either can be found if the other is known. The thickness parameter γ appears to have only a minor effect on the leading edge.

Rao (1974) described a method for the determination of all three

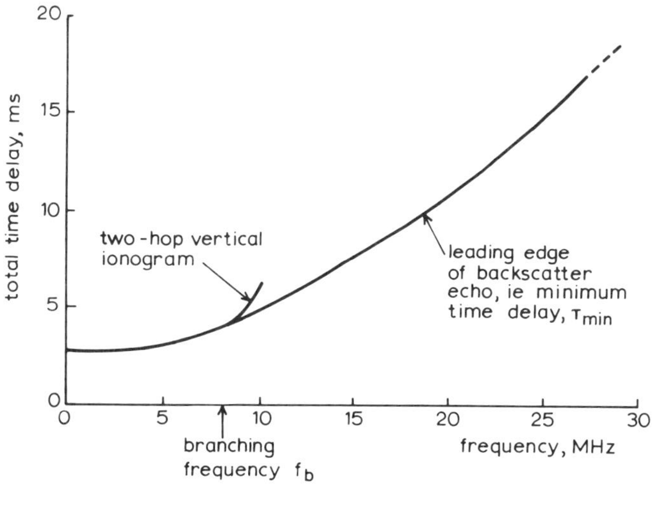

Fig. 15.5 Backscatter leading edge for a single parabolic layer layer parameters: f_c = 10 MHz, h_m = 300 km, γ = 0·4

parameters for a quasiparabolic layer, using as input the (f, τ_{min}) co-ordinates of three points on the leading edge. However, it is doubtful whether any such method is adequate for use with imperfect records, on which the effects of tilts might be present, because it can be shown that quite different ionospheric models lead to almost identical leading edges over a substantial frequency range (e.g. 9–17 MHz).

Additional sources of information must therefore be sought. The difficulty described above can be partly overcome if it is possible to make vertical angle of arrival measurements at various time settings in the backscatter echoes. George (1970) showed that his technique of direct step-by-step integration can be applied to BS data as well as oblique ionograms; the synthetic profile is calculated from measurements of elevation angle $\Delta(\tau)$ at various time delay settings τ in the echo received on a fixed frequency. However, he pointed out that the range of elevation angles necessary to probe the ionosphere in depth corresponds to a rather large spread in ground range and a correspondingly large area of the ionosphere. An alternative is to collect $\Delta(\tau)$ data over a range of frequencies, choosing values of τ and f so that a restricted area of the ionosphere is examined. Ideally, the points chosen would lie on a curve of constant ground range. Comparatively large antenna arrays are required for the angle measurements and the problems of resolving-power and masquerading discussed in Chapter 11 become more acute with the poor signal/noise ratios and complicated mode structure characteristic of BS returns. Nevertheless, some success has been reported by McCue (1956), Tveten and Hunsucker (1969) and Treharne (1969) in the measurement of elevation angles in backscatter.

With sophisticated BS equipment, enhanced echoes from ground features such as cities, islands and mountains are sometimes seen (Basler and Scott, 1972). In a swept-frequency BS ionogram, the enhanced return from a scatterer of limited area such as a city is very similar to an oblique bistatic ionogram from a transmitter at the same point, but with the time scale doubled. A synthetic profile can therefore be derived from such a curve by the methods described for oblique ionograms, *provided* the ground range of the feature responsible is known. In effect, the angle of arrival at a specified frequency can be calculated from the known value of D and the measured value of P', using Breit and Tuve's theorem; thus, identified ground features provide an angle calibration of the BS ionogram and hence add to the information available from the shape of the leading edge.

Lateral tilts in the ionosphere produce different leading edges to BS ionograms obtained on different sounding azimuths. TIDs can produce characteristic kinks in the shape of a leading edge; examples can be seen in Croft's review paper (1972). A longitudinal tilt along the sounding azimuth produces

(*a*) a change in the slope of the leading edge,
(*b*) a lack of simple proportionality between the time delays for various numbers of hops,

and, if vertical angle measurements are available,

(*c*) an increased spread in points transformed (using simple theory as for an untilted layer) from oblique to vertical incidence.

It seems possible that large-area tilts, i.e. tilts extending over distances of several hundred kilometres or more, could be detected and estimated if suitable analysis methods were developed. The resolving power of typical BS equipment is probably inadequate for the study of smaller tilt phenomena.

15.5 Summary

A full appreciation of the physical picture of tilts and TIDs that has emerged from experimental work, and of the power and limitations of various tilt-correction schemes, can be obtained only from a study of the extensive published literature on these topics. A comprehensive review of this literature has not been attempted here, but references have been given to some of the important papers.

For convenience, tilts are often divided into two or three rather arbitrary types. The average value of a tilt on a given path at (say) 11 a.m. each day over a period of (say) a month is often referred to as a systematic ionospheric tilt (SIT); it is important to bear in mind that such a tilt is visualised as systematic for a specified *time* and that observations on a single target tell us nothing about its *spatial* extent. Tilts that are 'systematic' over a comparatively large area (several hundred kilometres) at a given moment, but which are not necessarily systematic in time, are perhaps best described as large-area tilts (LATs). Those that change comparatively rapidly with both position and time are usually identified with travelling disturbances (TIDs).

Tilt correction is certainly not a simple or straightforward matter and only rather limited success has so far been achieved, on controlled experimental paths. A routine and reliable procedure for use under less favourable operational conditions, for long oblique paths, has yet to be devised; TIDs, which can lead to significant bearing errors, present a particularly difficult problem.

The chances of success with a tilt correction scheme would be greatly enhanced if the following goals could be achieved:

(*a*) mode resolution, so that a separate bearing can be taken on each arriving mode,
(*b*) mode identification, so that the geometrical factor linking tilts and bearing errors can be calculated.

In addition, detailed ionospheric information is required for several of the real-time correction schemes described above.

The following are tentatively suggested as the more promising lines of attack:

(*a*) For SITs: use of predictions, regression analysis or some other empirical method, (*b*) For LATs: use of several check targets, analysis of backscatter records,

(*c*) TIDs: Further development of Doppler techniques for experimental paths, but chances of success without the use of a special sounding network are not high.

Empirical SIT corrections are probably worth applying now to averaged ('consolidated') WADF azimuth measurements collected over many days at about the same time of day.

It must be emphasised, in conclusion, that a tilt of 1° is *not* the same thing as a bearing error of 1°; an unfortunate confusion sometimes arises between these two variables, which are often of about the same magnitude on low-angle mixed-mode paths. It is recommended that observed bearing errors should, wherever possible, be converted into corresponding tilts in the ionosphere; this interpretation helps to build up information about the real physical phenomena and facilitates the comparison of results obtained with different path geometries.

References

BASLER, R.P., and SCOTT, T.D. (1972): 'Ground backscatter observed with high resolution oblique sounders'. *Radio Sci.*, **7,** pp. 239–243

BRAMLEY, E.N. (1955): 'Some comparative directional measurements on short radio waves over different transmission paths'. *Proc. IEE.*, **102B,** pp. 544–549

CROFT, T.A. (1968): 'The influence of ionospheric irregularities on sweep-frequency backscatter'. *J. Atmos. & Terr. Phys.*, **30,** pp. 1051–1063

CROFT, T.A. (1972): 'Sky-wave backscatter: a means for observing our environment at great distances'. *Rev. Geophys. and Space Phys.*, **10,** pp. 73–155

CROFT, T.A., and FENWICK, R.B. (1963): 'Chart for determining the effects of ionospheric tilts using an idealised model'. *J. Res. Natl. Bur. Stand.*, **67D,** pp. 735–745

DAVIES, K. (1962): 'The measurement of ionospheric drifts by means of a Doppler shift technique'. *J. Geophys. Res.*, **67,** pp. 4904–4913

DAVIES, K. (1965): 'Ionospheric radio propagation'. NBS Monograph 80. (US Govt. Printing Office). See p. 17.

DAVIES, K. (1981): 'Review of recent progress in ionospheric predictions', *Radio Sci.*, **16,** pp. 1407–1430

FOLKESTAD, K. (1968): 'Exact ray computations in a tilted ionosphere with no magnetic field'. *Radio Sci.*, **3,** pp. 81–84

GEORGE, P.L. (1970): 'True height analysis of oblique incidence HF radio wave data', *J. Atmos. & Terr. Phys.*, **32,** pp. 905–916

GEORGE, P.L. (1972): 'HF ray tracing of gravity wave perturbed ionospheric profiles', Agard Conference Proceedings No. 115, Wiesbaden 1972, Paper 26

GEORGES, T.M., and STEPHENSON, J.J. (1969): 'HF radar signatures of travelling ionospheric irregularities, 3D ray-tracing simulation', *Radio Sci.*, **4,** pp. 679–696

GETHING, P.J.D. (1969): 'The calculation of electron density profiles from oblique ionograms', *J. Atmos. & Terr. Phys.*, **31,** pp. 347–354

GETHING, P.J.D., and MALIPHANT, R.G. (1967): 'Unz's application of Schlomilch's integral equation to oblique incidence observations', *J. Atmos. & Terr. Phys.*, **29,** pp. 599–600

HATFIELD, V.E. (1970): 'Derivation of ionospheric parameters from backscatter data', Agard Conference Proceedings No. 49 (Technical Editing and Reproduction Ltd, London), Paper 16

HAYDEN, E.C. (1961): 'Propagation studies using direction-finding techniques', *J. Res. Natl. Bur. Stand.*, **65D**, pp. 197–212

HUNSUCKER, R.D. (1971): 'Analysis of a backscatter signature obtained with a high-resolution HF radar system', *Radio Sci.*, **6**, pp. 763–768

HUNSUCKER, R.D., and TVETEN, L.H. (1967): 'Large travelling-ionospheric-disturbances observed at mid-latitudes utilising the high resolution HF backscatter technique', *J. Atmos. & Terr. Phys.*, **29**, pp. 909–916

JONES, T.B., KANTARIZIS, E., and MORGAN, A.D. (1975): 'Simultaneous measurements of the frequency and bearing of HF radio waves', *Proc. IEE.*, **122**, pp. 214–244

JONES, T.B., and REYNOLDS, J.S.B. (1975): 'Ionospheric perturbations and their effect on the accuracy of HF direction finders', *Radio and Electron. Eng.*, **45**, pp. 63–73

JONES, T.B, and SPRACKLEN, C.T. (1976): 'The correction of errors produced in H.F. direction finders by travelling ionospheric disturbances', AGARD Conf. Proc. No. 173, Athens 26–30 May 1975 (Technical Editing and Reproduction Ltd.)

JONES, W.B., and GALLET, R.M. (1962): 'Methods for applying numerical maps of ionospheric characteristics', *J. Res. Natl. Bur. Stand.*, **66D**, pp. 649–662

JONES, W.B., and STEWART, F.G. (1970): 'A numerical method for global mapping of plasma frequency', *Radio Sci.*, **5**, pp. 773–784

KOPKA, H., and MÖLLER, H.G. (1968): 'MUF calculations including the effect of the earth's magnetic field', *Radio Sci.*, **3**, pp. 53–56

LAMBERT, S. (1985): 'Travelling ionospheric disturbances and their effects on high frequency direction finding', *Trans. S. Afr. Inst. Electr. Eng.*, **76**, pp. 19–27

MALIPHANT, R.G. (1968): 'HF ground backscatter analysis with the aid of computer ray tracing', Agard Conference Proceedings No. 37, Part 1, pp. 7–1 to 7–16

MATHER, R.A., WILSON, F.C., and SCHNEIBLE, R. (1965): 'Determining ionospheric parameters over regions of denied access', *IEEE Trans.*, **AES-1**, pp. 104–107

McCUE, C.G. (1956): 'HF backscatter observations at Salisbury, South Australia', *Aust. J. of Phys.*, **9**, pp. 454–470

MORGAN, A.D. (1974): 'A qualitative examination of the effect of systematic tilts, in the ionosphere, on HF bearing measurements', *J. Atmos. & Terr. Phys.*, **36**, pp. 1675–1681

RANZI, I., and GIORGI, P. (1972): 'An important characteristic of some travelling ionospheric disturbances', Agard Conference Proceedings No. 115, Wiesbaden, pp. 25–1 to 25–2

RAO, N.N. (1968): 'Ray-tracing investigation of direction of arrival observations of HF radio waves', *Radio Sci.*, **3**, pp. 796–802

RAO, N.N. (1969): 'Direction of arrival of HF radio waves deduced from ionospheric electron content gradients', *IEEE. Trans.*, **AP-17**, pp. 111–113

RAO, N.N. (1974): 'Inversion of sweep-frequency sky-wave backscatter leading edge for quasiparabolic ionospheric layer parameters', *Radio Sci.*, **9**, pp. 845–847

REYNOLDS, J.S.B., and MORGAN, A.D. (1975): 'The effect of frequency separation on the correlation of bearing errors measured on ionospherically propagated HF signals from colocated transmitters', *J. Atmos. & Terr. Phys.*, **37**, pp. 545–551

SHEARMAN, E.D.R. (1956): 'The technique of ionospheric investigation using back-scatter', *Proc. IEE Pt. B.*, **103,** pp. 210–223

SMITH, M.S. (1970): 'The calculation of ionospheric profiles from data given on oblique incidence ionograms', *J. Atmos. & Terr. Phys.*, **32,** pp. 1047–1056

TITHERIDGE, J.E. (1958): 'Variations in the direction of arrival of high frequency radio waves', *J. Atmos. & Terr. Phys.*, **13,** pp. 17–25

TREHARNE, R.F. (1969): 'The measurement of the height of the distant ionosphere by backscatter observations', IREE Australia 12th National radio and electronics engineering convention, Sydney, pp. 162–163

TREHARNE, R.F., RAYMOND, M.L.J., and ELLARD, R.M. (1969): 'Ionospheric tilt observations in South Australia', IREE Australia 12th National radio and electronics engineering convention, Sydney, pp. 164–165

TUSHENTSOVA, L.A. (1969): 'Possibility of determining the slopes of the ionosphere by oblique sounding', *Geomag. and Aeron.*, **9,** pp. 82–85

TVETEN, L.H., and HUNSUCKER, R.D. (1969): 'Remote sensing of the terrestrial environment with an HF radio high-resolution azimuth and elevation scan system', *Proc. IEEE.* **57,** pp. 487–493

WHALE, H.A. (1959): 'The effects of ionospheric irregularities and the auroral zone on the bearings of short-wave radio signals', *J. Atmos & Terr. Phys.*, **13,** pp. 258–270

WRIGHT, J.W., and SMITH, G.H. (1967): 'Review of current methods of obtaining electron density profiles from ionograms', *Radio Sci.*, **2,** pp. 1119–1125

Chapter 16

Bearing accuracy and DF plots

16.1 Accuracy statistics for individual bearings

16.1.1 Definitions

Azimuth measurements taken on transmitters of known position, i.e. check targets, can be compared with the 'ideal' answer, namely the true g.c. bearing, so that accuracy statistics can be compiled for various purposes described in Section 16.1.2 below. The same procedure cannot be used for measurements of elevation angles, where the 'true' answer is generally unknown. We shall therefore confine our attention in this chapter to azimuth measurements, which will now be referred to for brevity as bearings. It should be noted that bearing errors can, and do, arise from propagation effects as well as from instrumental effects, and that we are not directly concerned here with the physical causes of instrumental errors, which were discussed in Chapter 6.

Suppose that a large number of bearings have been taken from one DF station on a number of different check targets. A histogram of bearing errors can then be constructed and a few parameters that summarise the properties of the histogram can be computed. The principal parameters are

(*a*) The percentage of 'wild' bearings, i.e. the percentage of bearing errors outside some arbitrary limit such as $\pm$ 10°,

(*b*) The systematic error (SE), i.e. the mean value of the bearing errors, excluding 'wilds',

(*c*) The standard deviation σ about the sample mean, with wilds excluded,

(*d*) DF variance V relative to zero bearing error, again excluding wilds.

The skewness and kurtosis of the distribution can also be computed (see Section 16.1.3). From these definitions it follows that

$$V = \sigma^2 + (\mathrm{SE})^2.$$

V is called the DF variance to distinguish it from the more usual definition of variance relative to the sample mean; it is a useful measure of the combined effect of systematic and random errors and is the correct quantity to use in weighting bearings and in computing probability areas.

V and σ depend partly on the DF site, partly on the instrument, partly on the averaging period used in taking a bearing and partly on the nature of the check target list (e.g. frequency range, distance range, transmitter power etc). In what follows, a single bearing should usually be interpreted as a bearing taken by an experienced operator who applies mental averaging to some form of bearing indicator over a period of about 2 minutes. The term 'snap bearing' sometimes used in published literature refers to a mental average over a much shorter period, usually a few seconds; it is *not* an instantaneous reading, such as would be obtained from a momentary bright-up of a visual display. The variance of single bearings is significantly less than that of snap bearings, which in turn is significantly less than that of instantaneous bearings.

The influence of the nature of the target list on accuracy statistics is sometimes overlooked. There is not a single variance figure that characterises the peformance of a given DF system on any type of target. Variances obtained on 'difficult' signals, e.g. weak transmissions heavily affected by interference, can easily be twice as great as those found for 'easy' signals. In comparison trials of two different DF instruments it is therefore most important that the same target list should be adopted for both. It must also be appreciated that a conscientious DF operator who tries to take bearings in even the most difficult conditions may appear to produce worse accuracy statistics than an operator who refuses to return bearings except under the most favourable conditions; the former will, however, produce much better figures for 'traffic pick-up'. In comparisons of equipments or operators the recommended procedure is to compute accuracy statistics from only a common sample of tasks.

16.1.2 Applications

The main applications for which statistics are collected and analysed are as follows:

(*a*) To monitor the performance of the direction finder and to detect instrumental faults,

(*b*) To study the nature and causes of errors arising on the propagation path,

(*c*) To study the variation of accuracy with parameters such as frequency, time of day, site etc.

(*d*) Hence, to assign weights to bearings from different DF stations on a given target under specified conditions,

(*e*) To use these weights in the calculation of a Best Point Estimate (BPE) for the position of the transmitter and a surrounding probability region.

In order to elucidate the physical sources of errors, for *a*, *b* and *c* above, it is often useful to sort bearings errors by factors such as frequency, azimuth sector etc. A division into subpopulations by one factor only, e.g. by 1 MHz frequency bands, is conveniently described as a primary sort; a further subdivision, e.g. by hour of day for each 1 MHz band, is conveniently described as a secondary sort. It is not usually sensible to go beyond secondary sorts, because sample sizes become too small to be useful. Factors such as frequency and time of day are often correlated and such a correlation reveals itself in very unequal sample sizes in secondary sorts; for example, most of the low-frequency bearings may be obtained at night and those on higher frequencies by day.

A Fourier analysis of average bearing errors (SEs) for (say) 10° azimuth sectors as a function of azimuth α may provide clues about the nature of instrumental errors. For example, a constant term can arise from a zero error of the azimuth scale or, with a spinning-goniometer system, from uncorrected time delay in the receiver. For a four-element Adcock, a significant ($\sin \alpha$) term suggests a phase inequality between the *N–S* feeders and a ($\cos \alpha$) term can arise from a similar inequality between the *E–W* feeds; a ($\sin 2\alpha$) term can arise from amplitude imbalance between the *N–S* and *E–W* pairs; a ($\sin 4\alpha$) term arises from uncorrected octantal error. However, random errors may mask small instrumental errors when small samples of sky-wave bearings are analysed and these effects are more easily studied by means of checks with test equipment and local ground-wave calibrations. In very large samples of bearings, even low-quality NADF bearings, comparatively small instrumental errors and the effects of systematic ionospheric tilts can sometimes be detected. Routine analysis of sky-wave bearings can also be used to detect any gross instrumental malfunction that arises between infrequent ground-wave calibrations.

With regard to weighting, it is relative weighting that determines the position of the BPE, but 'absolute' weights are required for the calculation of the probability region (see Section 16.3). When a variance can be assigned to a bearing (see Section 16.2.1) the weight is usually taken to be inversely proportional to the variance. Relative weighting is important because of the substantial differences in probable accuracy that can arise in bearings contributing to a fix; for example, a WADF bearing from a good site can easily have 10 times the weight of an NADF bearing from a poor site.

16.1.3 Distributions of bearing errors

Table 16.1 shows a fairly typical distribution of bearing errors obtained with a WADF on a variety of transmitters in the frequency range 1·7–17 MHz, the majority being below 10 MHz. The histogram between limits of ± 10° is plotted in Fig. 16.1. The moments of this distribution, relative to the sample mean and excluding wilds at ± 10°, are shown in Table 16.2.

Such a distribution is often treated in DF plotting programs as approximately normal. However, it is commonly found that real distributions are leptokurtic, i.e. more 'peaky' than normal. The number of wild bearings (50 in this example) is also very much larger than would be expected in a normal distribution.

Table 16.1 Bearing error distribution

Error	No. of bearings
$\leqslant -10{\cdot}0°$	26
$-9{\cdot}9$ to $-9{\cdot}1$	1
$-9{\cdot}0$ to $-8{\cdot}1$	1
$-8{\cdot}0$ to $-7{\cdot}1$	2
$-7{\cdot}0$ to $-6{\cdot}1$	4
$-6{\cdot}0$ to $-5{\cdot}1$	4
$-5{\cdot}0$ to $-4{\cdot}1$	7
$-4{\cdot}0$ to $-3{\cdot}1$	9
$-3{\cdot}0$ to $-2{\cdot}1$	30
$-2{\cdot}0$ to $-1{\cdot}1$	103
$-1{\cdot}0$ to $0{\cdot}0$*	409
$0{\cdot}0$ to $+1{\cdot}0$*	488
$+1{\cdot}1$ to $+2{\cdot}0$	195
$+2{\cdot}1$ to $+3{\cdot}0$	76
$+3{\cdot}1$ to $+4{\cdot}0$	63
$+4{\cdot}1$ to $+5{\cdot}0$	18
$+5{\cdot}1$ to $+6{\cdot}0$	11
$+6{\cdot}1$ to $+7{\cdot}0$	4
$+7{\cdot}1$ to $+8{\cdot}0$	1
$+8{\cdot}1$ to $+9{\cdot}0$	1
$+9{\cdot}1$ to $+9{\cdot}9$	1
$\geqslant +10{\cdot}0°$	24
	Total 1478

*Contains half of the zero-error bearings

Table 16.2 Properties of distribution shown in Table 16.1

First moment $\mu_1 = 0{\cdot}0$ deg
Second moment $\mu_2 = 2{\cdot}8$ deg^2
Third moment $\mu_3 = +\ 0{\cdot}3$ deg^3
Fourth moment $\mu_4 = 61{\cdot}8$ deg^4
Asymmetry $\beta_1 = \mu_3^2/\mu_2^3 = 0{\cdot}003$
Kurtosis $\beta_2 = \mu_4/\mu_2^2 = 7{\cdot}8$

For a normal distribution, β_2 would be $3{\cdot}0$.

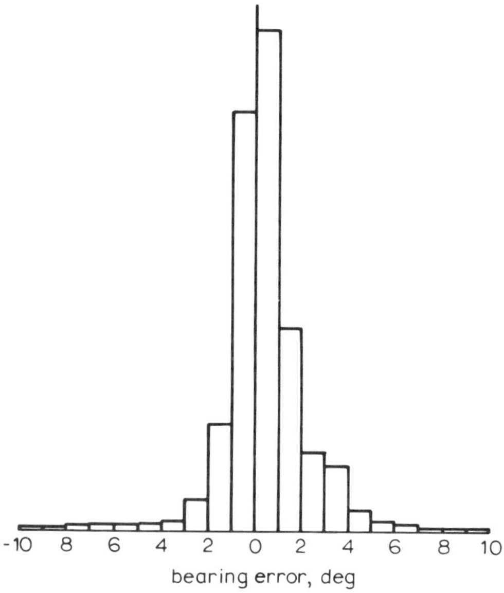

Fig. 16.1 Histogram of bearing errors for a WADF

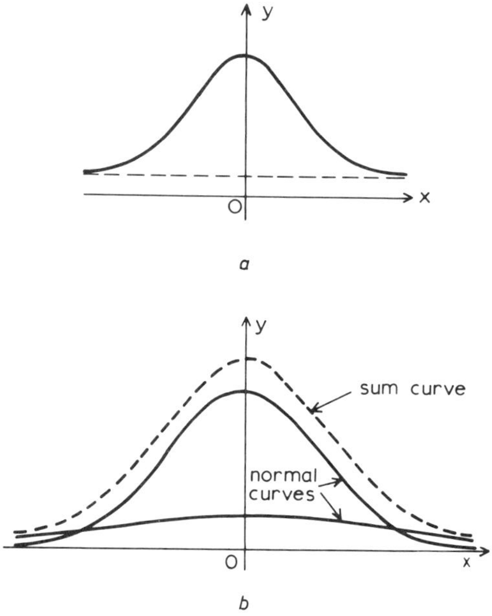

Fig. 16.2 Possible representations of bearing-error distributions
(*a*) Normal + plinth
(*b*) Sum of two normal curves

The total distribution, including wilds, is sometimes represented by a normal distribution plus a low-rectangular distribution, i.e. a normal distribution on a plinth (see Fig. 16.2*a*); the rectangular distribution might extend over the full 360°. An implication is that some small errors belong to the rectangular distribution of 'wilds', e.g. because they were taken on the wrong target. Other possibilities are to represent the total distribution as the sum of two normal distributions, one of which is comparatively low and has a large standard deviation (see Fig. 16.2.*b*) or as two normal distributions plus a plinth. Experience suggests that the two normal distributions plus a plinth give a more satisfactory representation than the other two models. Any observed distribution can, of course, be better fitted by a sum of two or more theoretical distributions rather than one, because of the extra degrees of freedom involved. It should be noted that the distribution of *rejected* bearings, with errors greater than ± 10°, is very far from normal, because there are no values in the range − 10° to + 10°.

It is often convenient to assume, as a simplification, that the truncated distribution of observed bearing errors can be approximated by a normal curve with the same standard deviation. However, if the aim is to estimate the error limits containing (say) 95% of the bearings, it would be better to find these limits directly from the observed distribution rather than infer them from the calculated standard deviation.

16.1.4 Wild bearings

Wild bearings arise from a variety of physical causes, of which misidentification of the transmission and side-scatter propagation are probably the most important. Misidentification can, of course, sometimes lead to bearing errors that are less than ± 10°. Errors outside these limits can be readily identified and rejected on check transmitters but not on a transmitter of unknown position. In a fix on an unknown transmitter, one or more bearings may sometimes be clearly inconsistent with the BPE obtained from the remainder; this question of rejection at the fixing stage is discussed briefly in Section 16.4.2. Wild bearings may also become apparent in a sequence of bearings collected on a fixed transmitter from a given DF site over a period of time.

Ross (1975) has suggested that it is legitimate to reject an apparently wild bearing in only two sets of circumstances, which he describes as follows: 'These are, when it is fairly clear that a case of misidentification has occurred, or when all the evidence as to propagation conditions points fairly certainly to no normal ray path having been possible'. His paper should be consulted for a fuller discussion of the philosophy of rejection.

16.2 Accuracy of bearings and sources of error

16.2.1 Estimation of variance of a single bearing

In the remainder of this chapter it will frequently be assumed that a variance figure can be associated with a single bearing. A variance can, in practice, be calculated only for a *population* of bearings. The variance of a single bearing should be interpreted as an estimate of the variance that would be found from a large sample of bearings taken on the same transmitter under identical conditions. The estimate is usually based on statistical analyses of previous samples of bearings taken on a variety of transmitters under a variety of propagation conditions. From such analyses the major factors affecting accuracy can be isolated by standard statistical methods; for example, a sort of bearing errors by frequency usually reveals that bearings taken on the higher frequencies are more accurate than those taken lower in the band; this is mainly a consequence of the increasing number of wavelengths represented by a given physical aperture at the higher frequencies.

As noted above, bearing accuracy depends not only on propagation conditions and the nature of the DF equipment and site, but also on parameters of the target list such as the distances and frequencies of the transmitters. Bowen (1955) analysed results obtained with U-Adcocks (i.e. NADFs) on fixed medium-power CW transmitters in the band 3–20 MHz and at ranges of 700–4000 km; snap bearings were averaged over a total period of about 30 s to produce each mean bearing. He was then able to estimate the variance components associated with various sources of error in a mean bearing as follows:

	Variance component, deg^2
Lateral deviation	2·0
Instrumental and distant site	1·2
Wave interference	2·5
Operator	2·5
Total	8·2

An average figure for an 'operator-error' component is included because operators vary in their ability to deal with the fluctuating bearing displays seen in conditions of wave interference. If all operators were identical in performance, it would be equally valid to classify their contribution to the variance as wave-interference error. The total variance of just over 8 deg^2 is probably typical of an Adcock installation on a fairly good site.

For a WADF and a similar target list, the lateral deviation component would have about the same magnitude, but all other variance components are greatly reduced. Wave interference errors and operator errors are almost entirely absent in bearings averaged over a minute or more. Under favourable conditions, e.g. a well-adjusted instrument on a very good site, a total variance of 2·2 deg^2 might be attained, but 3 deg^2 is probably a more realistic figure in practice.

For Adcock bearings, a variance estimation system was developed during the Second World War by Ross and Barfield and was later described in a paper by Ross (1947a). The system was subsequently modified and improved by Mr Norman Brooke and is now referred to as the Brooke system. A detailed discussion of the variance components that should be used, and the method of continuously up-dating them from the analysis of recent check-bearing data, can be found in a paper by Beale (1961a). It is sufficient to note here that the total Brooke variance B is assumed to be the sum of three components,

$$B = V_f + V_d + V_q.$$

Here V_f denotes a component depending on frequency and the DF site, representing instrumental and distant site error; V_d is a function of the distance of the transmitter, representing lateral deviation error; and V_q is a quality component, depending on the type of equipment and display, and on the stability of the display as reported by the operator.

For other types of DF system, e.g. a WADF, the Brooke system probably requires modification. The underlying *principle*, of determining the important factors and associated variance components from analyses of recent data should not, of course, be abandoned. The important *factors* might possibly be different, although V_f, V_d and V_q seem fairly natural choices on physical grounds. The component *values* would undoubtedly be different, and for any type of DF may require alteration from time to time.

The underlying assumption of any variance-estimation scheme is that variance components determined from an analysis of check targets can be applied to bearings taken on operational targets, i.e. transmitters of unknown position. It is therefore important that check targets should be broadly similar to the operational targets and that, in particular, the check targets should be neither 'easier' than the operational targets nor well known to the operators. An additional complication is the significant effect that rejection limits have on the variance of a sample of bearings taken on check targets. Unfortunately it is not possible to reject an operational bearing when it is more than $\pm$ 10° from the true bearing, because the latter is unknown. Some operational bearings can be rejected at the plotting stage as obviously inconsistent with the remainder of the bearings in the fix; a reasonable aim is to reject about the same percentages of operational bearings and check bearings. Variances of operational bearings can, in principle, be calculated from a sample of fixes, using the consistency of the bearings as the basis of the calculation (see Section 16.6) but in practice it is difficult to obtain reliable answers from samples of reasonable size.

16.2.2 Influence of propagation conditions on accuracy

Ionospheric tilts can lead to significant errors in what appear to the DF operator to be good, stable bearings. Ionospheric roughness, which could be regarded as tilts on a very small scale, can lead to what is variously described as blurring, broadening or scintillation of the observed bearing as a function of time. Thus it is clear that there will be correlations between bearing accuracy and

(*a*) mode or modes of propagation (1*E*, 1*F* etc),
(*b*) state of the ionosphere (calm, stormy etc).

These correlations become particularly significant with bearings from WADFs, for which other sources of error are much less important than in NADF. Both (*a*) and (*b*) above depend statistically on time of day, season etc; for example, normal *E* ionisation is absent at night in temperate latitudes, whereas spread-*F* is mainly a night-time phenomenon, which may be responsible for an observed broadening of bearing records at night (Morgan, 1972).

Evidence that single-hop modes give better bearings, on average, than multihop modes has already been mentioned in Section 15.1. In addition, it is found that low-angle rays are better than high-angle rays (for a given number of hops), that *E*-modes are better than *F*-modes and that single-mode conditions are to be preferred to multimode conditions. Note that it is theoretically possible to have multiple modes (1*E* and 1*F*) without multiple hops and multiple hops (2*F*) without multiple modes; these terms are sometimes confused. Sidescatter paths can lead to large bearing errors (e.g. 60°), particularly when the main great-circle path is not open. Propagation round the longer arc of the great circle through the transmitter and receiver can give rise to a 'reciprocal' bearing, approximately 180° in error. (A reciprocal bearing can also arise if the DF operator fails to obtain a correct 'sense' indication).

The author has shown (1961) that a degradation of fix accuracy can be detected during periods of ionospheric storms. Large-scale TIDs are usually associated with geomagnetic-storm activity (Georges, 1967) and there is some evidence that they are more commonly observed at night. Medium-scale TIDs are considerably more frequent in winter than in summer and the occurrence of such disturbances is a maximum near midday in both winter and summer (Jones and Reynolds, 1975). Ranzi and Giorgi (1972) found some evidence that intense winter TIDs occur on groups of days (three or more) and get earlier each day. These diurnal and seasonal trends should be present in the accuracy of bearings, but possibly partly obscured by the dependence of propagation mode on time and season. It is desirable to analyse a sample of bearings all taken on the same mode of propagation, if possible.

Medium- and large-scale TIDs are believed to result from the propagation of internal gravity waves through the atmopshere. Because such waves show a phase advance with increasing height, i.e. the phase fronts slope forward in the direction of propagation, a TID crossing a given transmitter-receiver path is seen first on the rays that penetrate furthest into the ionosphere. This phenomenon, which is most easily seen when a transmitter is active on more than one frequency (Reynolds and Morgan, 1975), should not be confused with the result embodied in eqn. 2 of Section 15.2.2, that the effective tilt of a layer increases rapidly as penetration conditions are approached.

Jones and Reynolds (1975) have studied ionospheric perturbations by means of Doppler techniques and have demonstrated a strong correlation between disturbances in the Doppler measurements and the bearing measurements taken on paths with a common reflection point.

A valuable source of information is provided by an 'atlas' compiled by Bailey *et al.* (1968) for a fixed path from Houston, Texas to the WADF at Urbana, Illinois. Observations of bearing and apparent elevation angle were made on five days (24 hours each) per month in the period March

1967 to June 1968. Also collected were oblique ionograms, vertical incidence ionograms and Doppler measurements; these data are also shown in the atlas. Near the midpoint of this path disturbances in the Doppler records are almost absent during daylight hours; a possible explanation is that *E*- or *E*s-modes are dominant and show much smaller effects that *F*-modes.

16.2.3 Consolidated bearings

It is often possible to take repeated bearings from a given DF site on a given fixed transmitter over a period of time. Provided the observations are separated by time intervals of at least 20 minutes, and preferably more, the influence of errors arising from TIDs can be greatly reduced when several observations are averaged. The weighted average of all the bearings is often referred to as the consolidated bearing; the simplest weighting scheme is to use the reciprocal of the estimated variance of each bearing as the weight.

If the transmitter changes frequency from time to time, or transmits on two or more frequencies simultaneously, the advantages of frequency diversity as well as time diversity are obtained from consolidation. Frequency diversity is particularly important with NADF systems, for which distant site errors are significant but vary comparatively rapidly with frequency and azimuth (Ross, 1947b).

If all errors were of a random nature, the variance of a consolidation of n bearings each with variance V would be V/n and the weight of the consolidated bearing would be n times that of a single bearing. However, when systematic errors are significant, consolidation is less effective than these simple results suggest. For NADF bearings a 'saturation point' is reached after (say) 20 or 30 bearings have been obtained on a given frequency; unless the transmitter is active on other frequencies, there is little point in collecting more observations. The consolidated variance on a typical target of the sort used by Bowen (1955) would then be about 1 deg^2, representing distant-site error; indeed, the consolidated variance provides one of the main pieces of evidence regarding the magnitude of this component. Because WADF bearings are less susceptible to site errors they do not exhibit an obvious saturation effect and it should be possible to obtain, from a large enough sample, a consolidated variance of 0·2 deg^2 or less, representing mainly the effect of systematic ionospheric tilts. That is to say, SITs probably represent the major source of error in consolidated bearings from good WADF sites.

16.3 DF plotting

16.3.1 The problem

Suppose that we have obtained a bearing from each of n separated DF stations of known position on a single transmitter of unknown position; we assume either that the bearings were obtained simultaneously or that the transmitter is of fixed position. The bearing lines, when plotted on a suitable chart, are not likely to intersect perfectly in a single point, nor should it be assumed that they are all of equal weight. The statistical problem to be discussed is that of determining the Best Point Estimate (BPE) for the position of the transmitter and a surrounding probability region (PR) with a specified probability (e.g. 90%) that it contains the true position of the transmitter.

It will be assumed that corrections for known instrumental errors have been applied to the bearings prior to the plotting stage. Before any corrections for deviations on the propagation path can be applied it is generally necessary to have an approximate location for the transmitter, so that the relevant areas of the ionosphere can be considered, the likely mode of propagation identified, and the conversion factor between tilt and bearing error calculated. It should also be noted that, in the Brooke classification system, the estimated variance of each bearing is a function of distance; V_d represents the effects of tilts. For these reasons it may be necessary to obtain a first approximation to the BPE, then adjust the corrections and weights of the bearings prior to the calculation of the second approximation. If one (or more) of the bearings now looks quite inconsistent with the remainder it might be rejected, and a new BPE calculated. Thus the process of arriving at the final BPE is often an iterative one in practice.

If only a few bearing lines are available and are drawn (or represented by strings) on a suitable chart, an experienced plotter can make a reasonable guess at the BPE. However, as the number of bearings increases it becomes more difficult for him to take proper account of all the information, including the weights. We shall not be concerned here with 'string plotting', or the crude methods that are sometimes used in order to arrive at a quick approximation to the BPE and the PR. The wide availability of digital computers means that more sophisticated and soundly-based methods can be used.

16.3.2 Method of Stansfield

Many modern plotting programs are derived from a least-squares-method first described by Stansfield (1947), which was probably used initially with desk calculators. The simplifying assumptions and basic theory will be presented first; then, in Section 16.4, an indication will be given of ways in which some of the assumptions can be removed in more sophisticated programs.

The simplifying assumptions are that

(*a*) the earth is flat near the true position of the transmitter,
(*b*) the bearing lines are straight lines,
(*c*) an error of observation displaces the bearing line parallel to itself,
(*d*) errors in separate bearings contributing to the fix are independent, i.e. there is no correlation,
(*e*) the errors are normally distributed about zero mean, and we have estimates of their variances.

We assume that conditions (*a*) to (*c*) are closely satisfied on a gnomonic chart, provided the bearing errors are comparatively small. No *a priori* knowledge about the position of the transmitter is assumed; thus, all elements of equal area on the chart are assumed initially to be equally likely to contain the true position of the transmitter. Because of the area distortion of the gnomonic projection usually used in plotting (in order that condition (*b*) above is satisfied), such elements do not represent equal areas on the earth's surface, except near the point of tangency between a given chart and a sphere representing the earth.

Calculation of the BPE

Suppose that we have a Cartesian co-ordinate system (x, y) on the chart and that we adopt some trial position (ξ, η) for the transmitter. For any bearing line there is an implied error in the observation, nonzero if the line does not pass through (ξ, η). Suppose that the implied error for the jth bearing line, of estimated variance V_j, is θ_j. We now calculate

$$E = \sum \frac{\theta_j^2}{V_j}$$

summed over all bearings and choose (ξ, η) to minimise E, i.e. we adopt a least-squares criterion.

A typical bearing line is

$$x \sin \theta_j - y \cos \theta_j = p_j$$

where p_j is the perpendicular distance from the origin to the line. A parallel line through (ξ, η) is

$$x \sin \theta_j - y \cos \theta_j = q_j$$

where q_j is given by

$$\xi \sin \theta_j - \eta \cos \theta_j = q_j$$

Hence the implied error in the bearing is proportional to the distance

$$\epsilon_j = p_j - q_j$$

which can be calculated for any assumed (ξ, η).

We now suppose that, for the optimum position $(\bar{\xi}, \bar{\eta})$ of the BPE, the errors ϵ_j are normally distributed with known variances σ_j^2, where

$$\sigma^2_j = V_j d_j^2$$

and d_j is the distance from the jth DF station to the BPE. The probability of errors falling in the ranges ϵ_1 to $\epsilon_1 + d\epsilon_1$, ϵ_2 to $\epsilon_2 + d\epsilon_2$ etc is given by the expression

$$dP = \frac{\exp(-\tfrac{1}{2}\Phi)}{(2\pi)^{n/2}(\sigma_1 \sigma_2 \ldots \sigma_n)} d\epsilon_1\, d\epsilon_2 \ldots d\epsilon_n \qquad (1)$$

where

$$\Phi = \left(\frac{\epsilon_1^2}{\sigma_1^2} + \frac{\epsilon_2^2}{\sigma_2^2} + \ldots + \frac{\epsilon_n^2}{\sigma_n^2}\right)$$

The sum-of-squares function Φ, whose minimum value defines the least-squares estimate, can be writen as

$$\Phi = \Sigma w_j(p_j - q_j)^2$$

where this and subsequent sums are taken over all bearings, and weight w_j is equal to $1/\sigma_j^2$. In terms of x and y, Φ is a quadratic function of the general form

$$\Phi = ax^2 + 2hxy + by^2 + 2gx + 2fy + c$$

where

$$a = \sum w_j \sin^2 \theta_j$$

$$b = \sum w_j \cos^2 \theta_j$$

$$c = \sum w_j p_j^2$$

$$f = \sum w_j p_j \cos \theta_j$$

$$g = -\sum w_j p_j \sin \theta_j$$

$$h = -\sum w_j \sin \theta_j \cos \theta_j$$

It is now convenient to eliminate the term in xy by rotating axes through an angle α such that the angles φ_j relative to the new x axis are $\varphi_j = \theta_j + \alpha$ and α is given by

$$\tan 2\alpha = \frac{2h}{(b - a)}$$

In the new co-ordinate system (X, Y)

$$\Phi = AX^2 + BY^2 + 2GX + 2FY + C$$

where

$$A = \sum w_j \sin^2 \varphi_j$$

$$B = \sum w_j \cos^2 \varphi_j$$

etc

The equation

$$\Phi = \text{constant}$$

therefore defines an ellipse; the minimum value of Φ, which corresponds to the maximum probability density in the X, Y plane, occurs at the centre of the family of ellipses, i.e. at

$$\bar{X} = -G/A$$

$$\bar{Y} = -F/B$$

These are the co-ordinates of the BPE in the new co-ordinate system and they can be calculated from the initial bearing data. If we now translate the origin of co-ordinates to the BPE, the probability density can be written as

$$dP = \frac{1}{2\pi r_1 r_2} \exp\left\{-\tfrac{1}{2}\left(\frac{X^2}{r_1^2} + \frac{Y^2}{r_2^2}\right)\right\} dXdY$$

where

$$r_1 = \frac{1}{\sqrt{A}}$$

$$r_2 = \frac{1}{\sqrt{B}}$$

and the constant term in front of the exponential term has been chosen so that the integral of dP over the X, Y plane is unity. r_1 and r_2 are roots of the equation

$$ABr^4 - (A + B)r^2 + 1 = 0$$

Calculation of probability region

We now consider the ellipse

$$\frac{X^2}{r_1^2} + \frac{Y^2}{r_2^2} = k$$

centred on the BPE, where k is a constant to be defined; a large value of k corresponds to a large ellipse and would therefore be expected to lead to a comparatively large total integrated probability within the ellipse.

At every point on the perimeter of the ellipse, the probability density has a constant value given by

$$dP = \frac{\exp\left(-\tfrac{1}{2}k\right)}{2\pi r_1 r_2} dxdy$$

At any point outside the perimeter, the density is lower than this value and at any point inside it is higher. It follows that the ellipse must be a region of *minimum area* associated with a given total probability, because the total probability is decreased if we exchange an element of area $\delta x \delta y$ outside the ellipse with the same area inside.

The area contained between two ellipses defined by k and $(k + dk)$ is $\pi r_1 r_2 dk$ and the total probability enclosed by the ellipse is therefore

$$P = \int_0^k \tfrac{1}{2} \exp\left(-\tfrac{1}{2}k\right) dk$$

$$= 1 - \exp\left(-\tfrac{1}{2}k\right)$$

Hence

$$k = -2 \log_e (1-P) \tag{2}$$

For a specified value of P, e.g. 0·9, we can therefore calculate $k(k = 4{\cdot}605$ for $P = 0{\cdot}9)$ and hence find the semi-axes $r_1 \sqrt{k}$, $r_2 \sqrt{k}$ of the ellipse. We now know the centre, orientation and size of the probability ellipse for a given probability value. The size of the ellipse depends, as expected, on the absolute values of the weights.

We could calculate a 10% ellipse in this way; the remainder of the earth's surface is then a 90% probability region. Many other regular or irregular figures could be found with a 90% integrated probability. It must again be stressed that the Stansfield calculation produces the region of *minimum area* for a given probability level, that this is what is required in practice and that the phrase 'probability region' should generally be regarded as an abbreviation for 'probability region of minimum area'.

Blachman (1989) has shown how the information in several ellipses and/or bearings obtained from independent fixes on the same target can be combined into one solution.

16.3.3 Confidence regions and probability regions

As noted in Section 16.2.1, the variance estimate assigned to a single bearing is really a prediction of the variance that would be obtained from a sample of repeated observations on a given transmitter made under identical conditions. From such a hypothetical sample it would be possible to calculate confidence limits within which (say) 50% of all future bearings are expected to fall, even though we make no statement about systematic errors, or about the probability of the true bearing of the transmitter lying between the confidence limits.

Similarly, there is a subtle distinction between confidence regions and probability regions. Because the variance estimates define confidence limits, the Stansfield type of calculation leads, strictly speaking, to a confidence region. The 90% confidence region is one in which we would expect 90% of the future BPEs for the same transmitter to lie, if bearings could be repeatedly taken under similar conditions. Now suppose that a large number of 90% confidence regions, obtained on a variety of check transmitters under a variety of conditions, are examined. We can reasonably expect 90% of these to contain the true transmitter position, provided our initial assumption that all positions are, *a priori*, equally likely is valid. The terms confidence region and probability region can then be used interchangeably. This topic is treated in more detail in a paper by Daniels (1951) and in the discussion of the paper at a meeting of the Royal Statistical Society.

For those who find the distinction confusing, it is perhaps helpful to point out that a confidence region could be tested without any knowledge of the true transmitter position ('How many BPEs fall within a specified ellipse in repeated tests?') whereas probability regions can be tested only when true positions are known ('How many true positions fall inside the corresponding PRs?').

It must be emphasised that the BPE and PR are calculated in practice from a very limited amount of information, e.g. five bearings with estimated weights. By normal statistical standards this must be regarded as a very small sample, with corresponding risks of a misleading answer. Fortunately, the bearing weights are usually based on a wealth of past experience.

16.4 Modifications to basic theory

16.4.1 Conversion from ellipse to rectangle

Although the PR of minimum area is strictly elliptical, it is more convenient to plot on a chart a rectangle of approximately the same elongation and area, with a specified probability level.

The semi-axes of the ellipse are $\sqrt{(k/A)}$ and $\sqrt{(k/B)}$ in the X and Y co-ordinates respectively, i.e. $2{\cdot}146/\sqrt{A}$ and $2{\cdot}146/\sqrt{B}$ for the 90% ellipse. Now suppose that the uncertainty in the position of the BPE is measured by 'standard deviations' σ_X and σ_Y respectively in the two co-ordinates. The meaning of these quantities becomes clearer if we define limits.

$$X_1 = -2\sigma_X$$

$$X_2 = 2\sigma_X$$

$$Y_1 = -2\sigma_Y$$

$$Y_2 = 2\sigma_Y$$

such that, for a normal distribution of BPE positions in repeated trials, 95·4% of X co-ordinates would lie between X_1 and X_2 and 95·4% of Y co-ordinates would lie between Y_1 and Y_2 (these percentages are obtained from standard tables for the normal distribution). The percentage probability associated with the rectangle is the joint probability that the BPE lies between the $\pm\ 2\sigma_X$ lines and between the $\pm\ 2\sigma_Y$ lines; it is therefore a 91% region, which can be referred to for practical purposes as a 90% region.

It can be shown (see Appendix 9) that

$$\sigma_X = \frac{1}{\sqrt{A}}$$

and

$$\sigma_Y = \frac{1}{\sqrt{B}}$$

Thus the total length of the rectangle is $4/\sqrt{A}$ and the total width is $4/\sqrt{B}$. The semilength and semiwidth of the rectangle are slightly smaller than the semi-axes of the ellipse, as would be expected. The rectangle is not, of course, strictly a PR of minimum area, but the distinction in shape between a long ellipse (for a narrow-baseline DF network) and a long rectangle is not of any great practical significance; the rectangle can be regarded simply as a graphical presentation of information about σ_X, σ_Y.

Suppose that, in addition to DF bearings, we have a range estimate $R \pm \sigma_R$ for the distance from the baseline to the transmitter. In order to make a fair comparison between the accuracy of this estimate and that provided by the DF network it is necessary either to compare σ_R and σ_X, or to compare $4\sigma_R$ with the total length of the rectangle.

16.4.2 Conversion from ellipse to circle

The elliptical probability region is sometimes replaced by a circle, which is again easier to draw on a chart. Isley (1980) showed that the circular error probability region (CEP) with a specified probability level can be calculated from a power series which converges efficiently for cases of practical interest.

16.4.3 Spherical geometry

In a computer program, spherical geometry can be adopted in order to avoid the assumptions of flat-earth geometry and of parallel displacements used in the basic Stansfield calculations. In this way the undesirable effects of the distortions associated with any flat projection of a portion of the surface of the sphere can be avoided. The distinction between a sphere and the true shape of the geoid produces errors that can be safely neglected in most practical applications; further information on this subject may be found in a paper by Houston and Nelson (1980).

Various trial positions for the BPE are first tested in an iterative search procedure. For each trial, the *angular* errors θ_j, in the bearings implied by that choice are calculated and a normalised sum of squared errors,

$$E = \sum \frac{\theta_j^2}{V_j}$$

is formed. The solution for the BPE is the position that gives a minimum sum E_0. The axis of the rectangle is then calculated as in the basic method. The end-points of the rectangle on this axis are found by searching for points at which

$$E = E_0 + C$$

where C is a constant related to the probability level. Similarly, two points on a line through the BPE perpendicular to the axis are located at which E reaches the same level. These four points define a rectangle with sides parallel and perpendicular to the axis.

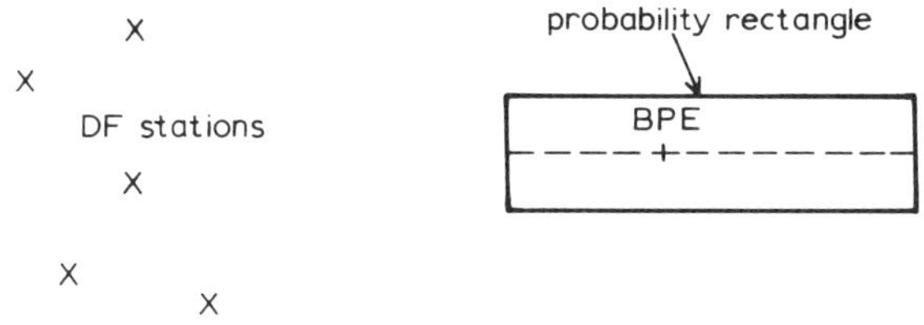

Fig. 16.3 Sketch showing position of BPE for a narrow-baseline DF network

The points corresponding to $\pm 2\sigma$ limits in the plane-earth case are those at which $C = 4$. When spherical geometry is used and angular errors are calculated, it is found that the BPE is nearer to the DF baseline (for a narrow baseline) than is the centre of the rectangle (see Fig. 16.3). The position of the BPE at the centre of an ellipse in elementary theory is a direct consequence of the assumption of parallel displacements on a flat chart. A recursive procedure for finding the BPE is given in a paper by Rabtsun (1987).

16.4.4 Correlated errors

In practice, bearing observations taken from separated DF stations are not likely to be strictly independent; large-scale ionospheric tilts, for example, can lead to positively-correlated errors. The magnitude of the correlation depends on the frequency and magnitude of tilt errors in comparison with random errors and on the geographical scale of a tilt compared with the station separation. Thus, correlation would be higher for closely-spaced WADFs than for widely-spaced NADFs. The effect of correlation on the size and shape of probability ellipses has been investigated by the author (1967) for some simple model situations. For a few DF stations on a comparatively narrow baseline the effect of positive correlation is to decrease the major axis and increase the minor axis of the ellipse of given probability level. For a two-station fix, Daniels (1951) showed that the area of the modified ellipse is always smaller than the original by a factor of $\sqrt{(1 - \rho^2)}$, where ρ is the correlation coefficient. With three or more stations the area becomes larger provided ρ (now the correlation between each *pair* of stations) is small. For large ρ the area can become smaller. If correlation is present, but is not taken into account, the stated probability corresponding to a given ellipse will be incorrect; it will certainly be too high in those cases in which the area of the modified ellipse is greater than that of the unmodified ellipse, since the former is a region of minimum area corresponding to the stated probability.

Cooper (1972) has given a more general treatment of position-fixing in three dimensions and of the effects of correlated errors.

16.4.5 Non-normal distributions and *a priori* information

For a normal distribution, the percentage of errors lying between specified limits can be found from standard tables; for example, 68·3% of observations should lie within $\pm\,\sigma$ limits, where σ is the standard deviation, and 95·4% within $\pm\,2\sigma$ limits. In order to calculate a 90% PR for (say) a two-station fix we are interested in the limits between which about 95% of the (repeated) bearings from each station would lie, because $(0{\cdot}95)^2 \simeq 0{\cdot}90$. Given a significantly non-normal distribution a possible approach is therefore to calculate directly the limits between which a desired percentage of bearings lie, rather than predict these limits from a calculation of σ.

It is convenient to consider this problem in conjunction with that of the use of any *a priori* information about the position of the transmitter. We suppose that the area of interest on the earth's surface is divided into a number of small cells, each perhaps 1 km square, and that a particular *a priori* probability can be associated with each cell on the basis of sources of information other than DF. For example, if a transmitter is known to be on a ship, a very low initial probability value would be assigned to each land cell. A standard Bayesian approach is now used to find the *a posteriori* probability. Stated in words, the probability associated with a specified cell is the *a priori* probability multiplied by the conditional probability of obtaining the given set of bearings if the transmitter is in that cell. Now the conditional probability for n bearings is itself a product of n terms; see, for example, eqn. 1 in section 16.3.2. It has therefore been suggested by A.D.V. Ridlington that the logarithms of probability should be used, so that they can be added for each cell.

One of the advantages of this procedure is that there is no restriction on the type of bearing distribution that can be assumed. For example, a normal distribution on a plinth could be used; the effect of a uniform plinth is simply to produce a small contribution of log (probability) in each cell. Each bearing line produces a maximum contribution to cells through which it passes, with a falling-off on each side. The total log (probability) in each cell, when all contributions have been added, will be referred to as the score. The cell with the highest score can be located and contours of constant score can be drawn; thus PRs with known probability levels are obtained.

This procedure, although rather expensive in terms of computer time, appears to be the only feasible way of dealing with awkward probability distributions. Depending on the nature of the assumed distribution we may find more than one significant probability 'hump' in the plane, and the 90% PR could consist of two or more distinct regions. Note that the process of adding log (probability) is essentially sequential; the scores obtained with n bearings can be easily modified for the effect of an $(n + 1)$th bearing. Relevant theory can be found in a paper by Butterly (1972).

16.4.6 Systematic errors

Known systematic errors in bearings arising from a known physical cause, for example, from a slight rotation of the azimuth scale of a DF instrument relative to due N, should naturally be removed before either the calculation of a variance or the plotting of a set of bearings. Some idea of the effect of unknown systematic errors on the calculation of PRs can be obtained as follows.

Consider a normal distribution with no systematic error and a standard deviation of 3°; then 95·44% of a sample of bearings should lie between the limits of ± 6°. Suppose now that the whole distribution is displaced by a systematic error of (say) −0·5° and suppose that the same value of σ is calculated relative to the sample mean; the percentage of errors lying between −5·5° and +6·5° from the sample mean can be found from standard tables and is 95·15%. From the combination of two such bearings we expect that the PR with a nominal probability of $(0{\cdot}9544)^2$ will become one with a nominal probability of $(0{\cdot}9515)^2$. There is, therefore, a loss of about 0·55% in the expected success rate for such PRs when tested against true positions, arising from the neglect of systematic errors. This quite trivial loss shows that small systematic errors are likely to have a negligible effect on success rates; note also that SEs are incorporated in the DF variance, which is usually used in practice to calculate the PR.

16.4.7 Combination of fixing systems

A number of authors have considered 'mixed mode' fixing systems, in which some combination of bearing measurement, range estimates and time-difference hyperbolic position lines is used to fix the position of the target transmitter.

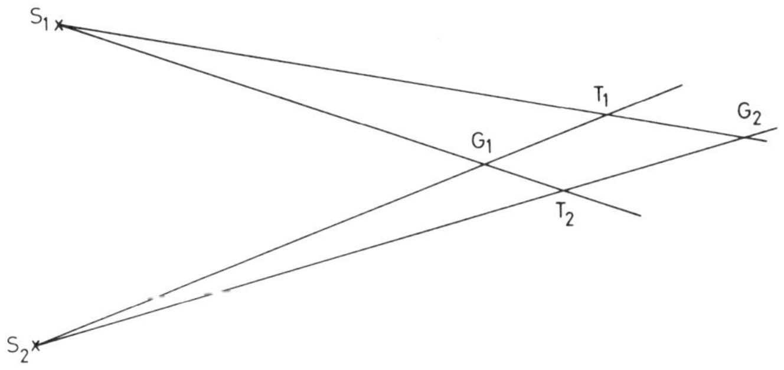

Fig. 16.4 Ghost targets
DF stations at S_1, S_2
True targets at T_1, T_2
Ghost targets (false association of bearings) at G_1, G_2

The situation can become rather complex in (say) an electronic warfare scenario when several forms of active and passive sensor are used against a number of target transmitters and jammers, not necessarily located on the surface of the earth. However, the underlying principle of finding the BPE that gives the best weighted least-squares fit to the observations remains valid and can be incorporated into numerical computer routines to refine an initial estimate by a process of successive approximations.

In three dimensions, lines of position become surfaces of position. The properties of tangent planes to the surfaces of position near some good initial approximation to the true position of the target were considered by Wax (1983). Unless a good initial estimate is available, a multiplicity of iterative solutions can be derived with different convergence properties and degrees of complexity. Wax's analysis included the effects of small uncertainties in the positions of the sensors, which might be carried on moving platforms. Expressions for the maximum likelihood estimator and its error statistics under a small error assumption were derived.

With two or more target transmitters active in the same frequency band, there may be a problem in associating bearing taken from different sites into sets taken on one transmitter: false associations can lead to the appearance of 'ghost' targets, as shown in Fig. 16.4. Berle (1986) described a mixed triangulation/trilateration technique to overcome the ghosting problem, by the combined measurement of DOA and time difference of arrival at the receiving sites.

16.5 Dispersion factor and the rejection of bearings

16.5.1 Dispersion factor

For a two-bearing fix, the BPE coincides with the intersection of the bearings lines. For three or more bearings we expect to find that some or all of the bearings do not pass exactly through the BPE. The value of the weighted-squared errors at the BPE,

$$E_0 = \sum \frac{\theta_j^2}{V_j},$$

is often referred to as the dispersion factor, and it is recommended that this number should form part of the output of any fixing program. The expected value for n bearings is $(n - 2)$, because two degrees of freedom have been used up in the calculation of the latitude and longitude of the BPE.

16.5.2 Bearing rejection

An abnormally large dispersion factor is a warning signal that one (or more) of the bearings may be wild, perhaps because it was taken on the wrong transmitter. A standard procedure with four or more bearings is then to drop each bearing in turn from the fix and re-compute the dispersion factor for the remainder. For example, we can form five separate 4-bearing fixes from a set of five by dropping each bearing in turn; if one of these fixes has a significantly smaller dispersion factor than the others, the suspect bearing is the one omitted from this fix.

The above remarks apply mainly to computer calculations, where a routine objective test is required. In hand-plotting, the existence of a wild bearing may be obvious, provided the number of bearings in the fix is sufficient. With two bearings there can be no evidence of inconsistency. An unusually large triangle formed by three bearings suggests that one at least is dubious, but it is not possible to say which. Only with four or more is there any real chance of applying rational rejection criteria.

16.6 Success rate of probability regions

A sample of probability regions, e.g. 90% rectangles, computed on a number of check targets can be compared with the true target positions and each labelled as a success if the true position falls inside the rectangle or a failure if it falls outside. We then hope to obtain a success rate of about 90%. In view of the approximations and assumptions that have to be made in a typical fixing program, it is sensible to apply this test from time to time.

Any significant departure from the nominal rate deserves further study, so that the causes of this result can be understood. For example, a lower success rate than 90% could be due to any of the following factors, or to a combination of factors:

(*a*) Systematic errors in the bearings,
(*b*) Correlated errors in the bearings,
(*c*) Errors in the 'true' positions adopted for the check targets,
(*d*) Under-estimates of bearing variances, leading to rectangles that are too small,
(*e*) Error distributions that differ from the distribution assumed,
(*f*) Wrong assumptions about *a priori* probabilities.

Systematic errors and errors in target position of a given magnitude, e.g. 1° or 10 km respectively, will have a larger effect on the success rate of comparatively small rectangles. The distribution of the true positions of the targets for failures should be examined with respect to the projected boundary lines of the rectangle (see Fig. 16.5). If the number of targets falling in regions 2 and 8 combined is greater than in 4 and 6 combined, positive correlation, perhaps arising from random tilts, may be present. If more fall in regions (1 + 2 + 3) than in (7 + 8 + 9), or vice versa, the influence of systematic tilts should be suspected.

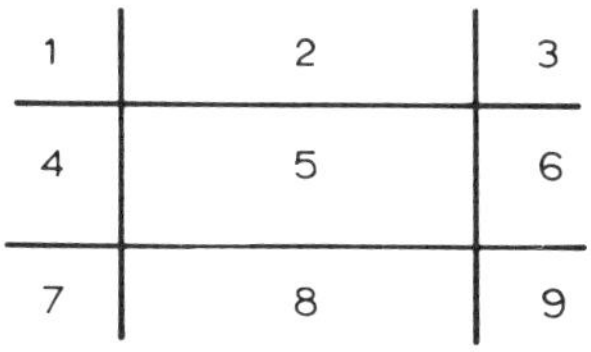

Fig. 16.5 Definition of regions for probability rectangle and surrounding area
Theoretical percentage probabilities for regions shown:
1, 3, 7, 9 each 0·05%. 2, 4, 6, 8 each 2·2%
Total 9% (failures). Region 5: 91% (success)

If variances of individual bearings are estimated from results obtained on a population of check targets and the same targets are used to test the success rate, it is unlikely that a wrong success rate can be due to incorrect variances; the trouble is more likely to have arisen in some way in the *combination* of bearings into a fix. However, we now consider the possibility that variances have been consistently underestimated or overestimated, and require multiplication by some factor f, greater or less than 1. It may be shown from eqn. 2 that the relationship between f and percentage success rate S of Stansfield ellipses can be written

$$f = -\log_{10}\left(\frac{100-S}{100}\right)$$

where $f = 1{\cdot}0$ for $S = 90$. Some values of S and f are shown in Table 16.3.

Table 16.3 Relationship between variance factor f and success rate S

f	S	$1/f$
1·30	95	0.77
1·00	90	1·00
0.82	85	1·22
0·70	80	1·43
0·60	75	1·67

The first column shows the factor by which the correct variances (for S = 90) must be multiplied in order to obtain other success rates. Conversely, incorrect variances must be multiplied by $1/f$, as shown in the 3rd column, to return to the correct success rate.

Another effect of incorrectly estimated variances would be on the average size of the dispersion factor. This factor can, of course, be calculated without any knowledge of true target positions. If, from a large sample of fixes, it were found that the success rate on check targets was too low and that dispersion factors were greater than $(n - 2)$ on average, the application of an empirical percentage increase to all estimated variances might well be justified. However, one would still prefer to know why such an increase is required.

16.7 Variance estimation from targets of unknown position

It is sometimes not possible to set up a system of check targets, or there may be a suspicion that the check targets are not representative of the unknown targets in some way, e.g. that the check targets are 'easier'. The possibility of basing a variance estimation scheme on fixes obtained on targets of unknown position (unknown targets for brevity) is therefore of considerable interest.

In any real situation we can safely assume that something is known about the performances of different types of DF equipment and sites, even if this information is derived only from strong commercial transmitters. We can further refine these initial estimates by examining

(*a*) the consistency of sets of bearings taken from each DF site on a fixed unknown target;
(*b*) the average value of the dispersion factors on unknown targets.

With regard to (*a*), it might reasonably be objected that, if the true position is unknown, we cannot be certain that the transmitter is stationary. However, there may be evidence (e.g. in the transmitted traffic) that it is.

It therefore seems reasonable to assume that the average level of variances on unknown targets can be adjusted to about the right value, but that small differential corrections might be required, i.e. that the variances for some stations need to be increased and those for others decreased. Two methods of calculating differential corrections will now be described. We assume that each variance v_i needs to be multiplied by a factor $(1 + \delta_i)$, where δ_i is the small differential correction. The true variance can then be written as $(1 + \delta_i)/w_i$, where the weight w_i is the reciprocal of the initial variance estimate v_i.

We now consider the miss distance d_k from a BPE calculated with n bearing lines to the kth line of this set. Beale (1961b) showed that the expected value of $w_k d_k^2$ is given by

$$E(w_k d_k^2) = \lambda_{kk} + \sum_{i=1}^{n} \delta_i \lambda_{ik}^2 \tag{3}$$

where λ_{kk} and λ_{ik} are defined by the equations

$$\lambda_{kk} = 1 - w_k \left(\frac{\sin^2 \varphi_k}{A} + \frac{\cos^2 \varphi_k}{B} \right)$$

$$\lambda_{ik} = -(w_i w_k)^{1/2} \left(\frac{\sin \varphi_i \sin \varphi_k}{A} + \frac{\cos \varphi_i \cos \varphi_k}{B} \right) \qquad (i \neq k)$$

Here φ_k is the angle between the kth line and the long axis of the probability rectangle: A and B are as defined in Section 16.3.2. Each fix gives n such equations, taking one position line at a time. From an analysis of a large number of fixes with various geometrical configurations it is possible, in principle, to obtain estimates of the δ_is. New BPEs and rectangles should be computed with the revised variances and the cycle repeated until the solution converges to stable values. Limited experiments with a scheme of this kind have suggested that a sample of several thousand fixes is required in practice if convergence is to be achieved.

Eqn. 3 can be simplified if we assume that, when the kth miss-distance is examined, only the kth variance requires a substantial correction. The equation can then be written

$$E(w_k d_k^2) = \lambda_{kk} + \delta_k \lambda_{kk}^2$$

from which,

$$\frac{1 + \delta}{w} = \frac{E(d^2) + ws^2 - s}{(1 - ws)^2} \tag{4}$$

where the subscript k has been dropped and

$$s = \frac{\sin^2\varphi}{A} + \frac{\cos^2\varphi}{B}$$

A further simplification is obtained if we omit the kth bearing line in the calculation of the BPE and probability rectangle. It is then legitimate to set $w = 0$ on the right-hand side of eqn. 4. The best estimate of the true variance from this one miss distance is then

$$\text{Variance} = d^2 - \left(\frac{\sin^2\varphi}{A} + \frac{\cos^2\varphi}{B}\right) \qquad (5)$$

Eqn. 5 is identical with an equation derived in Appendix 9 by a different route. It shows that the value of d^2 is, as might be expected, greater than the true variance; the 2nd term on the r.h.s. can be regarded as a contribution to d^2, to be added to the variance, arising from uncertainty in the position of the BPE. The greater complication of eqn. 3 as compared with eqn. 5 arises partly from the fact that the error in the position of the BPE is correlated with the error in the kth line if the kth line is used in the fix. A disadvantage of eqn. 5 is that n separate BPEs and rectangles must be computed for each set of n bearings on the same target, each bearing being omitted in turn.

Variances are sometimes calculated relative to BPEs of unknown targets, on the assumption that a BPE calculated from several bearings is likely to be very close to the true position. Such an assumption is both unsatisfactory and unnecessary; eqn. 3 shows that a simple allowance can be made in the calculations for the uncertainties in the BPE positions.

16.8 Concluding remarks on plotting

Plotting programs with various levels of sophistication can be written, but clearly the more complicated programs will be more expensive in terms of computer time. For some purposes a simple program based on the Stansfield equations may be perfectly adequate. In other situations, particularly where it is desirable to take account of nonuniform *a priori* probabilities arising, for example, from a knowledge of propagation conditions, the approach outlined in Section 16.4.5 becomes attractive.

Because azimuth measurements taken with different sorts of DF instruments and from different sites can differ greatly in accuracy, a rational system for assigning weights to bearing lines is important. Relative weights affect the position of the BPE and absolute weights determine the size of the probability rectangle. Although the problem of estimating variances has rightly attracted considerable interest in the past, it should not be inferred that high accuracy is required. Variances and weights correct to (say) ± 10% would be adequate; the error in the area of the rectangle with a stated probability level would then be 10% or less. Any plotting program should be tested for self-consistency in the sense that variances derived from check targets should lead to the correct success rate when PRs are tested against check targets.

References

BAILEY, A.D., ERNST, E.W., MILLER, L.J., and WOOD W.W. (1968): 'An atlas of reduced data obtained in a cooperative HF directional propagation experiment over a 1330 km path (March 1967–March 1968)'. RRL Publication no. 344, Dept. of Elec. Eng., University of Illinois

BEALE, E.M.L. (1961a): 'Brooke variance classification system for DF bearings'. *J. Res. Natl. Bur. Stand.*, **65D,** pp. 255–261

BEALE, E.M.L. (1961b): 'Estimation of variances of position lines from fixes with unknown target positions'. *J. Res. Natl. Bur. Stand.*, **65D,** pp. 263–273

BERLE, F.J. (1986): 'Mixed triangulation/trilateration technique for emitter location', *Proc. IEE.*, **133F,** pp. 638–641

BLACHMAN, N.M. (1989): 'On combining target-location ellipses', *IEEE. Trans.*, **AES-25,** pp. 284–287

BOWEN, K.C. (1955): 'Sources of error in U-Adcock high-frequency direction finding'. *Proc. IEE Pt. B.*, **102,** pp. 529–532

BUTTERLY, P.J. (1972): 'Position finding with empirical prior knowledge'. *IEEE. Trans.*, **AES-8,** pp. 142–146

COOPER, D.C. (1972): 'Statistical analysis of position-fixing: general theory for systems with Gaussian errors'. *Proc. IEE.*, **119,** pp. 637–640

DANIELS, H.E. (1951): 'The theory of position finding'. *J. R. Stat. Soc.*, **13B,** pp. 186–199

GEORGES, T.M. (1967): 'Ionospheric effects of atmospheric waves', ESSA technical report IER 57-ITSA 54

GETHING, P.J.D. (1961): 'Influence of ionospheric conditions on the accuracy of high frequency direction finding', *J. Res. Natl. Bur. Stand.*, **65D,** pp. 225–228

GETHING, P.J.D. (1967): 'Correlation effects on direction-finding probability regions', *Proc. IEE.*, **114,** pp. 199–200

HOUSTON, R.S., and NELSON, J. (1980): 'Model error and the direction-finder problem', *IEEE. Trans.*, **AES-16,** pp. 729–732

ISLEY, C.T. (1980): 'A procedure for directly calculating CEP in DF fixing', *IEEE. Trans.*, **AES-16,** pp. 867–870

JONES, T.B., and REYNOLDS, J.S.B. (1975): 'Ionospheric perturbations and their effect on the accuracy of HF direction finders', *Radio and Electron. Eng.*, **45,** pp. 63–73

MORGAN, A.D. (1972): 'The effects of ionospheric disturbances on the bearings of incoming sky waves', Agard Conference Proceedings no. 115, Wiesbaden, pp. 28–1 to 28–17

RABTSUN, A.V. (1987): 'Maximum likelihood estimate of signal source coordinates in multi-position direction finding', Radiotekhnika, **42,** no. 9, pp. 9–10. English translation in Telecom. Radio Eng., **42,** no. 10, pp. 34–35

RANZI, I., and GIORGI, P. (1972): 'An important characteristic of some travelling ionospheric disturbances', Agard Conference Proceedings no. 115, Wiesbaden, pp. 25–1 to 25–2

REYNOLDS, J.S.B., and MORGAN, A.D. (1975): 'The effect of frequency separation on the correlation of bearing errors measured on ionospherically propagated HF signals from colocated transmitters', *J. Atmos. & Terr. Phys.*, **37,** pp. 545–551

ROSS, W. (1947a): 'The estimation of the probable accuracy of HF radio direction-finding bearings', *J. IEE.*, **94,** Part IIIa, pp. 722–726

ROSS, W. (1947b): 'Site and path errors in short wave direction-finding', *J. IEE.*, **94,** Part III, pp. 108–114

ROSS, W. (1975): "Wild' bearings in high-frequency direction finding', *Proc. IEE.*, **122,** pp. 337–339

STANSFIELD, R.G. (1947): 'Statistical theory of DF fixing', *J. IEE.*, **94,** part IIIa, pp. 762–770

WAX, M. (1983): 'Position location from sensors with position uncertainty', *IEEE. Trans.*, **AES-19,** pp. 658–662

Chapter 17

Conclusions

The aim in this brief final chapter is to summarise the major conclusions reached in earlier pages and to suggest the likely trends in future research.

Until recently, the output of a DF system was usually thought of as a single number, the indicated bearing of the target transmitter. Traditionally, the wave-field, no matter how complex, was interpreted in terms of a one-ray field, with the ray arriving from a constant azimuth. Because of the inability of NADF's such as the Adcock to resolve modes closely spaced in azimuth, there was little incentive to use a more sophisticated model.

A new situation has arisen in recent years, as a result of the following factors:

(*a*) The more widespread use of various forms of WADF, which provide much greater resolving power,
(*b*) The development of techniques for the measurement of elevation angles,
(*c*) The reduction of instrumental and site errors to almost negligible proportions,
(*d*) The exploitation of multichannel receivers, digital recorders and computers to analyse the structure of wave-fields,
(*e*) The development of superresolution algorithms.

Improved resolving power has allowed the first steps to be taken along the path towards mode separation, mode identification and the application of appropriate corrections for deviations on the propagation path.

The need to achieve mode resolution with WADF's of reasonable aperture (a few wavelengths) implies that the benefits of time-sampling as well as space-sampling must be incorporated in the processing scheme. Even so, we cannot expect to obtain a correct and unique solution for all ray parameters in all circumstances: any analysis method is likely to break down if the underlying assumptions made about the wave-field are invalid or if the resolving power is inadequate to cope with the problem. With very close rays it will not be possible to resolve the rays successfully unless the signal/noise ratio is abnormally high (perhaps 75 dB or more). Given that we have no choice, in practical DF, about the occurrence of such problems, the important requirements are that they should be recognised and that the corresponding solutions should be regarded as unreliable.

The characteristics of wave-fields produced by a small number of rays have been described in Chapters 3 and 4. The well known 'kinked' form of the curves of constant phase for a two-ray wave-field can be related to the *U*-shaped p.d.f. If one ray is dominant, in the sense that its resolved amplitude is always greater than that of the other ray, the mean azimuth indicated by an orthodox direction finder is likely to be very close to that of the stronger ray. With three or more rays, more irregular field patterns occur. During any period in which one ray is dominant over all others combined, that ray will tend to 'capture' the bearing indication. When one ray is not dominant, the phenomenon of capture does not occur and the CPs are not continuous, but exhibit a 'line tearing' effect.

Possible techniques for the calculation of ray parameters were examined in Chapters 7–13. They can be summarised under the following headings:

(*a*) analysis of a single waveform from a spinning goniometer (feature extraction),
(*b*) analysis of a sequence of waveforms (intersection method, ratio intersection method),
(*c*) iterative search in azimuth when elevation angles have been separately determined,
(*d*) for the interferometer, interpretation of the polygon display,
(*e*) exploitation of quasiunimoded propagation, recognised by wavefront tests,
(*f*) various versions of wavefront analysis,
(*g*) use of superresolution algorithms such as MUSIC.

These methods provide a range of tools for the intelligent analysis of specific problems. In effect, we adopt some simplified model of the wave-field and then test whether the observations are consistent with this hypothesis. The need for such an approach is particularly clear in WFA where, for example, we may successively examine a one-ray hypothesis, a two-ray hypothesis etc., but initial assumptions are made, implicitly or explicitly, in other methods. The performance of a DF system in multiray conditions can be predicted only when the exact nature of the DF processor has been specified. One of the considerable advantages of WFA is that the processor can be varied after the data have been collected, by means of software changes.

Wavefront analysis combined with a suitable superresolution processing method seems to be the most attractive and powerful method of dealing with a limited number of rays. It is widely applicable, to almost any array configuration, and leads at worst to a two-parameter search in azimuth and elevation for nulls in the array radiation pattern. Elevation angles can be calculated correctly only if it is possible to determine accurately the position of nulls produced by a specified BFN; it is at this stage of the calculation that instrumental or site errors could produce an effect.

With realistic signal/noise ratios the number of rays that can be found may be rather less than the theoretical maximum for a given number of elements and receiving channels. In this connection, the phenomenon of masquerading has been described in Chapter 11. A sensible practical aim, given a signal/noise ratio of the order of 20–30 dB in the comparatively narrow bandwidth of a multichannel receiver, is to determine the parameters of the two or three strongest rays.

Many of the results presented here on wave-field structure and wavefront analysis are of general validity and application, independent of frequency band; wavefront analysis can be regarded as an extension of classical array theory. When the application is HF DF, we can also make use of the very large body of knowledge that exists about HF radio propagation and the ionosphere.

The importance, in experimental work, of finding out as much as possible about the state of the ionosphere, and the probable modes of propagation, cannot be over-stressed. Sequences of azimuth measurements alone, made in the absence of such knowledge, are now of very little value in the research context. Chapters 14 and 15 provide a review of those elements of propagation theory that are relevant to ray re-tracing, tilt correction and the interpretation of ionospheric data obtained with some forms of sounders. Oblique, backscatter or Doppler sounders are now often used in conjunction with WADF's in modern research and this part of the book should therefore be of particular interest to those involved in controlled experiments with co-operative transmitters.

A major objective of current DF research is to find new ways of attacking the difficult problem of tilt correction. Partial success has already been achieved, under suitable experimental conditions. Semi-empirical corrections for systematic tilts of the ionosphere, and theoretical corrections for magneto-ionic deviations, are probably worth applying now to high-quality WADF measurements. However, a more comprehensive system of tilt correction, applicable in less favourable operational conditions, remains a difficult target. As a first step, mode resolution, elevation angle measurements and mode identification must be developed into *routine* techniques, requiring a minimum of auxiliary equipment at the DF site. Such techniques would lead, at the very least, to an improved system of bearing weights based partly on the probable mode of propagation.

It seems almost inevitable that wavefront analysis methods, in some broad sense, will become more widely used in future, as the power of digital processing is applied to DF systems. The real-time control of nulls in radiation patterns will also be developed further for adaptive antenna arrays, designed to give improved reception rather than directional measurements. The control of patterns in elevation as well as azimuth will be an important part of this work, offering the prospect of the elimination of unwanted modes and the reduction of multipath distortion.

So far as DF equipment is concerned, further improvements in instrumental accuracy for single-ray conditions are of comparatively minor importance. Of much greater significance would be improvements and simplifications in the auxiliary equipment used for measurements on the 'remote' ionosphere. Improvements are needed because the quality of typical records places a limitation at present on the information that can be extracted. Simplifications of the equipment would make it cheaper to install, and easier to use, at operational DF stations. The basic requirement is a means of taking a rather broad look at a patch of the ionosphere, rather than a very detailed look at many small ionospheric features. Similarly, the fine structure of any one mode, although of interest to the ionospheric physicist, does not seem to be of great significance in practical DF with apertures of a few wavelengths.

Whereas a comparatively simple model of the current state of the ionosphere may be adequate for mode identification, and would be a great deal better than no model at all, rather more detailed information is probably required for tilt correction schemes. Ray-tracing programs, particularly three-dimensional programs, provide an important tool for the studies and simulations that should be undertaken, in order to decide on the best and simplest way of extracting the necessary data from records obtained with auxiliary sounding equipment. Numerical maps provide a convenient summary of averaged ionospheric parameters, for comparison with real measurements. From such comparisons, fresh knowledge about ionospheric morphology and physics may well be obtained. As in most fields of scientific endeavour, a combination of practical and theoretical research is likely to be more powerful than either alone.

Appendix 1

Circular WADF: Parameters used in pattern calculations

1 Array dimensions

Circular array of 96 equispaced doublets
Array radius = 163·83 m to centre of each doublet
Separation of elements in doublet = 4·88 m (radially)
Doublet delay line (attached to inner element) equivalent to 4·88 m

2 Goniometer

Low-angle delay lines cut for elevation angle of 15°
High-angle delay lines cut for elevation angle of 30°
Low-sum pattern formed from doublets (1 to 32)
Low-difference pattern formed from doublets (1 to 16)–(17 to 32)
High-sum pattern formed from doublets (49 to 80)
High-difference pattern formed from doublets (49 to 64)–(65 to 80)

3 Simplifying assumptions

(*a*) Response of single monopole assumed to be proportional to cos Δ, where Δ is elevation angle,
(*b*) All mutual coupling effects are neglected,
(*c*) Action of capacitor plates on rotor and stator neglected by assuming that goniometer rotation is equivalent to rotation of whole array, with delay networks permanently connected to the doublets specified above.

Appendix 2

The theory of multiplicative processing

In this appendix we consider the response of a particular multiplicative array to an incident field produced by one or two targets. The array can be scanned electronically or mechanically relative to the targets. The treatment here is based on a paper by Shaw and Davies (1964).

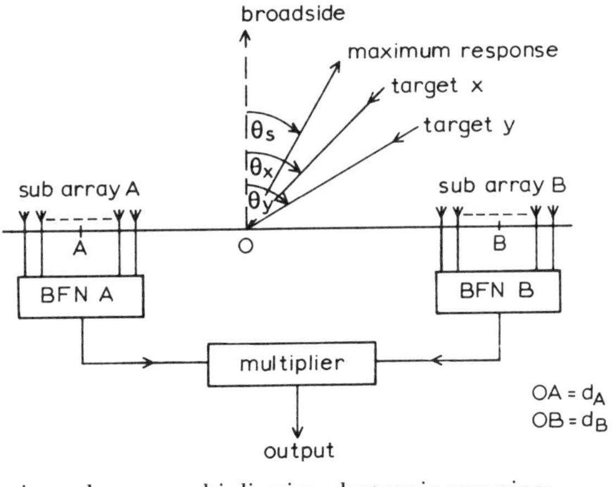

Fig. A2.1 Processing scheme, multiplicative electronic scanning

Case 1. Electronic scanning

In Fig. A2.1, A and B represent the phase centres of two additive sub-arrays. The two targets X and Y (in the same plane) are assumed to produce signals $\sin \omega t$ and $K \sin (\omega t + \psi)$ at O. It is important to note that the value of the phase difference depends on the choice of the point O.

For a spacing d between neighbouring pairs of elements, the progressive phase shift required to steer the direction of maximum response of each sub-array to a direction θ_s (measured from the broadside direction) is $(2\pi d \sin \theta_s)/\lambda$. This is the phase difference between neighbouring elements. In order to facilitate comparisons with (say) a full additive array, we assume that the phase shift changes progressively along the whole array.

For electronic scanning, it is convenient to replace angles by the equivalent value on a p scale defined by

$$p_s = \frac{\pi d \sin \theta_s}{\lambda}$$

$$p_x = \frac{\pi d \sin \theta_x}{\lambda}$$

$$p_y = \frac{\pi d \sin \theta_y}{\lambda}$$

where θ_x and θ_y are the azimuth angles of the two targets, measured from the broadside direction. It is also convenient to replace the distances d_A, d_B from O to the subarray phase centres by

$$\delta_A = \frac{\pi d_A}{\lambda}$$

$$\delta_B = \frac{\pi d_B}{\lambda}$$

and to write

$$r = \frac{(d_A + d_B)}{d}$$

The difference of phase shift applied in the BFN to neighbouring elements is $2p_s$ and the total difference between A and B is $2rp_s$.

The response of an additive subarray of n elements, phased to give a maximum at a 'scan angle' of θ_s, to a single target at angle θ is given by

$$D(\theta) = \frac{\sin \{n(p - p_s)\}}{\sqrt{n} \sin (p - p_s)}$$
$$= D(p - p_s)$$

If we represent the additive patterns of subarrays A and B by D_A, D_B respectively, we obtain the following expressions for the signals to be processed:

Signal at A due to $X = D_A\,(p_x - p_s) \sin (\omega t - 2\delta_A \sin \theta_x)$

Signal at A due to $Y = KD_A\,(p_y - p_s) \sin (\omega t - 2\delta_A \sin \theta_y + \psi)$

Signal at B due to $X = D_B\,(p_x - p_s) \sin (\omega t + 2\delta_B \sin \theta_x - 2rp_s)$

Signal at B due to $Y = KD_B\,(p_y - p_s)$
$\sin (\omega t + 2\delta_B \sin \theta_y + \psi - 2rp_s)$

The resultant signal at A has components $A_s \sin \omega t$ and $A_c \cos \omega t$, where

$$A_s = D_A(p_x - p_s) \cos(-2\delta_A \sin \theta_x) + KD_A(p_y - p_s) \cos(\psi - 2\delta_A \sin \theta_y)$$

$$A_c = D_A(p_x - p_s) \sin(-2\delta_A \sin \theta_x) + KD_A(p_y - p_s) \sin(\psi - 2\delta_A \sin \theta_y)$$

The corresponding components at B are $B_s \sin \omega t$, $B_c \cos \omega t$, where

$$B_s = D_B(p_x - p_s) \cos(2\delta_B \sin \theta_x - 2rp_s) + KD_B(p_y - p_s) \cos(\psi + 2\delta_B \sin \theta_y - 2rp_s)$$

$$B_c = D_B(p_x - p_s) \sin(2\delta_B \sin \theta_x - 2rp_s) + KD_B(p_y - p_s) \sin(\psi + 2\delta_B \sin \theta_y - 2rp_s)$$

The product of the resultant signals from A and B may be written as

$$\{A_sB_s \sin^2 \omega t + (A_sB_c + A_cB_s) \sin \omega t \cos \omega t + A_cB_c \cos^2 \omega t\}$$

When this signal is lowpass filtered (i.e. integrated over many RF cycles) the output is proportional to $(A_sB_s + A_cB_c)$. The multiplicative response, $D_m(p_s)$, of an array scanned electronically to p_s may therefore be written

$$\begin{aligned} D_m(p_s) = {} & e_1 D_A(p_x - p_s) D_B(p_x - p_s) \\ & + e_2 K D_A(p_x - p_s) D_B(p_y - p_s) \\ & + e_3 K D_A(p_y - p_s) D_B(p_x - p_s) \\ & + e_4 K^2 D_A(p_y - p_s) D_B(p_y - p_s) \end{aligned}$$

where

$$\begin{aligned} e_1 &= \cos\{2(\delta_A + \delta_B) \sin \theta_x - 2rp_s\} \\ e_2 &= \cos(2\delta_A \sin \theta_x + 2\delta_B \sin \theta_y + \psi - 2rp_s) \\ e_3 &= \cos(2\delta_A \sin \theta_y + 2\delta_B \sin \theta_x - \psi - 2rp_s) \\ e_4 &= \cos\{2(\delta_A + \delta_B) \sin \theta_y - 2rp_s\} \end{aligned}$$

The symbol e is used here to denote electronic scanning.

As might be expected, the coefficients e_1 and e_4, which represent the effects of target x alone and target y alone respectively, do not involve ψ and hence do not depend on the phase reference point O. The 'interaction' terms e_2 and e_3, on the other hand, depend on ψ and on the choice of O.

A slight simplification occurs when the subarrays are of equal length, i.e. $D_A = D_B$. We then have

$$D_m(p_s) = e_1 D_A^2(p_x - p_s) + (e_2 + e_3) K D_A(p_x - p_s) D_A(p_y - p_s) + e_4 K^2 D_A^2(p_y - p_s)$$

If the phase reference point O is now made to coincide with the centre of the whole array, δ_A is equal to δ_B and

$$(e_2 + e_3) = 2 \cos \{2\delta_A(\sin \theta_x + \sin \theta_y) - 2rp_s\} \cos \psi$$

Shaw and Davies, using what appears to be an unduly compressed notation, write the above coefficients as

$$e_1 = \cos rp$$
$$e_2 = \cos (rp + \psi)$$
$$e_3 = \cos (rp - \psi)$$
$$e_4 = \cos rp$$

and define rp as 'the phase difference between the signals arriving at the phase centres of the two parts of the array...'

The response of an array consisting of two four-element subarrays as it is scanned across two targets is shown in Fig. 5.6 in Chapter 5. The targets are at $\pi/16$ and $-\pi/16$ on the p scale and are of equal amplitude.

Consider an array consisting of two equal subarrays, with two targets at angles defined by the conditions

$$p_x = 0$$
$$p_y = a$$

Suppose that the array is scanned to an angle defined by

$$p_s = x'$$

If we write A in place of $2r$ and φ in place of ψ, the coefficients e_1 to e_4 become

$$e_1 = \cos (-Ax')$$

$$(e_2 + e_3) = 2 \cos \left\{ A\left(\frac{a}{2} - x'\right)\right\} \cos \varphi$$

$$e_4 = \cos \{A(a - x')\}$$

These are identical with the coefficients given in a paper by Ksienski (1965).

Case 2. Mechanical scanning

We now consider the response of an array with zero phase shifts in the BFNs as it is rotated about some fixed point. We refer to such a rotation relative to fixed targets as mechanical scanning. The same response curve could be obtained, of course, by rotating the two targets round a fixed array, provided the angular separation of the targets was held constant. The 'scan angle' of such an array, defining the direction of maximum response, is always in the broadside direction θ_s, measured from some fixed reference direction, OC in Fig. A2.2. (Note the slight change in the definition of θ_s). The phase difference ψ between the two target signals could be defined at any fixed point; the rotation centre O is the obvious choice and is used here.

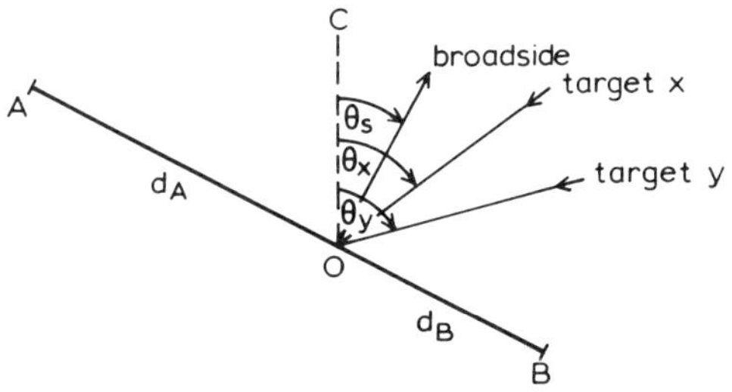

Fig. A2.2 Two-target situation, mechanical scanning

It is now more convenient to work in terms of angles θ_x, θ_y, θ_s rather than the sines of these angles. We write

$$x = \theta_x - \theta_s$$
$$y = \theta_y - \theta_s$$

The derivation of the expression for the array response is then very similar to that given for electronic scanning. The various steps are set out below.

$$\text{Signal at } A \text{ due to } X = D_A(x) \sin(\omega t - 2\delta_A \sin x)$$
$$\text{Signal at } A \text{ due to } Y = KD_A(y) \sin(\omega t - 2\delta_A \sin y + \psi)$$
$$\text{Signal at } B \text{ due to } X = D_B(x) \sin(\omega t + 2\delta_B \sin x)$$
$$\text{Signal at } B \text{ due to } Y = KD_B(y) \sin(\omega t + 2\delta_B \sin y + \psi)$$

From these expressions

$$A_s = D_A(x) \cos(-2\delta_A \sin x) + KD_A(y) \cos(\psi - 2\delta_A \sin y)$$
$$A_c = D_A(x) \sin(-2\delta_A \sin x) + KD_A(y) \sin(\psi - 2\delta_A \sin y)$$
$$B_s = D_B(x) \cos(2\delta_B \sin x) + KD_B(y) \cos(\psi + 2\delta_B \sin y)$$
$$B_c = D_B(x) \sin(2\delta_B \sin x) + KD_B(y) \sin(\psi + 2\delta_B \sin y)$$

The response is then given by

$$D_m(\theta_s) = m_1 D_A(x) D_B(x) + m_2 K D_A(x) D_B(y) + m_3 K D_A(y) D_B(x) + m_4 K^2 D_A(y) D_B(y)$$

where

$$m_1 = \cos\{2(\delta_A + \delta_B) \sin x\}$$
$$m_2 = \cos(2\delta_A \sin x + 2\delta_B \sin y + \psi)$$
$$m_3 = \cos(2\delta_A \sin y + 2\delta_B \sin x - \psi)$$
$$m_4 = \cos\{2(\delta_A + \delta_B) \sin y\}$$

These coefficients, m_1 to m_4, for mechanical scanning are very similar in form to the coefficients e_1 to e_4 for electronic scanning.

For small scan angles and targets near the broadside direction, θ_x, θ_y and θ_s are all small. We can then write

$$\sin x \simeq \sin \theta_x - \sin \theta_s$$

$$\sin y \simeq \sin \theta_y - \sin \theta_s$$

With these approximations, the values of m_1 to m_4 become approximately equal to e_1 to e_4 respectively. As might be expected, there is no significant difference between electronic and mechanical scanning near the broadside direction.

References

KSIENSKI, A. (1965): 'Multiplicative processing antenna system for radar applications', *Radio & Electron. Eng.*, **29,** pp. 53–67

SHAW, E., and DAVIES, D.E.N. (1964): 'Theoretical and experimental studies of the resolution performance of multiplicative and additive arrays'. *Radio & Electron. Eng.*, **28,** pp. 279–291

Appendix 3

Vertical array of loops: Parameters used in pattern calculations

1 Heights of loops

True heights of eight loops above ground: 9·25, 18·39, 27·60, 36·80, 46·00, 55·19, 64·69, 73·91 m. (Note: loops are often assumed to be uniformly spaced at intervals of 9·22 m).

2 Loops

Loops are circular, of twin construction, diameter approximately 1·6 m.

3 Simplifying assumptions

(*a*) Response of a single loop (in free space) is assumed to be proportional to cos Δ, where Δ is elevation angle,

(*b*) All mutual coupling effects are neglected,

(*c*) Each loop responds only to horizontal polarisation,

(*d*) Ground reflection is usually assumed to be 'perfect', i.e. reflection coefficient = − 1.

Appendix 4

An example of masquerading

A specific example will now be given, in which two distinct two-mode toy problems, *A* and *B*, give almost identical signals in the eight-loop vertical array. Data are shown for three frames, but the method of deriving the ray parameters for *B* from those for *A* is so general that the masquerading can in fact continue indefinitely. The frequency is 6 MHz.

The ray parameters for problem *A*, chosen arbitrarily, are shown in Table A4.1.

Table A4.1 Ray parameters for problem *A*

Angle	Frame 1		Frame 2		Frame 3	
	Amp	Phase	Amp	Phase	Amp	Phase
6·0°	100·0	0°	330·0	76·0°	260·0	168·0°
10·0°	210·0	64·0	170·0	217·0	50·0	19·0

The corresponding loop signals are shown in Table A4.2.

Table A4.2 Loop data, problem *A*

Loop	Frame 1		Frame 2		Frame 3	
	Amp	Phase	Amp	Phase	Amp	Phase
1	96	51°	50	313°	47	336°
2	187	51	99	312	93	336
3	273	51	147	310	139	336
4	348	50	192	308	183	336
5	411	50	235	305	227	337
6	460	49	275	302	268	338
7	493	48	315	297	309	339
8	507	47	352	293	346	340

Now it can be shown that rays at 6·0° and 10·0°, of equal amplitudes and of the same phase, produce almost the same loop signals as a single ray at 8·6°. Representing each ray by its amplitude and elevation angle, we can write

$$200\ (10{\cdot}0^\circ) + 200(6{\cdot}0^\circ)\ M\ 371\ (8{\cdot}6^\circ)$$

Similarly, if the rays are in anti-phase, the signals are very similar to those for a single mode at 15·2°

$$200(10{\cdot}0^\circ) - 200(6{\cdot}0^\circ)\ M\ 55(15{\cdot}2^\circ)$$

It follows that

$$400(10{\cdot}0^\circ)\ M\ 371(8{\cdot}6^\circ) + 55(15{\cdot}2^\circ)$$

and

$$400(6{\cdot}0^\circ)\ M\ 371\ (8{\cdot}6^\circ) - 55(15{\cdot}2^\circ)$$

Thus any combination of rays at 6·0° and 10·0°, in any amplitude/phase relationship, can be masqueraded by rays at 8·6° and 15·2° with amplitude components that are easily calculated from these equations. The ray parameters obtained from Table A4.1 are shown in Table A4.3 and the corresponding loop signals in Table A4.4.

Table A4.3 Ray parameters for problem *B*

Angle	Frame 1		Frame 2		Frame 3	
	Amp	Phase	Amp	Phase	Amp	Phase
8·6°	249·8	44·5°	208·6	104·4°	202·8	161·2°
15·2	26·0	92·4	65·2	243·0	41·8	352·9

Table A4.4 Loop data, problem *B*

Loop	Frame 1		Frame 2		Frame 3	
	Amp	Phase	Amp	Phase	Amp	Phase
1	96	51°	48	314°	46	335°
2	188	51	98	313	91	335
3	273	51	146	311	137	336
4	348	50	191	308	182	336
5	411	50	235	305	227	337
6	459	49	276	301	269	338
7	493	48	315	297	309	339
8	508	47	349	293	343	339

The agreement between Tables A4.2 and A4.4 is sufficiently close to demonstrate that the underlying theory is correct; the largest discrepancies are three units of amplitude and 1° of phase. It is also clear that the agreement could be improved slightly by means of minor adjustments, e.g. by increasing both ray amplitudes by about 1% on frames 2 and 3 of problem *B*. The figures quoted are for perfect equipment on a perfect site, with no noise or interference. With a signal/noise ratio of 40 dB or less, problem *B* seems quite indistinguishable from problem *A*.

In real problems, the possibility of small angle variations from frame to frame cannot be excluded. It therefore seems probable that the agreement between Tables A4.2 and A4.4 could be made even closer. Moreover, attention has been deliberately restricted to a two-mode problem; masquerading by three or more modes could be even more effective.

It should not be assumed that rays at 8·6° and 15·2° must be chosen in order to masquerade for rays at 6·0° and 10·0°. Many other equivalences can be found, for example

$$400\ (10{\cdot}0°) + 200(6{\cdot}0°)\ M\ 566(9{\cdot}2°)$$

$$200\ (10{\cdot}0°) + 400(6{\cdot}0°)\ M\ 550(8{\cdot}0°)\ \text{etc}$$

Continued manipulation of equations of this kind leads to the conviction that any two-mode problem involving angles below 15·2° (at least) at 6 MHz can be masqueraded by some other problem involving two or more modes. One mode can masquerade as two, but in this case the amplitude fluctuations of the two modes would show a strong correlation.

Masquerading occurs chiefly at low frequencies and low-elevation angles and is a manifestation of inadequate resolving power. On this physical interpretation, it cannot be avoided by alterations in the number and dispositions of the elements within a given aperture; it can arise with irregular, as well as regular, spacings.

Appendix 5

Covariance matrix of sensor outputs

Matrix notation

Each matrix is represented by a bold letter or by a typical element within square brackets. The superscript T is used for the transpose of a matrix, i.e.

$$[x_{ji}] = [x_{ij}]^{\mathrm{T}}$$

It is sometimes convenient to write column matrices, usually referred to as vectors, in the transposed row form. An asterisk is used as a superscript to denote the complex conjugate of an element or matrix, and superscript H for the transposed complex conjugate of a matrix. Note that a prime or dagger symbol is used instead of H in some papers on superresolution.

The operation represented by H can be performed on any matrix, whether square or not; it should not be confused with the condition for a square matrix $\boldsymbol{B}$ to be Hermitian, namely

$$\boldsymbol{B} = \boldsymbol{B}^H$$

The inverse of a square non-singular matrix $\boldsymbol{C}$ is written as $\boldsymbol{C}^{-1}$ and satisfies the relation

$$\boldsymbol{C}\,\boldsymbol{C}^{-1} = \boldsymbol{I}$$

Here $\boldsymbol{I}$ is the identity matrix (unitary matrix) of the same order as $\boldsymbol{C}$, with ones along the leading diagonal and zeros elsewhere. $\boldsymbol{C}^n$ represents $\boldsymbol{C}$ to the power n.

Matrix Pencil

Two matrices $\boldsymbol{A}$ and $\boldsymbol{B}$ of the same dimensions can be combined with an undefined multiplier α into $(\boldsymbol{A} - \alpha\boldsymbol{B})$. This combination is sometimes called the pencil of $\boldsymbol{A}$ and $\boldsymbol{B}$, denoted by $\{\boldsymbol{A}, \boldsymbol{B}\}$.

Projection

In Fig. A5.1, the sides of the triangle ABC might represent three vectors (such as forces, or velocity components, or phasors) in the *x–y* plane. The vector equation

$$\boldsymbol{AC} = \boldsymbol{AB} + \boldsymbol{BC}$$

applies because the triangle is a closed figure. The triangle, and hence the equation, can be projected on the x-axis, giving

$$DF = DE + EF$$

This is a scalar equation, or a vector equation with all the vectors aligned in the same direction. Similarly, the original vector equation can be projected on to the *y*-axis, or any other straight line in the *x–y* plane; equality is preserved provided the same operation is performed on both sides of the equation.

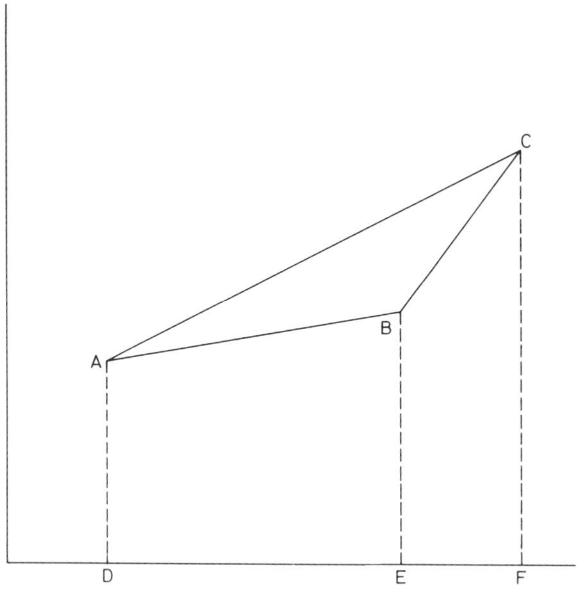

Fig. A5.1 Projection of vector triangle on to X-axis

The concept of 'projecting' an equation can be generalised to *n*-dimensional space, using matrix notation. If $\boldsymbol{A}$ and $\boldsymbol{B}$ are *n*-dimensional vectors, the product $\boldsymbol{A}^{\mathrm{T}}\ \boldsymbol{B}$ is a scalar (one-element matrix). When $\boldsymbol{A}$ is a vector of unit length, i.e. $\boldsymbol{A}^{\mathrm{H}}\boldsymbol{A} = 1$, the vector $(\boldsymbol{A}^{\mathrm{T}}\boldsymbol{B})\ \boldsymbol{A}$ is a projection of $\boldsymbol{B}$ along the direction of $\boldsymbol{A}$. When this vector represents a voltage or signal, the associated power is proportional to $(\boldsymbol{A}^{\mathrm{T}}\boldsymbol{B}\ \boldsymbol{A})^{\mathrm{H}}\ (\boldsymbol{A}^{\mathrm{T}}\boldsymbol{B}\ \boldsymbol{A})$. When the vector represents the deviation of a point from a plane, the same expression may be interpreted as a real squared distance.

If $\boldsymbol{B}$ is an $(n \times n)$ matrix, the product $\boldsymbol{A}^{\mathrm{T}}\boldsymbol{B}\ \boldsymbol{A}$ is a scalar, which may still be referred to as a projection of $\boldsymbol{B}$ on to $\boldsymbol{A}$. Note that $\boldsymbol{A}^{\mathrm{H}}\boldsymbol{B}\,\boldsymbol{A}$ is also a scalar and is another valid projection of $\boldsymbol{B}$ on to $\boldsymbol{A}$.

Data model

For r modes (rays) falling on n sensors, let m_{sk} be the contribution of mode k at unit amplitude to sensor s. Then if the amplitude of mode k in frame j is w_{kj}, the signal z_{sj} at sensor s in the absence of noise is found by summing over all modes:

$$z_{sj} = \sum_{k=1}^{r} m_{sk} w_{kj}$$

With noise components e_{sj} we may write

$$z_{sj} = \sum_{k=1}^{r} m_{sk} w_{kj} + e_{sj}$$

i.e.

$$\boldsymbol{Z}_j = \boldsymbol{M}\,\boldsymbol{W}_j + \boldsymbol{E}_j \tag{A5.1}$$

In this equation $\boldsymbol{Z}_j$, $\boldsymbol{W}_j$ and $\boldsymbol{E}_j$ vary from frame to frame. The matrix $\boldsymbol{M}$ is a function of the constant array geometry and of the angles of arrival, assumed constant; it is sometimes referred to as the matrix of array steering vectors.

A covariance matrix is now formed from the measured $\boldsymbol{Z}_j$. The sample covariance matrix for a total of f frames is

$$\hat{\boldsymbol{R}} = \frac{1}{f}\sum_{j=1}^{f} \boldsymbol{Z}_{\mathrm{j}}\,\boldsymbol{Z}_{\mathrm{j}}^{\mathrm{H}}$$

and is given by

$$\hat{\boldsymbol{R}} = \boldsymbol{M}\,\boldsymbol{S}\,\boldsymbol{M}^{\mathrm{H}} + \boldsymbol{U} \tag{A 5.2}$$

where

$$\boldsymbol{S} = \frac{1}{f}\sum_{j=1}^{f} w_{\mathrm{j}}\, w_{\mathrm{j}}^{\mathrm{H}}$$

$$\boldsymbol{U} = \frac{1}{f}\sum_{j=1}^{f} \boldsymbol{E}_{\mathrm{j}}\,\boldsymbol{E}_{\mathrm{j}}^{\mathrm{H}}$$

$\hat{\boldsymbol{R}}$, $\boldsymbol{S}$ and $\boldsymbol{U}$ are square matrices of order $(n \times n)$, $(r \times r)$ and $(n \times n)$ respectively. Each is bound to be Hermitian, because of the way in which it is calculated. $\boldsymbol{M}$ and $\boldsymbol{M}^{\mathrm{H}}$ are of order $(n \times r)$ and $(r \times n)$ respectively. The sample covariance matrix $\hat{\boldsymbol{R}}$ and the matrix of array steering vectors $\boldsymbol{M}$ play a key role in the theory of superresolution.

No noise

The case of no noise or measurements errors is of theoretical interest. Provided r is less than $(n - 1)$, one or more linear relationships will be satisfied exactly by the measured elements of $\boldsymbol{Z}_j$ on each frame and $\hat{\boldsymbol{R}}$ will be of reduced rank r. It is said to have nullity $(n - r)$.

Noise model

It is usually assumed that the noise is of uncorrelated Gaussian form, with components drawn from a population with standard deviation σ. Using the notation $E(x)$ for the expected value of any variable x, we then have

$$E\,(e_{sj}) = 0$$

$$E\,(e_{sj}\; e_{tj}) = 0, \; \mathrm{s} \neq \mathrm{t}$$

$$E\,(e^2{}_{sj}) = 0$$

Asymptotic form

With the above noise model applied to a large sample of frames, the off-diagonal elements of $\boldsymbol{U}$ approach zero and equation (A5.2) assumes the asymptotic form

$$\boldsymbol{R} = \boldsymbol{M}\,\boldsymbol{S}\,\boldsymbol{M}^{\mathrm{H}} + \sigma^2\boldsymbol{I} \qquad \text{(A 5.3)}$$

For completely uncorrelated fading of the arriving rays, the off-diagonal elements of $\boldsymbol{S}$ also tend to zero. The off-diagonal elements of $\boldsymbol{R}$ contain the directional information about the arriving rays and do not tend towards zero, but the leading diagonal elements are real and tend towards equality; the physical interpretation is that we expect all sensors to collect approximately the same power over a long sampling period.

Toeplitz matrix

The covariance matrix is of Toeplitz form when elements along each diagonal parallel to the leading diagonal are equal, i.e.

$$R_{12} = R_{23} = R_{34} = \text{etc.}$$

$$R_{13} = R_{24} = R_{35} = \text{etc.}$$

The Toeplitz property implies that the average mutual power for antenna elements of interval $(i - j)$ is the same for all pairs of elements with this interval. The condition is not likely to apply unless the elements are uniformly spaced, and may be only approximately satisfied by a finite sample of real data.

Appendix 6

Magneto-ionic deviations calculated with the Jones 3-dimensional ray-tracing program

Ionospheric model

The ionospheric model adopted was a single, untilted, quasiparabolic layer with the following parameters:

Critical frequency = 5·0 MHz

Height of maximum ionization, h_m = 280 km

Semi-thickness parameter, y_m = 80 km

$\gamma = y_m/(h_m - y_m)$

$= 0{\cdot}4$

Earth's magnetic field

The 'earth-centred' dipole model adopted in the Jones subprogram DIPOLY was used, with a gyrofrequency of 0·68228 MHz at the equator. The co-ordinates of the N magnetic pole were taken to be 290°E, 79°N.

Sign convention

Rays were traced from a fixed transmitter position in S. England. Bearing errors are measured at the receiver, i.e. the point where a ray strikes the ground after one hop, and are positive when the 'observed' azimuth angle, measured clockwise from N, is greater than the true g.c. bearing.

Key to column headings in Tables A6.1 and A6.2

f = signal frequency, MHz

Δ = elevation angle, deg

φ = azimuth angle (at transmitter) from magnetic N

D = ground range, km

P' = effective (group) path length, km

BE = bearing error, degrees, rounded to nearest 0·01°

Table A6.1 Variation of bearing errors with azimuth

	Launch conditions			Ordinary ray			Extraordinary ray		
Ray ref number	*f*	Δ	φ	*D*	*P′*	*BE*	*D*	*P′*	*BE*
1	8·0	25·0	0	1017·3	1166·6	0·00	986·4	1132·3	0·00
2			10	1017·3	1166·6	−0·01	986·7	1132·6	0·01
3			20	1016·6	1165·8	−0·02	986·8	1132·9	0·02
4			30	1016·3	1165·3	−0·02	987·2	1133·4	0·02
5			40	1014·9	1163·8	−0·03	988·0	1134·3	0·02
6			50	1014·2	1162·9	−0·03	988·6	1135·2	0·02
7			60	1013·4	1161·8	−0·02	989·7	1136·4	0·02
8			70	1012·6	1160·9	−0·02	990·4	1137·2	0·02
9			80	1012·3	1160·5	−0·00	990·3	1137·3	0·01
10			90	1012·1	1160·4	0·01	990·9	1137·8	0·00

Table A6.2 Bearing errors for various launch conditions

	Launch conditions			Ordinary ray			Extraordinary ray		
Ray ref number	*f*	Δ	φ	*D*	*P′*	*BE*	*D*	*P′*	*BE*
11	4·0	5·0	45	2230·4	2297·9	−0·00	2221·8	2289·2	0·00
12			135	2224·4	2292·0	0·00	2232·0	2299·3	−0·00
13			225	2224·4	2292·0	−0·00	2232·0	2299·3	0·00
14			315	2230·4	2297·9	0·00	2221·8	2289·2	−0·00
15		15·0	45	1249·9	1333·5	−0·00	1247·6	1330·8	0·00
16			135	1248·9	1332·5	0·00	1243·1	1326·5	−0·00
17			225	1248·9	1332·5	−0·00	1243·1	1326·5	0·00
18			315	1249·9	1333·5	0·00	1247·6	1330·8	−0·00
19		25·0	45	818·8	932·5	−0·01	811·1	924·3	0·01
20			135	820·5	934·2	0·01	809·0	922·4	−0·01
21			225	820·5	934·2	−0·01	809·0	922·4	0·01
22			315	818·7	932·5	0·01	811·1	924·3	−0·01
23		35·0	45	588·8	742·7	−0·03	575·5	728·4	0·02
24			135	589·2	743·1	0·04	575·0	728·0	−0·02
25			225	589·2	743·1	−0·03	575·0	728·0	0·02
26			315	588·8	742·7	0·03	575·5	728·4	−0·02

Table A6.2 (cont'd)

	Launch conditions			Ordinary ray			Extraordinary ray		
Ray ref number	*f*	Δ	φ	*D*	*P'*	*BE*	*D*	*P'*	*BE*
27	4.0	45·0	45	441·2	644·7	−0·07	421·4	622·8	0·04
28			135	441·3	644·8	0·08	421·9	623·2	−0·03
29			225	441·3	644·8	−0·07	421·9	623·2	0·04
30			315	441·2	644·7	0·07	421·3	622·8	−0·04
31	6·0	5·0	45	2260·9	2330·7	−0·00	2269·5	2339·1	0·00
32			135	2265·8	2335·8	0·00	2258·6	2328·2	−0·00
33			225	2265·8	2335·8	−0·00	2258·6	2328·2	0·00
34			315	2273·4	2343·3	0·00	2269·4	2339·1	−0·00
35		15·0	45	1298·3	1386·6	−0·01	1290·9	1378·9	0·00
36			135	1300·5	1388·9	0·01	1288·7	1376·7	−0·00
37			225	1300·5	1388·9	−0·01	1288·7	1376·7	0·01
38			315	1298·3	1386·6	0·00	1290·9	1378·9	−0·00
39		25·0	45	886·8	1012·4	−0·02	872·8	997·5	0·02
40			135	888·0	1013·7	0·02	872·5	997·1	−0·02
41			225	888·0	1013·7	−0·02	872·4	997·0	0·02
42			315	886·8	1012·4	0·02	872·8	997·5	−0·02
43		35·0	45	682·5	864·3	−0·05	655·8	835·1	0·04
44			135	683·1	865·1	0·06	656·7	835·6	−0·04
45			225	683·1	865·1	−0·06	656·6	835·6	0·04
46			315	628·5	864·3	0·05	655·8	835·1	−0·04
47		45·0	45	574·6	844·7	−0·13	522·1	781·2	0·09
48			135	575·3	845·9	0·16	523·3	782·1	−0·08
49			225	575·3	845·9	−0·15	523·2	782·1	0·08
50			315	574·6	844·7	0·13	522·1	781·2	−0·09
51	8·0	5·0	45	2323·0	2396·4	−0·00	2317·2	2390·4	0·00
52			135	2321·5	2395·1	0·00	2307·7	2380·8	−0·00
53			225	2321·5	2395·1	−0·00	2307·7	2380·8	0·00
54			315	2315·8	2389·3	0·00	2317·2	2390·4	−0·00
55		15·0	45	1374·4	1470·8	−0·01	1362·3	1458·0	0·01
56			135	1377·4	1473·9	0·01	1362·0	1457·5	−0·01
57			225	1377·4	1473·9	−0·01	1362·0	1457·5	0·01

Table A6.2 (cont'd)

Ray ref number	Launch conditions			Ordinary ray			Extraordinary ray		
	f	Δ	φ	*D*	*P′*	*BE*	*D*	*P′*	*BE*
58	8.0	15.0	315	1374·4	1470·8	0·01	1362·3	1458·0	−0·01
59		25·0	45	1014·6	1163·3	−0·03	988·3	1134·7	0·02
60			135	1016·9	1165·9	0·04	988·0	1134·2	−0·03
61			225	1016·9	1165·9	−0·04	988·0	1134·2	0·03
62			315	1014·6	1163·3	0·03	988·3	1134·7	−0·02
63		35·0	45	penetration			929·1	1200·3	0·10
64			135	penetration			924·5	1193·4	−0·11
65			225	penetration			924·5	1193·4	0·11
66			315	penetration			929·1	1200·3	−0·10
67		45·0	45	penetration			penetration		
68			135	penetration			penetration		
69			225	penetration			penetration		
70			315	penetration			penetration		

Appendix 7

Path integrals for parabolic layers

Basic expressions

The basic expressions for phase path P, group path P' and ground range D are

$$P = \int \mu ds$$

$$P' = \int \mu' ds$$

$$D = \int R d\theta$$

where μ is the phase refractive index,

μ' is the group refractive index,

ds is an element of path length,

R is the radius of the earth,

$d\theta$ is the angle subtended at the centre of the earth by ds

and each integral is taken over the total path from transmitter to receiver. In the absence of a magnetic field, $\mu' = 1/\mu$. Changing the variable of integration to r, the radial distance from the centre of the earth, and integrating from the surface of the earth ($r = R$) to the vertex ($r = r_v$) we obtain

$$P = 2\int_R^{r_v} \frac{\mu^2 r dr}{(\mu^2 r^2 - R^2 \cos^2 \Delta)^{1/2}}$$

$$P' = 2\int_R^{r_v} \frac{r dr}{(\mu^2 r^2 - R^2 \cos^2 \Delta)^{1/2}}$$

$$D = 2\int_{R}^{r_v} \frac{R^2 \cos \Delta dr}{r(\mu^2 r^2 - R^2 \cos^2 \Delta)^{1/2}}$$

Before these integrals can be evaluated, the way in which μ varies as a function of r must be specified. We now consider the particular forms of the function $\mu(r)$ within the ionosphere that define the parabolic and quasiparabolic layers.

Single parabolic layer

The appropriate expression for μ is

$$\mu(r) = \left\{1 - \frac{2}{x^2 y_m}(r - r_0) + \frac{1}{x^2 y_m^2}(r - r_0)^2\right\}^{1/2}$$

where $x = f/f_c$

f = signal frequency

f_c = critical frequency

y_m = semithickness of the layer

r_0 = radius of base of layer

It can be shown (Gething, 1965) that the total path integrals can be closely approximated by the following expressions; the subscript 1 is used to denote the part of the path within the layer and subscript 2 the part below the layer:

$$D_1 \simeq y_m\, x \frac{R}{r_0} \sin i_0 \{(1 - 2\epsilon) \ln F + 4\epsilon x \cos i_0\}$$

$$D_2 = 2R\left\{\frac{\pi}{2} - (\Delta + i_0)\right\}$$

$$P_1 \simeq \frac{y_m}{2}\left[x\{1 + (1 - 2\epsilon)\sin^2 i_0\} - \frac{1}{x}\right] \ln F$$

$$+\, y_m(1 + 3\epsilon x^2 \sin^2 i_0)\cos i_0$$

$$P_2 = 2R\cos(\Delta + i_0)/\sin i_0$$

$$P_1' \simeq x y_m \ln F$$

$$P'_2 = 2R \cos(\Delta + i_0)/\sin i_0$$

where

$$\epsilon = y_m/r_0 \quad \text{and is assumed to be small,}$$

$$F = \frac{1 - \epsilon x^2 \sin^2 i_0 + x \cos i_0}{1 - \epsilon x^2 \sin^2 i_0 - x \cos i_0}$$

i_0 =angle of incidence on base of layer.

Penetration through a parabolic layer

We now consider the case of penetration through one parabolic layer (subscript E) and reflection by a higher layer (subscript F). D_E, P_E and P'_E represent the increases in ground range, phase path and group path respectively for a single passage through layer E. Then

$$D_E \simeq \frac{y_E x_E R^2 \cos \Delta}{r_E^2} \{(1 - 2\epsilon) \ln F_E - 2\epsilon x_E (\sqrt{X} - \cos i_E)\}$$

$$P_E \simeq y_E \sqrt{X} - \frac{y_E}{2}(1 + 3\epsilon x_E^2 \sin^2 i_E)(\sqrt{X} - \cos i_E)$$

$$+ \frac{y_E}{2}\left[x_E\{1 + (1 - 2\epsilon) \sin^2 i_E\} - \frac{1}{x_E}\right] \ln F_E$$

$$P'_E \simeq x_E y_E \ln F_E$$

where

$$i_E = \text{angle of incidence on base of lower layer, etc.}$$

$$F_E = \frac{x_E \sqrt{X} + \epsilon x_E^2 \sin^2 i_E + 1}{x_E \cos i_E + \epsilon x_E^2 \sin^2 i_E - 1}$$

$$X = \cos^2 i_E + 4\epsilon \sin^2 i_E$$

Once D_E, P_E and P'_E have been found, the total path lengths can be readily calculated. For example, the total ground range D is given by

$$D = D_F + 2D_E - 2D_S$$

where D_F represents the ground range for a single reflection from the upper layer in the absence of the lower layer and D_S is the ground range corresponding to a single undeviated passage through the lower layer.

Quasi-parabolic layer

The following results are taken from p. I–70 of a paper by Davies (1968). The refractive index is defined by the usual equation

$$\mu^2 = 1 - \frac{f_N^2}{f^2} = 1 - \frac{1}{x^2}\left(\frac{f_N}{f_c}\right)^2$$

and now

$$\left(\frac{f_N}{f_c}\right)^2 = 1 - \left(\frac{r - r_m}{y_m}\right)^2\left(\frac{r_0}{r}\right)^2$$

for $r_0 < r < r_0 + 2y_m$

where r_m is the radius to the level of maximum ionisation. Then

$$D = 2R\left\{\frac{\pi}{2} - (i_0 + \Delta) - \tfrac{1}{2}RJ\cos\Delta\right\}$$

$$P = 2\{-R\sin\Delta + K + \tfrac{1}{4}Br_mJ\}$$

$$P' = 2\left\{r_0\cos i_0 - R\sin\Delta - \frac{1}{A}(r_0\cos i_0 + K)\right\}$$

where

$$A = 1 - \left(\frac{f_c}{f}\right)^2 + \left(\frac{f_c r_0}{f y_m}\right)^2$$

$$B = \frac{-2r_0^2 r_m}{y_m^2}\left(\frac{f_c}{f}\right)^2$$

$$C = \left(\frac{f_c r_0 r_m}{f y_m}\right)^2 - R^2\cos^2\Delta$$

$$J = \frac{1}{\sqrt{C}}\ln\left\{\frac{B^2 - 4AC}{4C\left(\cos i_0 + \frac{\sqrt{C}}{r_0} + \frac{B}{2\sqrt{C}}\right)^2}\right\}$$

$$K = \frac{B}{4\sqrt{A}}\ln\left[\frac{B^2 - 4AC}{\{2r_0(A + \sqrt{A}\cos i_0) + B\}^2}\right]$$

The minus sign in the expression for B is incorrectly omitted by Davies (1968).

References

DAVIES, K. (1968): 'Magneto-ionic theory of radio propagation'. AGARD lecture series XXIX: Radio wave propagation

GETHING, P.J.D. (1965): 'Relationship between phase path and effective path for oblique ionospheric propagation'. *J. Atmos. & Terr. Phys.*, **27,** pp. 57–66

Appendix 8

The effective tilt of a parabolic layer

For simplicity, earth-curvature effects are neglected throughout this Appendix. The appropriate integrals for a flat earth and horizontally-stratified ionosphere are

$$D = y_m x \sin i_0 \ln \left(\frac{1 + x \cos i_0}{1 - x \cos i_0}\right) + 2h_0 \tan i_0$$

$$P = \frac{y_m x}{2} \left\{(1 + \sin^2 i_0) - \frac{1}{x^2}\right\} \ln \left(\frac{1 + x \cos i_0}{1 - x \cos i_0}\right)$$

$$+ y_m \cos i_0 + 2h_0 \sec i_0$$

$$P' = D/\sin i_0 = y_m x \ln \left(\frac{1 + x \cos i_0}{1 - x \cos i_0}\right) + 2h_0 \sec i_0$$

The bearing error $\delta\beta_1$ on a one-hop path is, in terms of the notation used in Section 15.2.1,

$$\delta\beta_1 = \tfrac{1}{2} \operatorname{cosec} i_0 \left(\frac{dP}{dI}\right)\left(\frac{dI}{d\xi}\right)$$

where the differential of phase-path P with respect to any ionospheric parameter I is calculated subject to the condition that D is constant. The spatial gradient of I transverse to the path, $dI/d\xi$, is assumed known, perhaps from a numerical map of the ionosphere.

The basic equation for dP/dI is

$$\frac{dP}{dI} = \frac{\partial P}{\partial I} - \sin i_0 \frac{\partial D}{\partial I}$$

We now write S for $\sin i_0$

C for $\cos i_0$

and

F for $\left(\frac{1 + xC}{1 - xC}\right)$

Gradient in critical frequency

By direct differentiation

$$\frac{\partial P}{\partial f_c} = \frac{-y_m x}{2f_c} \left\{\left(1 + S^2 + \frac{1}{x^2}\right) \ln (F) + \frac{2xC}{1 - x^2C^2} \left(1 + S^2 - \frac{1}{x^2}\right)\right\}$$

$$\frac{\partial D}{\partial f_c} = \frac{-y_m x S}{f_c} \left\{\ln (F) + \frac{2xC}{1 - x^2C^2}\right\}$$

Hence

$$\frac{dP}{df_c} = \frac{-y_m C}{2f_c} \left(xC + \frac{1}{xC}\right) \ln (F) + \frac{y_m C}{f_c}$$

which is equal to $-(P'-P)/f_c$, as expected from Appleton's theorem.

$$\delta\beta_1 = \frac{1}{2S} \left(\frac{df_c}{d\xi}\right) \left(\frac{dP}{df_c}\right)$$

$$= \frac{y_m \cot i_0}{2f_c} \left(\frac{df_c}{d\xi}\right) \left\{1 - \tfrac{1}{2} \left(z + \frac{1}{z}\right) \ln \left(\frac{1 + z}{1 - z}\right)\right\}$$

where $z = xC$. This equation was first derived by Titheridge (1958).

Gradient of base height

Differentiating the path integrals with respect to h_0 and holding all other layer parameters constant, we obtain

$$\delta\beta_1 = \cot i_0 \frac{dh_0}{d\xi}$$

which is the standard equation for a tilted mirror.

Gradient of height of maximum ionisation

Similary, we can find the effect of a gradient of h_m. We clearly cannot hold both y_m and h_0 constant, because $h_m = (h_0 + y_m)$. If we hold y_m constant, a gradient of h_m is exactly the same as a gradient of h_0 and produces the same bearing error as that found above. Holding h_0 constant, we obtain

$$\frac{dh_m}{d\xi} = \frac{dy_m}{d\xi}$$

and

$$\frac{dP}{dy_m} = \frac{C}{2}\left(z - \frac{1}{z}\right) \ln\left(\frac{1+z}{1-z}\right) + C$$

$$\therefore \delta\beta_1 = \frac{\cot i_0}{2}\left(\frac{dh_m}{d\xi}\right)\left\{1 + \frac{1}{2}\left(z - \frac{1}{z}\right) \ln\left(\frac{1+z}{1-z}\right)\right\}$$

Gradients of h_0 and h_m with γ constant

A case that may be closer to a real physical situation is one in which the thickness parameter γ (equal to y_m/h_0) remains constant while h_0 and h_m vary with geographical position. The path lengths can be written

$$D = h_0\left\{\gamma x S \ln\left(\frac{1+xC}{1-xC}\right) + \frac{2S}{C}\right\}$$

$$P = h_0\left\{\frac{\gamma x}{2}\left(1 + S^2 - \frac{1}{x^2}\right) \ln\left(\frac{1+xC}{1-xC}\right) + \frac{2}{C} + \gamma C\right\}$$

Expressing the final bearing error in terms of the gradient of h_m (the quantity likely to be readily available from numerical maps) we obtain

$$\delta\beta_1 = \frac{\cot i_0}{2}\left(\frac{dh_m}{d\xi}\right)\left\{\left(\frac{2+\gamma}{1+\gamma}\right) + \frac{\gamma\left(z - \frac{1}{z}\right)}{2(1+\gamma)} \ln\left(\frac{1+z}{1-z}\right)\right\}$$

Two layers

For penetration through one layer, denoted by subscript E, and reflection at a higher layer, subscript F, it may be shown that the path integrals are

$$D = 2y_E x_E S \ln\left(\frac{x_E C + 1}{x_E C - 1}\right) + \frac{2h_1 S}{C}$$

$$+ y_F x_F S \ln\left(\frac{1 + x_F C}{1 - x_F C}\right) + \frac{2h_2 S}{C}$$

$$P = 2y_E C + y_E x_E \left(1 + S^2 - \frac{1}{x_E^2}\right) \ln\left(\frac{x_E C + 1}{x_E C - 1}\right) + \frac{2h_1}{C}$$

$$+ y_F C + \tfrac{1}{2} y_F x_F \left(1 + S^2 - \frac{1}{x_F^2}\right) \ln\left(\frac{1 + x_F C}{1 - x_F C}\right) + \frac{2h_2}{C}$$

where h_1 is the height of the base of layer E and h_2 is the difference in height between the upper boundary of layer E and the lower boundary of layer F.

If both critical frequencies vary with position and all other layer parameters are held constant, the total bearing error is found by the usual procedure to be

$$\delta\beta_1 = \frac{y_E \cot i_0}{f_E}\left(\frac{df_E}{d\xi}\right)\left\{1 - \tfrac{1}{2}\left(z_E + \frac{1}{z_E}\right)\ln\left(\frac{z_E + 1}{z_E - 1}\right)\right\}$$

$$+ \frac{y_F \cot i_0}{2f_F}\left(\frac{df_F}{d\xi}\right)\left\{1 - \tfrac{1}{2}\left(z_F + \frac{1}{z_F}\right)\ln\left(\frac{1 + z_F}{1 - z_F}\right)\right\}$$

As noted by Titheridge, an overall tilt in the lower (refracting) layer has no effect; mathematically, a change in h_1 has no effect on P when $(h_1 + h_2)$ remains constant, i.e. when the upper layer is not tilted.

If a layer acts as a prism, it can be assigned an effective tilt ψ and, for a penetrated layer, this tilt is encountered on both the upward and downward paths. Titheridge (1958) stated:

> 'The changes in direction when a ray penetrates several layers 1, 2,...and is reflected from a layer r are thus described by an overall effective tilt:
>
> $$\psi = 2\psi_1 + 2\psi_2 + \ldots + \psi_r + \psi_{or}$$
>
> where ψ_{or} is the tilt of the base of the reflecting layer...any overall tilt of a penetrated layer having no net effect.'

For more complex situations such as multihop paths, M- and N-type modes etc., and in cases where the effective tilt on the downward path is not necessarily the same as on the upward path, a more general equation is required. The total bearing error must then be obtained from the basic result

$$\delta\beta = (\delta P \operatorname{cosec} i_0)/a$$

where δP is a total change in phase path arising from many contributions, subject to the condition of constant ground range, and a is the aperture of the DF. If there are n reflection or refraction points,

$$\delta P = \delta P_1 + \delta P_2 + \ldots + \delta P_n$$

where

$$\delta P_r = \frac{d_r a}{D} \sum_i \frac{dP}{dI_i} \frac{dI_i}{d\xi}$$

d_r is the ground distance from the transmitter to the point vertically below the rth point of reflection or refraction; Σ_i denotes a summation over all possible ionospheric parameters i.

Reference

TITHERIDGE, J.E. (1958): 'Variations in the direction of arrival of high frequency radio waves'. *J. Atmos. & Terr. Phys.*, **13,** pp. 17–25

Appendix 9

Standard deviations of the best point estimate

We suppose that a Best Point Estimate (BPE) has been calculated from n bearing lines by means of the standard Stansfield theory given in Section 16.3.2. The position of the BPE relative to an origin at the (unknown) true position of the transmitter is $(\bar{X}, \bar{Y})$, where

$$\bar{X} = -\frac{G}{A}$$

$$\bar{Y} = -\frac{F}{B}$$

We now wish to calculate the standard deviations σ_X, σ_Y of $\bar{X}$, $\bar{Y}$ respectively. These values can be derived from the standard theory rather simply by considering an $(n + 1)$th bearing and its 'miss distance' d from the BPE. If p is the 'miss distance' from the origin, we have

$$d = p + \frac{G}{A}\sin\varphi - \frac{F}{B}\cos\varphi$$

where φ is the angle between the bearing line and the X axis. This is equation (6.7) in Beale's paper (1961), but with his subscript k omitted because we are now considering an *additional* bearing, not used in the calculation of the BPE. Hence.

$$E(d^2) = E(p^2) + E\left\{\left(\frac{G}{A}\sin\varphi - \frac{F}{B}\cos\varphi\right)^2\right\}$$

where E denotes the expected value and the bearings are assumed to be independent. Noting that, for the n bearings used in the BPE,

$$E(p_j^2) = \sigma_j^2 = \frac{1}{w_j}$$

$$E(p_i p_j) = 0, \qquad i \neq j$$

and that A, B do not involve the ps, we obtain

$$E(d^2) - E(p^2) = \left(\frac{\Sigma w_j \sin^2 \varphi_j}{A^2}\right) \sin^2 \varphi$$

$$+ \left(\frac{2\Sigma w_j \sin \varphi_j \cos \varphi_j}{AB}\right) \sin \varphi \cos \varphi$$

$$+ \left(\frac{\Sigma w_j \cos^2 \varphi_j}{B^2}\right) \cos^2 \varphi$$

$$= \frac{\sin^2 \varphi}{A} + \frac{\cos^2 \varphi}{B}$$

Now, if the test bearing has a very low variance, i.e. $E(p^2)$ is very small, the major contribution to $E(d^2)$ must come from the uncertainty in the position of the BPE. By considering a test bearing parallel to the Y-axis ($\varphi = \pi/2$) we see that

$$\sigma_X = \frac{1}{\sqrt{A}}$$

and similarly, (with $\varphi = 0$),

$$\sigma_Y = \frac{1}{\sqrt{B}}$$

These are the required standard deviations; they apply for any position of the origin, but depend on the assumption that the X and Y axes are parallel to the axes of the rectangle.

Reference

BEALE, E.M.L. (1961): 'Estimation of variances of position lines from fixes with unknown target positions', *J. Res. Natl. Bur. Stand.*, **65D**, pp. 263–273

Index